The Spatial Context of Technological Development

Edited by

RICCARDO CAPPELLIN

Bocconi University
Milan

and

PETER NIJKAMP

Free University
Amsterdam

Avebury

Aldershot · Brookfield USA · Hong Kong · Singapore · Sydney

Published by
Avebury
Gower Publishing Company Limited
Gower House
Croft Road
Aldershot
Hants GU11 3HR
England

Gower Publishing Company
Old Post Road
Brookfield
Vermont 05036
USA

British Library Cataloguing in Publication Data
The spatial context of technological development
1. Great Britain. Regional economic development.
Effects of technological change.
I. Cappellin, Riccardo II. Nijkamp, Peter 1946-
338.06

ISBN 0 566 07149 5

Printed and bound in Great Britain by
Billing & Sons Ltd, Worcester

Contents

List of contributors

N. Alderman
Centre for Urban & Regional Development
University of Newcastle upon Tyne
Newcastle upon Tyne NE1 7RU
UK

F. Arcangeli
Contrada I Biceghi
37020 San Rocco di Piegara (Verona)
Italy

A. Boitani
Facoltà di Economia
Università Cattolica
Largo Gemelli, 1
20123 Milano
Italy

R. Camagni
Via Sismondi, 9
20133 Milano
Italy

R. Cappellin
Istituto di Economia Politica
Università Commerciale L. Bocconi
Via Sarfatti, 25
20136 Milano
Italy

E. Ciciotti
Facoltà di Economia
Università Cattolica
Largo Gemelli, 1
20123 Milano
Italy

J.R. Cuadrado Roura
Fundación Fondo para la Investigación
Económica y Social
Juan Hurtado de Mendoza 14
28036 Madrid
Spain

M.M. Fischer
Institute of Geography
University of Vienna
Universitaetsstrasse 7/V
A-1010 Vienna
Austria

R. Funck
Institut fur Wirtschaftspolitik
und Wirtschaftsforschung
University of Karlsruhe
Postfach 6380
D-7500 Karlsruhe 1
West Germany

G. Garofoli
Facoltà di Economia e Commercio
Università degli Studi di Pavia
Strada Nuova, 65
27100 Pavia
Italy

M. Giaoutzi
Dept. of Geography and Regional Planning
National Technical University
Zographou Campus
Athens 157 10
Greece

G. Inoue
Department of Planning and Economics
Toyohashi University of Technology
Toyohashi Aichi 440
Japan

D.J.F. Kamann
State University of Groningen
Economics Faculty
P.O. Box 800
NL-9700 AV Groningen
The Netherlands

J.S. Kowalski
Institut fur Wirtschaftspolitik
und Wirtschaftsforschung
University of Karlsruhe
Postfach 6380
D-7500 Karlsruhe 1
West Germany

C. Lacour
Institut Economie Régionale Sud-Ouest
Université Bordeaux 1
F-33604 Pessac
France

T.R. Lakshmanan
Center for Energy and Environmental Studies
University of Boston
675 Commonwealth Avenue
Boston, Mass. 02215
USA

D. Maillat
Institut de Recherches Economiques et Régionales
Université de Neuchatel
Pierre-à-Mazel 7
CH-2000 Neuchatel
Switzerland

E. Malecki
Department of Geography
University of Florida
3141 Turlington Hall
Gainesville, Fla. 32611
USA

J.N. Marshall
Department of Geography
University of Birmingham
P.O. Box 363
Birmingham B15 2TT
UK

W. Molle
Netherlands Economic Institute
Burgemeester Oudlaan 50
NL-3062 PA Rotterdam
The Netherlands

P. Nijkamp
Department of Economics
Free University
P.O. Box 7161
1007 MC Amsterdam
The Netherlands

I. Orishimo
Department of Planning and Economics
Toyohashi University of Technology
Toyohashi Aichi 440
Japan

R. Premus
Wright State's Center
for Industrial Studies
Wright State University
Dayton, Ohio
USA

G. Scimemi
OECD
2, rue Andre Pascal
F-75775 Paris Cedex 16
France

A. Thwaites
Centre for Urban & Regional Development Studies
University of Newcastle upon Tyne
Newcastle upon Tyne NE1 7RU
UK

Preface

Preface

The triangular relationship between technological change, local and regional development, and public policy is an issue which deserves much attention in the light of the current restructuring of all regions and cities of our world economy. Analytically, however, the current state of the art is by no means satisfactory from both a scholarly and a political viewpoint. The wealth of research efforts is very much dispersed and heterogeneous, and has insufficient policy impact.

The increasing interest in the problems of regional economic restructuring and spatial imbalances in the adoption of new technologies has fostered the idea to organise the European Advanced Summer Institute of the Regional Science Assocation on the topic: 'Theories and policies of technological development at the local level', in Arco, Lake of Garda, Italy, from 17-23 July 1988. This book presents the main findings of this venue.

The contributors to this volume were invited to present a survey paper of the literature on the following fields, which were the major topics of the workshop: theories and models of technology diffusion, the role of technology in metropolitan and peripheral areas, the economics of research and development, the relationship between producer services and technological change, the role of small and medium size firms, the effects on local labour markets, the international and interregional transfers of technology, and policies to promote and to adjust to technological change.

This volume contains most of the overview papers presented at the above mentioned Summer Institute. An accompanying volume contains a selection of specific research papers in this area (see E. Ciciotti, A. Thwaites and N. Alderman, (eds.) Technological Change in a Spatial Context, theory, empirical evidence and policy, Springer Verlag, Berlin, 1989).

The autors wish to thank the L. Bocconi University of Milan, the University of Trento and the Catholic University of Milan for having sponsored this workshop; in particular, they thank Prof. Carlo Borzaga and Prof. Enrico Ciciotti, who - together with the advisory committee - have organised the scientific programme of this Summer Institute. Our special appreciation goes to the Provincia Autonoma di Trento and to various local institutions of Arco, which have financially supported this Summer Institute.

In the final stage of preparing this book, the editors received indispensable support from Winny Browne (who was responsible for linguistic editing) and Rita Hittema (who did the painstaking keyboarding work).

September 1989

Riccardo Cappellin
Peter Nijkamp

1 The role of space in technological change

R. CAPPELLIN AND P. NIJKAMP

1. Prologue

Technological change and spatial transformation are increasingly regarded as dual phenomena with closely intertwined relationships. Technological and regional revitalisation is at present often considered to be the main vehicle for new growth in our economies. Especially in recent years there has been considerable reflection on the impact and effectiveness of public policy (including regional policy) on technological innovation. This interest has centered around two major questions, viz. the conditions for a technological take-off (the **technogenesis** question) and the degree of diffusion and pervasiveness of new technologies (the **technological impact** question). However, a comprehensive view on these issues and on the ways policy may intervene is still lacking.

Given this observation, this book is addressing these two important issues, with a special emphasis on the perspective offered by regional science studies.

The first issue thus concerns the role of technology in the regional development process. In fact, traditional contributions on the problem of regional disparities have often emphasized the role of spatial mobility of production factors, such as capital and labour, whereas they assumed the evolution of technoloy as given and mainly uniform in the various regions. In contrast, more recent contributions indicate that the capability to timely introduce innovations is a crucial factor in explaining the growth and the birth of firms. In fact, in a process of dynamic competition, the economic development of the regions is not only affected by the level of the production costs but also by their regional receptivity toward technological change.

The second issue is related to the specific role which regional or spatial factors have in both technology generation and technology diffusion. Clearly, spatial factors play a critical role in affecting the transfer of technology at the interregional and international level. In fact, the high spatial concentration of most modern and innovative economic activities and the very existence of wide technology gaps between the various countries (or between different areas within these countries) point at the existence of spatial factors, which either facilitate or hinder the generation and adoption of new technologies.

A closer emphasis on the role of the spatial factors, which affect the generation and diffusion of technology, implies that it is misleading to consider technology as an a-spatial phenomenon which is the result only of large firms having an internal organisation capable to overcome national and regional boundaries.

On the contrary, according to the 'endogenous' or 'bottom-up' approach to regional development (Aydalot, 1985; Molle and Cappellin, 1988), the development process and the evolution of technology at the national level may be considered as the result of the interdependence among the various activities, which are localised in the different regions and tightly related to their local environment.

In this context, it is necessary to pay more attention to the spatial dimensions of technological change (Section 2), to the role of innovations in the evolution of the economic structure (Section 3) and of the spatial structure (Section 4) of various regions, and to the specific scope of innovation policy at the local level (Section 5).

2. The Role of Spatial Factors in Technology Diffusion

The generation of new technologies is not exclusively the result of external forces; there is an increasing awareness that the emergence and direction of technological change can be manipulated by means of, for instance, well focussed R & D efforts (see also Davelaar, 1989). A stimulation of innovations may thus be an important means of gaining a competitive edge over one's competitors. In a dynamic competitive situation the notion of a spatial product life cycle may be very important. On the basis of a distinction into various phases (take-off, slow growth, rapid growth, saturation and decline), it is plausible to assume that different regions are actively operating in different phases of a product life cycle. For instance, the design or invention of new products often takes place in central regions due to their agglomeration economies, their flexibility concerning input use, and the socio-cultural climate. Standardized products however are less in need of centripetal forces and may be manufactured outside of the major centres, especially if labour costs are lower there. Thus spatial differences in production costs may lead to a geographical dispersion, either to peripheral areas or even to Third World countries.

Clearly, process innovation may still play an important role in standardization of new products (cf. Taylor, 1987).

Besides, the geographical pattern of new technological regimes is co-determined by demand pull factors and bottlenecks (e.g., rising marginal costs or decreasing marginal productivities of R & D investments; see also Morphet, 1987).

According to the seminal contribution of Nelson and Winter (1982), the generation of new technologies or the adoption of existing technologies is affected by the 'selection environment', in which the firms are integrated. In fact, the adoption of innovations depends on various environmental factors (such as social factors which may stimulate the existence of entrepreneurial capabilities in the various regions), economic factors (which may determine the fear of bankrupcy and stimulate radical changes), or institutional factors (which increase the dependence of local firms and of local institutions on external decision makers by reconverting local, unused resources toward new production modes).

Also, the productivity of research is affected by local factors. For example, the generation of new technology highly depends on the coordination of research efforts by existing firms. Therefore, success depends on organisational factors, such as the capability to maintain the cohesion among a group of firms and researchers, who are jointly working toward a specific aim during a rather long period.

Clearly, as illustrated by the 'epidemic disease model' of the diffusion of innovations, the adoption of innovations is critically dependent on the availability of information. Therefore, the existence of cooperative agreements in research among various firms, belonging to the same local environment, may stimulate innovations. However, even more important may be the access to international scientific networks.

This increased mobility of technology at the interregional and at the international level may be related to the increased immaterial nature of technology, which is requiring relatively lower capital investments and higher human captial investments. Thus, the role of local or immobile factors is becoming more important in redirecting these technological flows toward specific regions. In particular, the speed in adopting new technology is related to the quality of the local labour force, as well as to the geographical mobility of researchers and technicians, who may be considered as 'gatekeepers' of the international scientific networks. Thus, regional policy should improve the quality of the local environment in order to be capable of attracting and maintaining these qualified workers.

Various contributions (Momigliano, 1984; Dosi, 1984) underline the role of dynamic processes in the diffusion of new technologies. In particular, the diffusion of technology implies a slow but important improvement and modification of the original techniques or products. This change is clearly the result of specific efforts in research and development undertaken in the various regions and also of the effort to comply with the specific characteristics of local demand. The increased customisation of production is creating new opportunities for the small and medium size firms which usually have tighter relationships with the respective local markets than larger firms, thus stimulating these firms to introduce product innovations or to differentiate among existing products.

The productivity of research efforts at the local level may be affected by the existence of dynamic scale economies, or by learning processes, which are tightly related to the existence of tacit factors and to the easy accessibility to specialized complementary factors, which characterize a particular local environment and may hinder the transfer of particular technologies to other areas.

Another important dynamic factor in the diffusion of technology is represented by the existence of adjustment costs, related to the reallocation of resources from more traditional to more modern production modes (Cappellin, 1983). These changes may be related to organisational changes within the firms, as well as to the closure of some firms and birth of new firms. Various local specific factors may affect the flexibility of regional economies in adjusting themselves to new technologies. The gap between the technology already in use and the new technologies and the speed of the reconversion process increase the level of the adjustment costs or the resources, which have to be employed in order to overcome the various obstacles which hinder the adoption of a particular technology in a particular region.

It should be underlined that the technology frontier becomes more differentiated, and countries and regions compete on this frontier. Thus it is becoming impossible to detain an overall leadership in technology, while the various countries and regions may become sectoral leaders in particular technologies, by exploiting specialization economies and the availability of local comparative advantages.

3. The Role of Technology in Regional Economic Development

The role of technology in the explanation of regional disparities is directly related to the fact that disparities in the growth rate of productivity are the most important components in the evolution of per capita income disparities (Camagni and Cappellin, 1985).

Moreover, technology affects economic development indirectly, through the impact on the competition of firms, as process and product innovations allow a decrease of production costs or an improvement of the quality of products. The resulting increase of exports induces a multiplier effect on the growth of regional product (see Molle; this volume).

Clearly, technology has an impact on both the level and the structure of employment, as it modifies the professional profile of the labour demand, although direct and indirect effects are rather complex and difficult to estimate (see Fischer; this volume).

Technology also changes the sectoral structure of a regional economy, as it determines the development of new sectors and induces a crisis in other sectors. In particular, technological change is the major factor in the recent growth of service sector employment (see Lakshmanan, Cuadrado Roura and Marshall; this volume).

In fact, services represent a process innovation for the firms using them. Moreover, services are a product innovation for the tertiary firms from which they are produced. Finally, also service firms adopt frequent process and organizational innovations. The relationship between services and industry has a complementary character, as an adequate use of services is crucial in promoting the competition of the exports of manufacturing products. However, also services may be exported, especially through the establishment of subsidiaries in other regions and through interregional networks linking the service firms of various regions (Cappellin, 1988). Therefore, the growth of services is tightly related to the characteristics of the local environment and in particular to the exploitation of a local know-how, as is also reflected in the intense mobility of qualified workers among tertiary firms localized in the same area. In this respect, service branches using the fruits of modern information technology (including telecommunication) may be the most favourable candidates for a successful innovative regional policy, provided such initiatives are adequately supported by means of proper labour force stimuli and entrepreneurial spirits. Such local and regional policies would be a clear representation of the 'third wave' society (a society marked by a highly diversified service and information orientation), advocated by Toffler (1983). It is clear that the emergent new technologies offer a window of rich opportunities for regional revitalisation. Whether such opportunities will be grasped is a matter of entrepreneural spirit and policy creativeness.

A further effect of technological change is on the spatial organisation of production, as here the vertical disintegration trends determine the creation of new spatial hierarchies (see Scott, 1986). Two factors are of decisive importance her, viz. an **inter-firm** process and an **intra-firm** process.

At the inter-firm level we observe the increasing significance of **networks** as vehicles supporting the creation of innovative, dynamic and active milieus (see Kamann, 1988). Networks provide the potential information infrastructure which is a sine qua non for creating the atmosphere and attitude for development. Networks presuppose competent actors, socio-economic resources and appropriate organisational environments. It goes without saying that the above mentioned development towards a world city system is also mirrored in the dynamics of network configurations.

The policy implications of a network economy are far reaching: instead of focussing attention on nodes (e.g., locations, actors), one has to look at edges (e.g., interaction patterns or transfer mechanisms for knowledge). For instance, a stimulation of knowledge centres, of telecommunication infrastructure, or of new information technology may be more adequate in a network society. Policies should then - from a spatial angle - be directed toward a balanced development of metropolitan and non-metropolitan areas. Conventional growth pole concepts would then have to be rephrased in terms of network interactions. Such interactions would need a much better and more cohesive consideration of public and private sector initiatives, focussing on both the necessary informatics infrastructure (see Giaoutzi and Nijkamp, 1988) and the diffusion of relevant know-how via appropriate communication and contact channels. Information management is here an important issue and may be reinforced and supported inter alia via a system of information brokers, local information centres etc. In such networks of chains of interactions the transaction costs (at both the inter-firm and the intra-firm level) are of crucial importance (see Porter, 1980).

At the intra-firm level we observe an increasing tendency toward **flexible specialisation.** Large firms with standardised and bureaucratic procedures appeared to be less flexible during the past period of economic stagnation, whilst small firms were able to respond much faster to the changing demand for more specialized products. This flexible specialisation of SMEs may generate a competitive edge for less central areas. The increased importance of such flexible production complexes marks the end of the conventional Fordist mass production system (see Piore and Sabel, 1984).

The spatial consequences of this development are also far reaching: many examples of flexible specialisation can be found outside the traditional industrial heartlands, so that evidently also intra-firm dynamism has clear spatial implications.

The previous transformation to a post-Fordist production system has been analysed in many countries, although surprisingly the policy implications have not yet been addressed very well. But it is clear that in general local and regional gentrification is a necessary condition for inducing flexible production systems. Specific vehicles in this setting may be the stimulation of new information technologies or of spatially varied branch networks in order to ensure flexible adjustment (cf. Lazerson 1988).

In the light of the previous observations on technological organizational and geographical dynamism, it is evident that a region- or city-specific technology policy needs a focussed orientation towards firm-specific and location-specific circumstances. Variety and flexibility of policy seems to be more important in this context than comprehensiveness.

4. The Role of Technology in the Spatial Structure of Regions

From a spatial angle, the discontinuous emergence of new technological regimes is an important background for spatial shifts in economic heartlands. Technogenesis and Schumpeterian 'swarming' processes create a high degree of dynamics in a space economy. However, adoption of new technologies is never a 'carbon-copy' process, but is the result of a 'creative' diffusion process, so that the spatial mapping of new technological trajectories may bear resemblance to a fractal-theoretic pattern.

In particular, urban areas perform three major functions in the process of regional development. First of all, they are the major locations of specialised service activities and of modern industrial activities. In fact, these activities have played a crucial role in the recent process of revitalisation of the large metropolitan areas in European countries. Secondly, cities represent the 'incubator' of innovations and of the birth of new firms specialised in modern technologies. Finally, a third major function of urban centers is their function as 'gateway' in the relationships between the regional and the national and international economy.

Local and industrial 'dynamism' is also reflected in contact and network patterns of firms. From a geographical perspective we observe a new hierarchy of cities in our world system, which seems to reflect a global

Christaller model (see Friedmann, 1986 and Moss, 1987). Such a world city system is likely to embody a socio-economic dominance of large agglomerations. Such a dominance pattern may be repeated - in a way comparable to fractal theory - at lower spatial levels (e.g., nations, regions, cities), so that also local and regional economic systems are influenced by global evolutions. Especially the producer service sector appears to play a strong role in this evolutionary process (see also Bailly, 1985 and Marshall et al., 1988), a role which is reinforced by the de-industrialisation trend (see Bluestone and Harrison, 1982).

Thus cities represent the nodes of networks through which information and services circulate at the intraregional and at the interregional and international level (Cappellin, 1988). In particular, after a process of relative homogenisation took place in the 1970s, the regional space is becoming more differentiated and organised according to networks of urban poles, specialised in different functions and linked by hierarchical relationships (see Kamann and Nijkamp; this volume).

Technology affects also the internal structure of metropolitan areas. Modern information technologies have contradictory effects, as they may support a further decentralisation of industrial and service activities. On the other hand, they may also have very powerful effects on concentration. However, an unambiguous effect of new technologies and of the development of service activities is the increase in mobility, as is also reflected by the increased transport and communication flows and by a faster change in the location of the individual firms and in the functions which characterise the various areas within a metropolitan region. Certainly, the urban orientation of modern technologies supports the need for a better integration of the urban dimension within regional policies.

5. The Scope of Regional Innovation Policy

Policies focussing on both technogenesis and technological impacts are necessarily broad and general in nature (witness the current popularity of science park and information centre concepts), but cannot certainly boost a high degree of achievement (see also Malecki and Nijkamp, 1988). Although it has become a widespread belief that technological innovation is not a 'deus ex machina' or 'manna from heaven', but can be induced by policy intervention, at the same time the question has emerged: is it possible to develop well focussed policies (in close cooperation with private initiatives) that may act as potentially successful vehicles for technological revitalisation? More precisely, the question is: which type of policy is needed for which type of technology and in which area? The micro-electronics technology, the telecommunications technology, the bio-technology, the agro-technology or the offshore technology, for instance, have all different seedbed conditions. Therefore, it is most unlikely that a uniform technology policy can be developed which is able to generate all kinds of new technologies. It is thus increasingly realized that technology policy is essentially a tailor-made or customized endeavour to favour the creation of specific innovative activities in specific sectors and in specific regions. This also implies that technology policy cannot be separated from other fields of public policy such as socio-economic policy and physical planning (cf. Rosenberg, 1982). Seen from this perspective, much more emphasis would have to be placed on the self-organizing potential of regions (including private initiatives) and to packaging of policy measures.

Given these observations, technology policy should explicitly address the question of different phases in spatio-economic evolution, viz. incubation, catching-up or competition, and stagnation.

In the incubation phase, the attention should be focussed upon favourable conditions for new Schumpeterian entrepreneurs (e.g., by removing entry barriers to new firms). Important conditions in this respect are: agglomeration economies (e.g., accessibility), skilled labour, specialized information and communications infrastructures, and appropriate social overhead capital (cf. Moss, 1985).

In the catching-up or competition phase policies favouring market expansion are needed, so that attempts at stimulating the relocation of firms (or in general branch plant economies and endogenous new firm formation) are crucial at this stage. In general, much attention has to be given to product improvement by means of process technologies. Thus local initiatives in non-metropolitan areas are critical in attracting a certain share of the innovative performance of an economy.

In the stagnation phase we encounter a situation of market saturation. Usually under these cirumstances price competition is becoming an important instrument, so that there will be an intensive effort at lowering labour costs (and in general all productions costs).

In the light of the phase-wise interpretation of spatial dynamics, it is obvious that those policies tend to be successful which try to assist regions to adjust themselves flexibly and quickly to new circumstances. On the other hand, protective policies that try to cope with emerging new phases are probably less successful and may fail in the long run, to the detriment of the regions protected. Also R & D should be clearly oriented toward the adoption of new technological regimes in order to ensure that a region is gaining a competitive edge in a highly competitive international economic situation.

Regions may be regarded as the seedbeds for new entrepreneurial initiatives and at the same time as workfloors of innovation policies. Since however a fine tuning of technology development and regional policy is hard to achieve, we observe many failures in the field of regional technology policies. In many cases such policies are over-ambitious and hence almost bound to fail. Regional technology policy should not be considered to be a panacea for all regional evils, but should provide a focussed support to feasible regional development initiatives. Snell (1988) distinguishes three different stages in local government strategies:

- selling the local area to non-local manufacturing industries (e.g., by means of a system of local industrial officers);
- promoting growth and export in local SMEs, amongst others by developing information and training programmes;
- integration of all local policies and efforts in order to develop the local (or regional) economy; examples of such endeavours are: stimulation of information technology, or integration of job training and labour market policy.

In general, the focal points of regional policy seem to have shifted from direct invention towards the mobilisation of indigenous growth potential and creativity of areas.

In this context, new firm formation is but one of the many strategies for favouring the regional development potential. Although incubation policies stimulating the birth of new firms should not be underrated (see also Keeble and Wever 1986), it is at the same time clear that creative development policies should have a much broader focus: they should seek to induce and support all initiatives aiming at an efficiency increasing technological transformation. Industrial 'dynamism' at the local level needs of course a creative entrepreneurial climate and a favourable selection environment.

This book on all above questions of technology, space and policy tries to provide a broad spectrum for the above mentioned issues. Most contributions aim at providing a critical overview and synthesis of various thoughts and results in this field, rather than an indepth treatment of a given case study. Thus the focus is more on generality than on specificity, so that this book may be used as a source of reference for those involved in research in technology, space and policy.

References

Aydalot, P. (1985), Economie Régionale et Urbaine, Economica, Paris.

Bailly, A.S. (1985), The Service Sector as a Motor for Endogenous Development, Self-Reliant Development in Europe (M. Bassand et al., eds), Gower, Aldershot, pp. 101-122.

Bluestone, B. and B. Harrison (1982), The De-industrialisation of America, Basic Books, New York.

Camagni, R., and R. Cappellin (1985), Sectoral Productivity and Regional Policy, Office for Official Publications of the European Communities, Luxembourg, Document, ISBN 92-825-5535-6.

Cappellin, R. (1983), Productivity Growth and Technological Change in a Regional Perspective, Giornale degli Economisti, n. 7/8, pp. 459-482.

Cappellin, R. (1988), Transaction Costs and Urban Agglomerations, Revue d'Economie Régionale et Urbaine, no. 2, pp. 261-278.

Davelaar, E.J. (1989), Incubation and Innovation, Ph.D. Thesis, Dept. of Economics, Free University, Amsterdam.

Dosi, G. (1984), Technical Change and Industrial Transformation, MacMillan, London.

Friedmann, J. (1986), The World City Hypothesis, Development and Change, vol. 17, pp. 69-83.

Giaoutzi, M., and P. Nijkamp (eds.) (1988), Informatics and Regional Development, Avebury, Aldershot.

Kamann, D.J. (1988), Spatial Differentiation in the Impact of Technology on Society, Avebury, Aldershot.

Keeble, D., and E. Wever (eds.) (1986), New Firms and Regional Development in Europe, Croom Helm, London.

Lazerson, M.H. (1988), Organizational Growth of Small Firms, American Sociological Review, vol. 53, pp. 330-342.

Malecki, E., and P. Nijkamp (1988), Technology and Regional Development, Some Thoughts on Policy, Environment and Planning C, vol. 6, pp. 383-399.

Marshall, J.N., P. Damesick, and P. Wood (1987), Understanding the Location and Role of Producer Services in the UK, Environment and Planning A, vol. 19, pp. 575-595.

Molle, W., and R. Cappellin (eds.) (1988), Regional Impact of Community Policies in Europe, Avebury, Aldershot.

Momigliano, F. (1984), Revisione dei modelli interpretativi delle determinantie ed effetti dell'attivita' innovativa e della diffusione intraindustriale e territoriale delle innovazioni tecnologiche, in: R. Camagni, R. Cappellin and G. Garofoli (eds.), Cambiamento Tecnologico e Diffusione Territoriale, Milano, Franco Angeli, pp. 19-57.

Morphet, C.S. (1987), Research, Development and Innovation in the Segmented Economy: Spatial Implications, in: G.A. van der Knaap and E. Wever (eds.), New Technology and Regional Development, Croom Helm, London, pp. 45-62.

Moss, M.L. (1985), Telecommunications and the Future of Cities, Paper presented at the Conference on Landtronics, London.

Moss, M. (1987), Telecommunications, World Cities and Urban Policy, Urban Studies, vol. 24, pp. 534-546.

Nelson, R.R., and S.G. Winter (1982), An Evolutionary Theory of Economic Change, Harvard University Press, Harvard.

Piore, M., and C. Sabel (1984), The Second Industrial Divide, Basic Books, New York.

Porter, M.E. (1980), Competitive Strategy, MacMillan, London.

Rosenberg, N. (1982), Inside the Black Box: Technology and Economics, Cambridge University Press, Cambridge.

Scott, A.J. (1986), Industrial Organization and Location, Economic Geography, vol. 62, pp. 215-231.

Snell, H. (1988), Local Government Promotion of Economic Activities, Local Economic Development in Denmark (S. Illeris, ed.), AKF, Copenhagen, pp. 26-34.

Taylor, M. (1987), Enterprise and the Product-Cycle Model, in: G.A. van der Knaap and E. Wever (eds.), New Technology and Regional Development, Croom Helm, London, pp. 75-93.

Toffler, A. (1983), The Third Wave, Pan Books, London.

PART A
INDUSTRY

2 The location of industrial R & D: retrospect and prospect

A. THWAITES AND N. ALDERMAN

1. Introduction

Technological change has always been important to economic advance, releasing men in general from the limits to growth imposed by the simple expansion of other factors (e.g., capital, labour) and the exhaustion of easily extractable natural resources (Bacon, 1625; Defoe, 1797; Mill, 1862; Arndt, 1975). It is claimed by some to be the key factor underlying the waves of economic growth which characterise the economies of most advanced industrial nations (Schumpeter, 1934; Mensch, 1979; Freeman, 1985). More recently it has been recognised that there is a spatial dimension to the generation and adoption of new technologies which works to the advantage of some regions and countries of the world and to the disadvantage of others (Thwaites et al., 1981, 1982; Rees et al., 1983; Aydalot, 1986; Camagni, 1986; Brugger et al., 1987). As a consequence, it is argued that technological change may also be crucial to the conversion and rejuvenation of economically depressed regions in Europe and elsewhere.

However, technological change is not an homogeneous activity, but covers a spectrum of activities which have various interactive facets to them, ranging from the generation of scientific knowledge to the development and exploitation of new products, processes and systems in the market place. One such factor is research and development (R & D) activity; an activity which has become an increasingly professionalised input to the whole technological process as the twentieth century has progressed (Freeman, 1982). Substantial expenditures and employment in R & D are seen as one of the keys to participation in the new high technology industries and to the accompanying benefits, couched in terms of local employment, income multipliers, and so forth.

Denison and Poullier (1968), following the pioneering studies of Solow (1957), found that 'advances in knowledge' made a substantial contribution to economic growth when the contribution of other factors had been taken into account. More recently, evidence suggests that the richest industrial nations are those expending considerable sums on R & D effort, in both relative and absolute terms, although the direction of causation remains open to debate (OECD, 1986a, 1986b). Disaggregation of the national and international economies also suggests that this positive association between economic growth and R & D also exists at the industry level (Freeman, 1982) where recent OECD evidence has shown that R & D intensive industries grew fastest within the OECD manufacturing industrial system over 1970-80 period (OECD, 1986a, 1986b). Similarly, Odagiri (1985) has concluded that R & D activity is positively associated with sales growth in Japanese industry. R & D is also found to be associated with export success (Rothwell, 1979, 1980). Moreover, as early as 1968, Mansfield showed that growth of productivity was significantly related to the rate of growth of cumulated R & D expenditures made by firms or industries (Mansfield, 1968) and more recent work by the OECD (1986a, 1986b) has confirmed the validity of this association for individual industrial sectors. Evidence also shows that R & D advances are a source of profits and associated over time with their scale (e.g., Baily, 1972; Nolan et al., 1980). Thus, investment in the mechanisms for the generation or adoption of new technologies appears on average to bring about increased prosperity in innovative firms and industries, providing a basis for maintaining a competitive lead in relevant markets.

Whilst it is clearly the opinion of many industrialists and policy makers that the expansion of R & D expenditures and employment, most notably in the private sector, brings competitive advantages in the market places of the world, there remains sufficient uncertainty surrounding the conclusions of past academic research to suggest there is room for further study. Moreover, the changing economic conditions of the 1980s suggest that existing research and current approaches to our understanding of the link between R & D and local economic development may need to be revised, together with the methods of enquiry. The remainder of this chapter is devoted to raising some of these issues for debate. While industrial R & D is a subject of study in its own right, in this chapter we discuss it in the context of its role in regional or local economic development and particularly development based on manufacturing industry.

The next section discusses problems of definition and measurement, which, we argue, hamper our understanding and interpretation of the spatial dimension to R & D. This is followed by a brief examination of the national, regional and sectoral distribution of R & D. Our current understanding of factors influential in determining R & D location is reviewed in section 4. In section 5 we consider some of the likely impacts on R & D of major structural economic changes being experienced by the advanced industrial nations and the mediating role played by information and communications technologies, while section 6 considers the specific case of industrial R & D within small and medium size enterprises. In section 7 the policy response to these developments by both the public and private sector is explored. The concluding section draws out the implications for the location of R & D in different types of region and identifies a number of research needs.

2. Definition and Measurement of Industrial R & D

R & D is regarded as an important input to the process of technological change. There is, however, a high degree of uncertainty about the quantity, quality and spatial impact of research outputs in terms of subsequent economic processes. At the national and regional levels this technological potential is often judged on the number of R & D workers employed locally and/or R & D expenditure. The difficulty with this is that it only reflects R & D imputs that are formally recognised as such. As Charles states, our concerns in understanding the role of R & D in local economic development lie

> "... with those processes and mechanisms whereby the existing state of technology is redefined, developed, altered, added to and subsequently diffused and brought into use. In this definition technology includes knowledge of a practical and theoretical nature, physical devices embodying that knowledge and disembodied expertise in the implementation of knowledge into the production of physical devices and other applications. This intangible collection of knowledge is the product of R & D and therefore the reason for definitional problems. If someone contributes to this knowledge is he an R & D worker?" (Charles, 1988a, p. 4).

If the answer is "yes" to the last question then we need to take account of other personnel and expenditures in our traditional compilation of R & D figures. In a period of expanding interest in the small firms sector we must also be conscious that many owners/executives find great difficulty in providing accurate figures for their R & D efforts. Of more fundamental importance perhaps is the fact that the nature of research itself is changing. Not only is R & D funded specifically to support existing facets of a business, but often developments elsewhere in the organisation give rise to new types of product or service. The most notable example of this is the development of computer software, which may arise from the R & D department, or may be developed in other areas, such as production engineering or data processing. Such software, in providing solutions to specific manufacturing problems, is increasingly being recognised as marketable in its own right, but such developments go largely unrecognised in normal R & D accounting frameworks.

The assembled figures are left as only crude indicators of activity or potential, precisely because they fail to capture the nature of the process of technological advance in enterprises and locations. We need to understnad the sources, flows and costs of the different types and qualities of knowledge required by establishments and enterprises in specific locations and the impact these have on the competitive situation of the firm. This is particularly relevant in a period of corporate and institutional restructuring. The key measurement therefore is the degree of access and the terms of access to technology-related information, and the way it is incorporated into local economic activity, rather than simple counts of R & D workers and expenditures. Therefore, whilst efforts to improve these statistics should be made, they need to be preceded by studies which contribute more to our understanding of process. Such studies would perhaps allow us to identify those key elements of the process we need to measure, and how this could be achieved. Similarly, measures of R & D output, such as patents, or numbers of innovations need to be re-evaluated, because in many cases they provide only imperfect surrogates for actual R & D activities (Parker, 1974; Pavitt, 1982).

3. The Location of R & D

The previous section indicates that the precise definition of R & D and its appropriate measurement remains problematic. Nevertheless, the best available statistical indicators suggest wide national and regional variations in the location and extent of R & D. In this section three indicators are used: Gross Expenditure on R & D (GERD), Business Expenditure on R & D (BERD) and employment in R & D.

Green (1986) suggests that R & D effort grew quite strongly in the industrial world in the 1950s and 1960s with some retardation in subsequent time periods. Table 1 demonstrates that in absolute terms the United States dominates gross world R & D expenditure and most noticeably in terms of share of private investment in R & D. In 1983 the US spent $88bn on GERD, $62bn on BERD and employed nearly 700,000 researchers in all fields of science (OECD, 1986a). Other countries engaging in relatively, and absolutely, high levels of R & D include Japan and Germany. In Europe the research field is dominated by Germany, France and the UK who between them employ 75% of R & D manpower recorded in all member states of the European Community (EEC, 1987b). Table 1 also suggests that GERD, BERD and R & D

TABLE 1 Gross Domestic Expenditure on R & D (GERD), Business Expenditure R & D (BERD) and R & D Employment in Selected OECD Countries

	GERD			BERD			R & D Employment	
	1981		1983	1981		1983	1981	
	Million $	Percentage of total[b]	Million $	Million $	Percentage of total[b]	Million $	Full-time equivalent	Percentage of total[b]
United States	73 678.0	46.4	88 329.0	51 810.0	50.2	62 816.0	683 700	41.8
Japan[c]	25 574.5	16.1	33 493.7	15 517.3	15.0	21 270.0	392 625	24.0
Germany	15 644.8	9.9	18 130.2	10 686.3	10.4	12 649.0	128 162	7.8
France	10 700.8	6.7	13 134.4	6 304.4	6.1	7 461.1	85 500	5.2
United Kingdom[a]	11 369.8	7.2	12 552.8	7 029.7	6.8	7 662.9	104 445	6.4
Italy	4 546.6	2.9	5 568.0	2 563.1	2.5	3 179.0	52 060	3.2
Canada	3 877.3	2.4	4 619.2	1 902.9	1.8	2 168.0	29 976	1.8
Spain	908.0	0.6		442.7	0.4		14 376	0.9
Netherlands	2 508.4	1.6	2 992.4	1 336.0	1.3	1 607.2	19 436	1.2
Sweden (n.s.c.)	2 166.7	1.4	2 776.8	1 442.1	1.4	1 874.6	15 035	0.9
Belgium[d]	1 065.7	0.9		950.1	0.9	1 148.7	10 943	0.8
Switzerland	1 785.4	1.1	1 979.8	1 324.8	1.3	1 471.0	10 720	0.6
Austria	765.2	0.5	932.9	427.4	0.4		6 712	0.4
Denmark	539.8	0.3		273.8	0.3	372.3	5 988	0.4
Norway	593.0	0.4	757.6	309.1	0.3	414.2	7 496	0.5
Greece	102.1	0.1		22.9	0.0		2 634	0.2
Portugal[e]	154.5	0.1		48.3	0.0		3 019	0.2
Ireland	155.3	0.1		67.7	0.1		2 636	0.2
Total OECD	158 720	100		103 120	100	1 634 880		100
of which								
EEC[a]	46 940	29.5		29 230	28.5	414 640		25.5

Source : OECD/STIIU data bank - December 1985.
a) OECD estimates; b) Secretariat estimates are provided where 1981 figures are not available.
c) Data adjusted by OECD would be 23 408 million $ in 1981 or 15% of the OECD total, and in 1983, 30,750 million $.
d) 1979; e) 1982.

employment have been growing over the 1970s for most nations. This is particularly true of Japan, which has thus increased its standing in the research world. Countries such as France, Italy and Sweden have also devoted more resources of late to R & D, whereas the UK and Switzerland show slight declines in R & D expenditure and employment over recent years (see also Bosworth, 1979; Buswell, 1983). Smaller countries have also devoted greater proportions of resources to R & D, but as they start from a relatively low base their impact at the international level may be restricted. Indeed, a major problem for smaller countries may be their inability to overcome barriers to entry in research fields dominated by at least one of the larger and higher spending countries.

A further difference between nations lies with the relative shares of government and industry research effort. In Greece and Portugal, for example, 60% of research is performed in the government sector and as much as 80% of funds are received from the public sector to undertake this work. The business sectors in these countries are undertaking and funding very low levels of research. At the other end of the scale, in Switzerland, Belgium, Japan and Sweden, industry undertakes and finances nearly two-thirds of all research. The US lies between these extremes with some 50% of total R & D funding coming from government, but over 70% of research carried out by industry, implying that government funds are very important to the level of industrially based R & D in that country.

Any data and measurement difficulties at the national level are exacerbated at the regional or local level. In some countries, such as the UK, regional figures for GERD, BERD and R & D employment are not available. However, a recent study for the European Commission (EEC, 1987a, 1987b) attempted to collect data on GERD, BERD and R & D employment for the Level II regions within Member States. While there are many comparability and missing data problems, these data showed, against a Community-of-twelve average, the dominance of regions such as Ile de France, Midi Pyrenees, Auvergne, Koeln, Darmstadt, Oberbayern, Mittelfranken and Utrecht. At the other end of the scale there are numerous regions in Spain, the whole of Portugal with the exception of the Lisbon area, Greece, Luxemburg (B), Sardegna, Calabria, Puglia, Molise, Abruzzi, Marche and Trentino - Alto Adige with an indicator of 20 or less against an Euro 12 average of 100 in terms of BERD/GVA (Gross Value Added) and/or GERD/GDP.

Notwithstanding the poor position of the less favoured regions (LFRs) in R & D terms, some recent evidence for the mechanical engineering industry in the UK suggests that some of the LFRs might be marginally improving their relative position, at least in terms of the proportion of establishments undertaking some form of R & D (Alderman and Thwaites, 1987). Moreover, in other countries, such as France, strategies to decentralise R & D may be having some effect (EEC, 1987a). What seems noticeable is that, despite variations in the scale of research effort, there is a tendency in most nations for R & D effort to concentrate around major cities, e.g. London, Athens, Paris, Lisbon, etc. (EEC, 1987b). Similar conclusions for the location of R & D near major cities were also reported in the United States (Malecki, 1979).

4. Factors Influencing the Location of Industrial R & D

To some degree the paucity of data at the local level reflects the fact that until the 1970s relatively little work was undertaken on the location of R & D within countries and the reasons for its pattern of distribution. Until that time,

> "Invention was therefore still essentially seen as an autonomous process associated with the spontaneous, almost random, contributions of individuals. As such, general 'supply' and 'demand' factors were seen as the primary determinants behind the pattern of inventive activity. Population size was seen as a key (supply) influence, since the larger the urban or regional population the more geniuses and therefore the greater the creative or innovative potential. Demand factors were associated with the initial demand for new inventions being greater in larger cities and more particularly with the process of urbanisation itself generating a new set of technological demands. Additional factors were also put forward. In particular the pattern of interaction and information flows which were seen as reinforcing the existing urban and regional distribution of inventive activity, centred on the most densely populated areas. The distribution of manufacturing activity was also recognised as a factor influencing the developing pattern of technical knowledge and innovation, although it was seen essentially as a demand factor. Thus a high concentration of manufacturing activity would serve to encourage the innovative process in a locality" (Howells, 1984, p. 15).

More recent work on location by Malecki (1981), Howells (1984) and Buswell et al. (1985), concentrating on manufacturing industry, suggests that the location of R & D is determined largely by an interaction of numerous factors internal and external to the enterprise.

The corporate organisation of R & D

Buswell et al. (1985) remind us that the largest proportions of privately funded research are conducted in large firms. In the UK at least, this situation is exacerbated by the proportion of government funding of research which goes to a few large corporations (Table 2). Research effort

TABLE 2 THE 100 ENTERPRISES WITH THE LARGEST R & D EXPENDITURE IN THE UK IN 1985

	R & D EMPLOYMENT	INTRAMURAL EXPENDITURE ON R & D £ MILLION	GOVERNMENT FUNDS FOR R & D £ MILLION
	AS A PERCENTAGE OF THE TOTAL IN THE SIZE ANALYSIS		
Enterprises with the largest expenditure on R & D			
First 5	29	32	58
First 10	43	48	71
First 15	50	57	79
First 20	60	64	83
First 50	75	80	92
First 100	84	89	95

Source : Business Monitor M014 (1985).

Note : The size analysis covers 98.4% of total R & D employment, 98.9% of intramural expenditure and 99.7% of government funds for R & D.

in world terms is dominated by a few multinationals (MNCs). "In 1982 the top 10 US-based MNCs spent more than $13bn on R & D, more than the GNP of many nations" (Grant, 1983, p. 249). While these companies directly invest in many countries, their engagement in R & D in those recipient countries is lagged, relative to warehousing and production activities, with a priority given to adapting existing MNC products to the local market rather than undertaking research into new products (Vickery, 1986). In countries like Australia, Canada and Ireland this type of R & D activity makes up 40 to 50% of the national total Business Expenditure on R & D. Comparison between investing and foreign direct investment (FDI)-recipient countries suggests that the proportion of total employment in R & D remains inferior to that of the investing country. Further research is required if we are to determine the contribution this makes to local economic development in recipient countries; however, there is less likely to be benefit to LFRs if MNCs increasingly invest in core rather than peripheral areas (Smith, 1986).

Research suggests this tendency to undertake larger proportions of total research effort near to home base is a characteristic of the national as well as the multinational corporation (Howells, 1984; Malecki, 1979, 1985). It is suggested that factors internal to the corporation are the key to understanding the location of R & D within a country and its distribution between regions. Many large corporations centralise research on one or a number of sites, depending on whether they have relatively simple organisational structures and product lines or are more complex and multidivisional, with multi-product portfolios (Howells, 1984). Examination shows that many major R & D sites are the result of the historical development of the enterprise and its research activities. After natural growth, or through takeover activity, merger and rationalisation procedures, these firms have ended up with a limited number of major R & D sites and perhaps some distribution of R & D in selected operating divisions. Executives claim that centralisation is justified on grounds of economies of scale in equipment and specialised research personnel, and because of the formal and informal synergetic relationships which emerge within large research units, resulting in economies of scope and "interesting ideas", both planned and serendipitous (Gibbs et al., 1985). Such units usually tend to undertake both basic and applied elements of research for the company, servicing both total company needs and the specific needs of separate divisions or plants.

On the other hand, development work is freqently more widely, but not necessarily evenly, distributed between group manufacturing establishments.

In the UK, the corporate organisation of R & D is reflected in the method by which R & D is funded, whereby funding of R & D appears to come from internally generated funds rather than external sources of capital (perhaps because of the inherently risky nature of such activity). The research at centralised units is usually funded from central corporate sources, while product or process related work carried out by the centralised unit for individual plants is funded from the plant's own profit and loss account (Gibbs et al., 1985). The nature of on-site R & D at production plant level is to a large extent governed by the content of that plant's product range (Morphet, 1987). This in turn is governed to a large extent by corporate policy. Where this involves standardised products then R & D is more akin to customisation or process (productivity) improvement. In the Thwaites et al. (1981) survey, establishments located in peripheral regions were more dependent on centralised corporate resources than plants located in core regions.

There is also a dynamic in this historical arrangement of types of production and R & D. One possible model suggests that new products will tend to be located within the plant that originates the ideas and has the R & D **and production capability** to exploit them. "Free products" (i.e. those emerging directly from central R & D or through licensing) go to the most appropriate plant for production, which again must satisfy the research and production criteria (Gibbs et al., 1985). Only if an existing site is unavailable will a greenfield site be considered. In a time of growing internationalisation of production, joint ventures and collaborative research agreements, the less technologically capable plants of the LFRs may well miss out altogether.

To a large degree, the organisation of research until the 1980s would appear to have been relatively stable for most companies. Corporate allocation of resources to R & D were freqently made on a vague notion of "resources allocated as necessary". Some resources appear to be decided historically in that some projects (e.g. in pharmaceuticals) last for many years (Howells, 1984). A considerable amount of expenditure is already decided upon through existing staff levels and required specialisations (indivisibilities) which introduces a certain amount of inertia if a change in research direction is required. However, without substantial change brought about through corporate decisions to alter the technological and related capacities of individual plants, existing patterns of R & D activity can be expected to continue. Mobility of R & D is usually desired by policy makers and even supported by individual executives, but is really only feasible in specific circumstances. As Morphet (1987) states, there is little likelihood of locating R & D in peripheral areas in which the corporation has no production, or only production which cannot use the output of the research.

However, the 1980s have brought rationalisation of many functions in manufacturing industry, including R & D. Rationalisation and the breaking up of traditional work practices and procedures is sometimes easier on a new site and encourages relocation (Clark, 1980). A second reason for movement of R & D is expansion of research activity (e.g. in electronics), especially if the existing site is constrained and, more notably, valuable (e.g. a capital city location). A third reason for movement arises if the company needs to move into a new area of research which does not fit comfortably into its existing research framework. It will first try to locate on an existing corporate site, which could be in a LFR if there is a centre of excellence in the particular research field already located in that region (e.g. government or HEI centre). There have also been some movements of R & D staff in response to a demand for closer co-operation between R & D and marketing (and in some cases production); for example, in in the UK there is the case of TMC moving from Swindon to Glasgow.

Howells (1984) also examined the external and environmental factors influencing the location of R & D. He suggests these are less important than internal factors in explaining existing patterns, at least in Britain. He found the availability of suitable sites and premises important to a large proportion of firms (60%), but only the main location factor for 13% of his sample. Surprisingly, the communications factor was of little importance, which other research (Gibbs et al., 1985) suggests is because key R & D staff are used to travelling in any case. Howells' work also confirms that industrialists show relatively little interest in locations close to other research centres.

Personnel and the location of R & D

On environmental grounds, peripheral or LFR locations for R & D do not appear to be necessarily disadvantageous. However, one stumbling block to participating in R & D in the LFRs is the quality of the local labour market and the ability to attract key research workers to reside there.

R & D is essentially a human activity and the quality of R & D output will be highly dependent upon the quality of R & D personnel. In many industries labour costs form the largest component of R & D expenditures: often more than 50 per cent. It is not surprising, then, to find labour cited as a key location factor in studies of high technology industries, which are usually research intensive by definition (for example, Oakey, 1979; Malecki, 1980; Premus, 1982; Rees et al., 1983; Scott, 1986).

The personal preferences of R & D workers are believed by academics and industrialists alike to be a major locational constraint upon R & D activity (Buswell, 1983; Gibbs et al., 1985), although this is not true of all sectors (see Charles, 1988a). Important components of these preferences appear to be access to amenities, alternative job opportunities, expecially for spouses and siblings, and access to the professional institutions (see also Herzog et al., 1986). These particular conditions thus imply urban (Malecki, 1979) or core region locations.

It is not just the locational preferences of R & D workers themselves that act as a powerful constraint on R & D location. It is estimated that scientists and engineers spend between 50 and 75 per cent of their time in verbal communication, an activity that assumes greater importance in the context of R & D (Tushman, 1979). There is a significantly large literature to indicate that, amongst the people within an R & D unit, there are key actors, often referred to as "technological gatekeepers", who are responsible for a large proportion of the information that enters a research unit from its external environment (Allen, 1977; Epton, 1981). It is to be expected that these technological gatekeepers will be more effective in locations where R & D activities are concentrated and where the opportunities for face-to-face contact are greater.

In practice, Gibbs et al. (1985) found that the importance of key research staff to the organisation was a primary deterrent to the relocation of R & D units. This was not due to the resultant increased costs of interpersonal interaction, which peripherally located R & D units appeared willing to bear, but because of the fear of losing the personnel involved. Such key staff can not be instantly hired elsewhere (Myers, 1983).

Core locations, such as the South East of England, whilst offering advantages for R & D activities through well-developed interpersonal contact networks, have disadvantages in that there are, by the same token, plenty of competing job opportunities amongst rival firms and a considerable amount of labour poaching goes on (Macdonald, 1986). This, however, tends to be tolerated, no doubt partly due to the perceived benefits of transferring knowledge "on the hoof" (Ettlie, 1980).

Corporate R & D employment policy, frequently forsaking possible long term gains for the avoidance of short term disruption, thus tends to perpetuate existing patterns of R & D location. Empirical findings from research in the Centre for Urban and Regional Development Studies at the University of Newcaste upon Tyne, UK, suggest that distinctions in R & D employment exist between centralised (core region) R & D units and production plant (often peripheral) locations. This reflects the particular "mission" of each unit. Typically, staff at centralised units are given more freedom to pursue "academic" activities appropriate to purer scientific research and they receive better fringe benefits. At production plant level staff grades are often lower, reflecting the more directly commercially oriented type of R & D prevalent there.

Peripheral production plants are therefore likely to be less attractive to key research personnel for a variety of reasons. This is in addition to the oft-quoted difficulties of attracting research workers into a manufacturing environment and into older and less "glamorous" industries, such as mechanical engineering, in particular. Again, however, this is not true of all disciplines. Locational choices for some researchers such as metallurgists or chemists, are constrained by the location of the leading companies [e.g. ICI on Teesside (Charles, 1988a)].

5. Structural Economic Change and the Location of R & D

The need for a review of the characteristics and direction of research into R & D and regional economic development is prompted in part by the changing nature of the industrialised economies, and the implications this has for R & D activity. A significant feature of the 1980s has been the world recession, particularly affecting manufacturing industries in the advanced industrial economies where there has been considerable restructuring, and the increasing attention being paid to competitiveness based upon technological change. A parallel feature of the world economy over the past two decades has been the growing internationalisation of production and service activities in multinational corporations and increased flows of commodities and services across national boundaries, both within and external to the corporation, including research and business services.

Moreover, one of the principal dimensions to this currently unfolding structural economic change within advanced capitalist nations is what is regarded by some as a change from a mode of accumulation based on mass production towards a mode of "flexible accumulation" (Harvey, 1985). Under the mass production regime known as Fordism the fragmentation of the production process in accordance with Taylorist principles reached its peak. In particular, conceptual and physical tasks became dissociated, such that R

& D and production were separately located according to their different labour requirements (Massey, 1984).

In response to the current perceived crisis in Fordism (Roobeek, 1987), brought on through market saturation (Piore and Sabel, 1984), market fragmentation (Anderson, 1985) and increasingly volatile demand conditions (Mensch, 1979), a greater flexibility is required on the part of many manufacturers. R & D and technological change are key factors in facilitating this movement towards flexibility of response. Closer relationships are necessary between suppliers and customers, with greater attention to market response times if other competitive advantages are not to be lost. It has also been suggested that the life cycle of products in certain industries is decreasing (e.g. Goldman, 1982), challenging R & D staff to provide the basis for a flow of new or improved products onto the market. The notion of modularisation may provide one possible design answer to this problem of maintaining a flexible response.

These changing circumstances, we would argue, mean that it is necessary to understand the changing role and nature of R & D in this context and to examine the extent to which new market conditions are impinging upon the process of R & D in individual establishments and enterprises. It is conceivable, but by no means inevitable, that the requirements of increased flexibility will lead to a much more integrated role for R & D within individual production facilities and an erosion of current spatial divisions of labour. Alternatively, we may see an increasing polarisation between basic (centralised) research activities and development (production-related) activities geared towards flexible response to customer demands.

As already stated, R & D patterns have appeared relatively stable, but the economic shocks of the last decade have resulted, in part, from the emergence of new technology, notably the integration of information and communication technologies (ICT) which permit the extended control over, and reorganisation of, existing economic activities and lie at the heart of many totally new products and services. As a largely information-based activity it would be unlikely for R & D to be untouched by these activities (Charles, 1988b).

Hepworth (1986) (see also Williams and Gillespie, 1987; Taylor and Williams, 1987; Gillespie and Williams, 1988a, 1988b) argues that corporate restructuring is taking place at an accelerating rate as a result of the opportunties provided by Local Area Networks (LANs), Wide Area Networks (WANs) and the increasing availability of useful commercially and technologically relevant on-line information systems (see Howells, 1988a,

1988b). The integration of these activities will allow corporations on a national and increasingly global scale to control

> "... the entire process by which goods and services are produced and distributed" (Hepworth, 1986, p. 422).

Hepworth forecasts:

> "... computer networking will figure as the key process technologies through which re-industrialisation of manufacturing economies and 'automation' of office economies will come about";

and that

> "... Spatial analysis of these transformations...will produce a new geography of economic opportunity" (Hepworth, 1986, p. 422).

It is these opportunities which permit the globalisation of activities and the spatial divorce of capital and information sources from complementary labour inputs. This provides the opportunity for a new geography of R & D effort, where specialisation and perhaps economies of scale and scope are made available 'at a distance' and no longer need to be replicated or concentrated in a few locations. In theory at least, there are growing opportunities to access the inputs and outputs of R & D on a national and international scale. Such access could benefit LFRs. However, this may not work out in practice as formal (Single European Markets) and informal (CoCom) regulation affects access and information flows (Macdonald, 1988). There is obviously a need for research to establish how the present nature and patterns of R & D will be affected by ICT and its related activities and how these changes will impact on local economic developments.

A case study of Bell Northern Research (BNR) of Canada illustrates some of the issues for research and debate (Hepworth, 1986). BNR concentrates on telephone switches, circuit-based design and office automation equipment. Major customer needs are satisfied from a widely distributed R & D workforce which maintains close links with major clients. The establishment of a computer based WAN and a layered approach to product development allows each laboratory to specialise in only one part of the product, but to integrate the work of the other units to provide the final product to the customer. This system is labour saving, secure, and avoids the international border customs checks usual on physical goods. Nevertheless, it has allowed the concentration of R & D to remain in Ottawa, Canada, despite the increasing value and complexity of the research activity. Other laboratories are supported by the Ottawa "office" over the network. The effect has been to centralise the major core of new product development with distributed customisation in a rapidly changing world.

This particular case study provides some evidence to suggest that the mediation of ICT in the organisation of R & D could lead to both of the locational scenarios postulated above. Specifically, whilst on the one hand there is increasing local specialisation in research activity to provide efficient and flexible market response, there is at the same time increasing centralisation of core technologies. Clearly the spatial outcomes of increasingly networked R & D activities are going to depend upon a great many corporate organisational decisions and the result of different individual and company approaches to R & D.

6. Small Firms and the Location of R & D

The discussion in the foregoing section has essentially been concerned with the corporate organisational and geographic patterns of R & D in large, multi-regional or multi-national enterprises. To a large degree this reflects the general paucity of evidence surrounding the research activities and linkages of the small or medium size independent enterprise (SME).

The concentration on assembling data on professionalised and specialised research usually found within identifiable units, or on individual employees occupied on research 100% of their time, has tended to undervalue the contribution made to research by the small firm sector. While many employ full-time research staff, others frequently combine the research task with other responsibilities within the enterprise: management, engineering, etc., and as a result go unrecorded in official statistics or even survey data (Oakey et al., 1987). This perhaps helps to explain - in part - the finding of the Bolton Committee (HMSO, 1971) that small firms obtained a better return per unit of input than did larger firms. It also, perhaps in part, accounts for the low incidence of R & D frequently recorded in SMEs in comparison with their larger counterparts within the same industries, but does not account for the spatial variations in R & D activity in SMEs within the same industries (Thwaites et al., 1981, 1982; Oakey, 1984; Alderman, 1986). This evidence suggests that SMEs located in areas peripheral to core economies may be less inclined to invest in R & D (Thwaites et al., 1981). They may of course be more inclined than their larger counterparts to substitute internal research by using external sources of knowledge useful in their innovative or adoptive processes.

Oakey (1984) however, in his survey of high tech SMEs located in the UK and the United States, found that few firms maintained what they considered to be important technical links with outside sources and only 14 per cent thought that severance of such links would seriously damage their innovative performance, while Thwaites et al. (1981) found SMEs in peripheral areas of the UK tended to have fewer external links than their core located counterparts in the same industries. Contrary to the generally held opinion, Oakey found that the highest proportions of technical information contacts were with institutions of higher education; a particularly strong feature of Scottish and San Francisco Bay located enterprises. Sweeney (1987) argues that small firms frequently have information linkages on an informal basis and between a few individuals within the firm and the external environment, which tends to lead to under-estimation of their involvement in technological networks. These networks, which are more about information transfer between individuals rather than between organisations, are heavily dependent upon personal and face-to-face contact. If these personal linkages are broken the network tends to collapse. While stressing the individuality of firms, Sweeney suggests that the higher the technical intensity of the firm the greater will be its external contacts and that a very high use is made of information gleaned from supplier and, in particular, from customer sources (see also Rothwell and Zegveld, 1985). As Sweeney states: "The heaviest flow of information is that from firm to firm" (p. 154). In the case of the SME the supplier and customer are frequently large firms. While Taylor and Thrift (1980) have provided a conceptual framework for exploring the links, relationships and dependencies between large and small firms, there have been surprisingly few empirical tests undertaken of their hypotheses and of the directions of flows of technological information in particular.

On the evidence provided, the conclusion is reached that the research effort of SMEs is probably under-estimated, that they might rely on less obvious and informal systems of learning to gain technological expertise and that a high proportion of this may come from other enterprises, particularly large enterprises, acting as suppliers or customers.

We have little knowledge of whether or not SME research is similar to or different from that undertaken in large firms, has the same or alternative goals, employs similarly qualified staff, etc. The research outlined above does suggest, however, that we will not discover the answers to these questions, or the reasons for the level and nature of SME research organisation and technical information networks, without understanding the wider process of technological change which links SMEs and other firms and in turn links the technological outputs to economic outputs, at a variety of corporate industrial and geographic levels.

7. Public and Private Sector Policy Responses

The responses to the economic and technological challenges of the 1980s in both the public and private sectors of economies have been varied, with largely speculative implications for the structure of R & D and its locational possibilities, in particular from the perspective of existing disadvantaged areas. Observation would suggest that, in the absence of strong ameliorative policy, new processes of technological change will only include LFRs as recipients of their output (for example, Williams et al., 1986).

An obvious response from the public sector, either at national or local level, has been the identification of specific areas of technology for special encouragement and support. These include electronics, biotechnology, new materials, etc. They are promoted across the world from Japan to Ireland, from France to the USA. Targeting is also extended to cover anything regarded as high technology (and by implication research intensive) and therefore, virtually by definition, found only to a limited extent in LFRs. Such programmes therefore tend to mitigate against participation by LFRs, where different industries or even service activity may be dominant. What is perhaps under researched is the link between the locations of producers and users and the extent to which a dynamic user sector is necessary to drive forward 'high-tech' sectors (see Carter and Williams, 1957; Pred, 1976; Hippel, 1977).

The public and private sectors have come together in Europe to offset technological challenges from the US and Japan by setting up collaborative ventures in R & D, such as ESPRIT, encouraging the participation of both large and small firms from research intensive and research constrained regions of the Community. What remains largely untested is whether or not such activities divert scarce resources in LFRs away from other research projects of more fundamental importance to, and long term impact on, the local areas; moreover, whether or not there is a natural loss of sovereignty over research outputs because there is no local capability to exploit any output that emerges. Anecdotal evidence suggests that this may be the case in certain areas of the UK and elsewhere (EEC, 1987a).

In many of these research areas, and perhaps in part because of public sector interest, competition is intense. The costs of participating in certain 'high-tech' areas are rising rapidly, to the extent that most enterprises and even some countries cannot support the necessary expenditure. LFR-based enterprises may therefore be barred from entry to large parts of the new industries. As a result, we have witnessed the establishment and extension of joint research ventures and agreement, outside of policy initiated agreements, (Williams et al, 1986; Charles, 1988b) linking enterprises around the world. The majority of enterprises and individual establishments party to these agreements appear not to be located in LFRs. Research is needed to investigate the forms of these agreeements between large firms and their implications for small firms and local area development (see Charles, 1988b), particularly in the light of the Single European Act (EEC, 1987c).

A feature of recent economic change has been the pressure in Europe, and in particular the UK, to improve HEI to industry links in the belief that this will improve the competitive position of industry nationally and internationally, whilst reorienting HEI research towards more practical outcomes, and providing the latter with much needed revenue. Indeed, within the UK there is a move to privatise elements of government research with a view not only to increasing competition, but also to release into the market place scarce manpower resources currently locked up in publicly funded institutions. However, some nations, regions, enterprises and institutions will be able to respond more positively and more effectively than others with the result that the effects are likely to be spatially differentiated. Although considerable research has been carried out to show the extent of (or limits to) HEI relevance to industry in the innovation process (Oakey, 1984; Howells, 1986), there is a need for research to establish who, what and where benefits from such collaboration and what are the possible dangers of vigorously following such a route. For example, Flood (1984) suggests that such links forced on HEIs for economic circumstances may jeopardise their long term future in research, because they are likely to be unstable short term, rather than stable long term, programmes of work. Any selectivity in HEI research programmes, or in any other public sector organisation, is always open to the debate of who makes the selections and on what criteria. There is always the danger that this will be left to the "great and the good" and that they will recommend research in intensely competitive fields with higher associated risks. In this respect organisations such as Japan's MITI appear to have been quite successful in their selection strategies.

8. Conclusions

R & D has been identified as an important component of the process of technological change, which much evidence suggests plays a part in economic development. Notwithstanding the problems of definition and measurement we have discussed, past research has clearly demonstrated the uneven spatial distribution of R & D activity, both in terms of quality and of quantity, to the apparent detriment of some nations and regions. Whilst the R & D activities of some sectors, and of small and medium size enterprises in particular, are probably underestimated, current regional imbalances would seem to be well entrenched.

The future prospects for different types of region are likely to depend upon their position within increasingly globalised corporate structures of organisation and control. Thus, existing spatial concentrations of R & D, reinforced by favourable labour market conditions for R & D workers, current patterns of Government R & D contracting (particularly in the military sphere) and inertia within organisations, can be expected to assume greater significance, at least in the short to medium term. Current evidence suggests that these patterns may also be influential on the scale and nature of technical information flows towards the SME sector. Without additional research we cannot establish the impact changes in SME-based research can have on national or local technological and economic development.

For the LFRs, much will depend upon the choices made over the implementation of ICT and the networking of R & D between institutions and within corporations. Whilst there is a window of opportunity that could allow LFRs to tap into R & D information currently denied them, there is also a danger of increasing centralisation of the research function, with peripheral locations continuing to be locked out of the system.

An understanding of corporate policy and the ways in which R & D is used as a means of effecting swift market response and achieving increased requirements for flexibility, in the face of market fragmentation and shortened product life cycles, is necessary if we are to understand the role R & D in large and small firms does (or should) play in relation to the production process in different types of location.

To a large extent, past studies have been concerned with pattern rather than process; content to tally numbers of workers and sums of money, rather than attempting to establish precisely how R & D takes place and what impact this has on the economic development of the individual establishment and its local area. We would suggest that attention now needs to be directed to research into the process of R & D as the pace of technological change quickens. Remote observation of static patterns needs to be replaced with participant observation of dynamic processes.

Such research needs to identify the nature and flows of information into the R & D process, the context of crucial decisions and their impacts upon technical, organisational, economic and, ultimately, locational outcomes. Key issues that need addressing are the impacts of ICT; the targeting of specific technologies; moves towards deverticalisation and specialisation; the interface between small, medium and large enterprises in technological terms; the reorganisation of corporate functions and their interface with the market; and the rising costs of R & D and the importance of joint ventures and technology transfer.

All these issues are intrinsically dynamic and will have different implications for different locations. These implications will only be identified and understood if we can develop appropriate methodologies that allow us to get to grips with the dynamic processes of R & D and technological change.

Acknowledgements

We should like to acknowledge the assistance we have received in the preparation of this paper from all staff and secretaries in CURDS and in particular from David Charles, Jeremy Howells, Chris Pywell and Denise Weites. All errors remain the responsibility of the authors.

The financial support of the UK Economic and Social Research Council (ESRC) during the preparation of this paper is also gratefully acknowledged.

References

Allen, T.J. (1977), Managing the Flow of Technology, M.I.T. Press, Cambridge, MA.

Alderman, N. (1968), A case study of the application of log-linear and logit models in industrial geography. Ph.D. dissertation, Department of Geography, University of Newcastle upon Tyne, UK.

Alderman, N. and Thwaites, A.T., (1987), Research and development and technical change in traditional industries: a regional perspective. Paper delivered at the 27th European Congress of the Regional Science Association, Athens.

Anderson, C.G. (1985), Planning for restructured competition, Long Range Planning, 18:21-29.

Arndt, H.W. (1975), Limits to development, Aust Q 47:78-89.

Aydalot, P. ed. (1986), Milieux Innovateurs en Europe, GREMI, Paris.

Bacon, F. (1625), The Essays On Councils Civil and Morale, London.

Baily, M. (1972), Research and development costs and returns, Journal of Political Economy, 80:70-85.

Bosworth, D.L. (1979), Recent Trends in R & D In The United Kingdom, Research Policy 8:164-185.

Brugger, E. and Stuckey, B., (1987), Regional Economic Structure and Innovation Behaviour in Switzerland, Regional Studies 21:241-254.

Buswell, R.J., (1983), Research and development and regional development: a review, in: Gillespie, A.E. (ed.), Technological Change and Regional Development, London Papers in Regional Science 2:9-22, Pion, London.

Buswell, R.J., Easterbrook, R.P. and Morphet, C.S. (1985), Geography, regions and research and development activity, in: Thwaites, A.T. and Oakey, R.P. (eds.), The Regional Economic Impact of Technological Change, Frances Pinter, London.

Charles, D. (1988a), R & D Activity and Local Authorities - A Report for Cleveland County Council. Newcastle upon Tyne, UK: CURDS, University of Newcastle upon Tyne.

Charles, D. (1988b), The Globalisation of R & D: Technology and the Formation of International Strategic Alliances. Paper delivered at the ESRC, Urban and Regional Economic Seminar Group, Newcastle upon Tyne, UK.

Clark, G. (1980), The Employment Relations and Spatial Divisions of Labour. Paper delivered at NSF/ESRC Anglo-American Workshop, North Carolina, USA.

Camagni, R. (1986), The Need for a Regional Dimension to Research and Technology Development (RTD) Policies and Programmes with Particular Reference to the LFRs of the Community. Milan: University of Luigi Bocconi. Final Report to CURDS, University of Newcastle upon Tyne.

Carter, C.F. and Wiliams, B.R. (1957), Industry and Technical Progress,

Oxford University Press, London.

Defoe, D. (1797). An Essay on Projects. London.

Denison, F.E. and Poullier, J. (1968), Why Growth Rates Differ, Allen & Unwin, London.

EEC (1987a), Research and Technological Development in the Less Favoured Regions of the Community (STRIDE), Commission of the European Communities, Brussels.

EEC (1987b), Science and Technology for Regional Innovation and Development in Europe, Commission of the European Communities, Brussels.

EEC (1986c), Single European Act, Commission of the European Communities, Brussels.

Epton, S.R. (1981), Ten Years of R & D Management - some major themes: the role of communication in R & D, R & D Management 11:165-170.

Ettlie, J.E. (1980), Manpower Flows and the Innovation Process, Management Science 26:1086-1095.

Flood, J. (1984), The Advent of Strategic Management in CSIRO: a History of Change, Prometheus 2:38-72.

Freeman. C. (1982), The Economics of Industrial Innovation (2nd Edition). Francis Pinter, London.

Freeman, C. (1985), The role of technical change in national economic development, in: Amin, A. and Goddard, J.B. (eds.), Technological Change, Industrial Restructuring and Regional Development, Allen & Unwin, London.

Gibbs, D.C., Alderman, N., Oakey, R. and Thwaites, A.T. (1985), The Location of R & D in Great Britain. Newcastle upon Tyne: Final Report to DTI/EEC, CURDS, University of Newcastle upon Tyne.

Gillespie, A.E. and Williams, H. (1988a), Telematics and the Reoganisation of Corporate Space. Paper delivered at the International Symposium on Telematics, Transportation and Spatial Development, The Hague, Netherlands.

Gillespie, A.E. and Williams, H. (1988b), A Small Firms Perspective on the Liberalisation of Telecommunications Services. Paper delivered at the CPR 1988 Conference on European Telecom Policy Research, Windsor Great Park, London.

Goldman, A. (1982), Short product life cycles: implications for the marketing activities of small high-technology companies, R & D Management 12:81-89.

Grant, P. (1983), Technological Sovereignty: forgotten location in the "Hi Tech" razzamatazz, Prometheus 1:239-270.

Green, K. (1986), Research Funding in Australia: A View from the North, Prometheus 4:68-92.

Harvey, D. (1985), The Geopolitics of Capitalism, in: Gregory, D. and Urry, J. (eds.), Social Relations and Spatial Structures, Macmillan, Basingstoke.

Hepworth, M. (1986), The Geography of Technological Change in the Information Economy, *Regional Studies* 20:407-424.

Herzog, H.W., Schlottmann, A.M. and Johnson, D.L. (1986), High technology jobs and worker mobility, *Journal of Regional Science* 26:445-459.

Hippel, E. von (1977), The dominant role of the user in semiconductors and electronic sub assembly process innovation, *IEEE Transactions on Engineering Management*, EM-24:60-71.

HMSO (1971), Report of the Committee of Inquiry on Small Firms (Bolton Report), HMSO, London, Cmnd.4811.

Howells, J. (1984), The location of research and development: some observations and evidence from Britain, *Regional Studies* 18: 13-29.

Howells, J. (1986), Industry-academic links in research and innovation: a national and regional development perspective, *Regional Studies* 20:472-476.

Howells, J. (1988a), *Economic, Technological and Locational Trends in European Services*, Avebury-Gower, Aldershot.

Howells, J. (1988b), The Location and Organisation of Research and Development, Newcastle upon Tyne: CURDS, University of Newcastle upon Tyne, unpublished paper.

Macdonald, S. (1986), Headhunting in high technology, *Technovation* 4:233-245.

Macdonald, S. (1988), United States Export Controls and Technological Information. Paper delivered at the Conference on International Information Flows between Eastern and Western Europe, Arnaldsheim, West Germany.

Malecki, E.J. (1979), Locational Trends in R & D by Large U.S. Corporations, 1965-1977, *Economic Geography* 55:309-323.

Malecki, E.J. (1980), Corporate organisation of R & D and the Location of Technological Activities, *Regional Studies* 14:219-234.

Malecki, E.J. (1981), Recent Trends in the Location of Industrial Research and Development: Regional Development Implications for the United States, in: Rees, J., Hewings, G.J.D. and Stafford, H.A. (eds.), *Industrial Location and Regional Systems*, Croom Helm, London.

Malecki, E.J. (1985), Industrial and Corporate Organisation in High Technology Industries, *Economic Geography* 61:345-369.

Mansfield, E. (1968), *The Economics of Technical Change*, Longmans, London.

Massey, D. (1984), *Spatial Divisions of Labour*, Macmillan, London.

Mensch, G. (1979), *Stalemate in Technology*, Ballinger, Cambridge, MA.

Mill, J.S. (1862), *Principals of Political Economy* Book II. London.

Morphet, C. (1987), Research, development and innovation in the segmented economy: spatial implications, in: Van der Knaap, G.A. and Wever, E. (eds.),

New Technology and Regional Development, Croom Helm, Beckenham, Kent:

Myers, L.A.J. (1983), Information systems in Research and Development: the Technological Gatekeeper Reconsidered, R & D Management 13:201-206.

Nolan, M.P., Oppenheim, C. and Withers, K.A. (1980), Patenting, profitability and marketing characteristics of the pharmaceutical industry, World Patent Information 2:169-176.

Oakey, R.P. (1979), Labour and the Location of Mobile Industry: Observations from the Instruments Industry, Environment and Planning A 11:1231-1240.

Oakey, R.P. (1984), High Technology Small Firms, Frances Pinter, London.

Oakey, R.P., Rothwell, R. and Cooper, S. (1987), Management of Innovation in High Technology Small Firms, Pinter Publishers, London.

Odagiri, H. (1985), Research activity, output growth and productivity increase in Japanese manufacturing industries, Research Policy 14:117-130.

OECD (1986a), R & D Invention and Competitiveness, Paris: Science and Technology Indicators no. 2.

OECD (1986b), STI Indicators Newsletter no. 9, Paris: Scientific, Technological and Industrial Indicators Division.

Parker, J.E.S. (1974), The Economics of Innovation, Longman, London.

Pavitt, K. (1982), Patent statistics as indicators of innovation activities: possibilities and problems, Scientometrics 7:77-99.

Piore, M.J. and Sabel, C. (1984), The Second Industrial Divide: Possibilities for Prosperity, Basic Books, New York.

Pred, A. (1977), City Systems in Advanced Economies, Hutchinson, London.

Premus, R. (1982), Location of High Technology Firms and Regional Development. Washington DC: Joint Economic Committee, Congress of the United States.

Rees, J., Stafford, H., Briggs, R. and Oakey, R. (1983), Technology and Regional Development in the American Context. Laxenburg, Austria: Collaborative paper, International Institute for Applied Systems Analysis.

Roobeek, A.J.M. (1987), The crisis in Fordism and the rise of a new technological paradigm, Futures: April.

Rothwell, R. (1979), The Relationship between Technical Change and Economic Performance in Mechanical Engineering, in: Baker, M.J. (ed.), Industrial Innovation: Technology, Policy, Diffusion, London, Methuen: 36-59.

Rothwell,l R. (1980), Innovation in Textile Machinery, in: Pavitt, K. (ed.), Technical Innovation and British Performance, London, Methuen: 126-141.

Rothwell, R. and Zegveld, W. (1985), Reindustrialisation and Technology, Longman, Harlow.

Schumpeter, J.A. (1934), The Theory of Economic Development, translated by

R. Opie, Oxford University Press, London.

Scott, A.J. (1986), Industrial Organisation and Location: Division of Labour, the Firm and Spatial Process, Economic Geography 62:215-223.

Solow, R.M. (1957), Technical Change and the Aggregate Production Function, Review of Economics and Statistics 39:312-320.

Smith, I.J. (1986), Takeovers, Rationalisation and the Northern Region Economy, Northern Economic Review 12:30-38.

Sweeney, G. (1987), Innovation, Entrepreneurs and Regional Development. Frances Pinter, London.

Taylor, M. and Thrift, N. (1980), Organisation, Location and Political Economy: towards a Geography of Business Organisations. Paper delivered at the IGU Commission on Industrial Systems. Fujiyoshida, Japan.

Taylor, J. and Williams, H. (1987), Regulatory Change and Organisational Response. University of Newcastle upon Tyne: CURDS unpublished paper.

Thwaites, A.T., Oakey, R.P. and Nash, P.A. (1981), Industrial Innovation and Regional Development. United Kingdom, Newcastle upon Tyne, UK: CURDS Final Report to the Department of the Environment.

Thwaites, A.T., Edwards, A. and Gibbs, D.C. (1982), The Inter-regional Diffusion of Production Innovations in Great Britain. Newcastle upon Tyne, UK: CURDS Final Report to Department of Trade and Industry/EEC.

Tushman, M.L. (1982), Managing Communication Network in R & D Laboratories, in: Tushman, M.L. and Moore, W.L. (eds.), Readings in the Management of Innovation, Pitman, London.

Vickery, G. (1986), Technology Transfer Revisited; Recent Trends and Developments, Prometheus 4:25-49.

Williams, H. Gillespie, A.E., Howells, J.R., Pywell, C. and Thwaites, A.T. (1986), New Information Technology Production in the Regions of Europe, Newcastle upon Tyne, UK: CURDS Final Report to EEC.

Williams, H., and Gillespie, A.E. (1987), Telecommunications and the Reconstruction of Regional Comparative Advantage. Paper delivered at the 5th World Telecommunications Conference, Geneva.

3 The micro-electronics revolution and its impact on labour and employment

M. M. FISCHER

1. Introduction

The relationship between technological change and employment has received increasing attention in the most recent period of time when fears of renewed inflationary pressures, intensive international competition and budgetary constraints have limited the scope for macro-economic demand management on the one side and when the rise in structural unemployment since the mid 1970s has been more or less dramatic in most industrialised countries on the other (see Fischer and Nijkamp 1987). Great concern has been expressed that job displacing types of technological change might outstrip the creation of new job opportunities over a fairly long period of time, involving much higher levels of structural unemployment in the 1980s and 1990s than in the past. These fears have been accentuated by the emergence of a new cluster of new technologies associated with the combination of micro-electronics, telecommunications and information technologies (see Soete and Freeman 1985) which may be loosely termed micro-electronics based technologies.

This new technological system satisfies all the requirements for a Schumpeterian revolution in the economic system. The revolutionary nature manifests itself in several features. First, it is pervasive in generating a whole range of new applications over a wide range of sectors. There is virtually no field in manufacturing and service industries which can fail to be influenced. The OECD (1987), for example, suggests that some 80 per cent of today's and medium future technological advances are micro-electronics based. Second, the technological changes induced already were rather dramatic and sudden and of sufficient magnitude and consequence up to now. There has been never such a powerful cluster of new technologies advanced so rapidly in such a short period of time. Not only the capabilities and the performance of the microprocessor have increased significantly, but also the costs have been reduced remarkably. Third, micro-electronics based technologies provide a large potential for productivity increased in relation to all factor inputs, ranging from labour, capital, materials and energy to other more intangible factor inputs such as stock control, organisation, quality control, management and marketing; and, moreover, large potentials for flexibility in manufacturing and services as well as for more interactive decision making and organisation. All these characteristics point to the radical change potential associated with the adoption of micro-electronics based technologies.

This new technological system seems to affect every branch of the economy, both in terms of its present and future employment and skill requirements and its labour market prospects. There is a wide range of studies in various OECD countries by a variety of research groups in industry and academic life which have attempted to assess and forecast the impacts of the new technologies on employment and occupations.

This chapter discusses some of the labour and employment implications the micro-electronics revolution is likely to have. It draws largely on findings of a series of case studies in different countries. In order to provide a better understanding of how labour and jobs are changed we first describe the nature of these micro-electronics based technologies (see section 2). It is not a straightforward task to identify and measure the employment effects of new technologies. There are different types of employment effects and serious methodological problems involved in this task. This issue is discussed in section 3 in some more detail. One of the key questions concerning the employment implications of technological change is whether the initial impact in term of job displacement may be compensated for by endogeneous and exogeneous factors which tend to increase employment in other ways at the macro-economic level. A discussion of this question is followed by a consideration of present and projected employment effects in manufacturing and service industries (see section 4). Regardless of whether these new technologies have a net positive or negative effect on aggregate employment they will affect job tasks, job skills and the work environment. Such changes in the quality of labour will be described in section 5. Finally some policy implications for labour market adjustments are considered.

2. The Nature of Micro-Electronics Based Technologies

There is no established definition of micro-electronics based technologies (METs). METs may be roughly defined as comprising all those new technologies which use microprocessors or their electronic equivalents (such as custom or semi-custom integrated circuits), normally with large-scale or very large-scale integration circuitry, either in the form of single integrated circuit devices or in small groups of linked devices. The major distinguishing feature between microprocessors and custom integrated circuits refers to the fact that microprocessors can be programmed by the user for different applications while the programs of custom integrated circuits are fixed at the time of manufacture. Thus, custom integrated circuits are much less flexible, but might be considerably cheaper for large volume applications than the use of microprocessors. Semi-custom integrated circuits are a compromise between these two alternatives and offer varying degrees of user flexibility and cost savings for middle volume applications.

Potential applications of METs are innumerable in manufacturing, but METs are also applied increasingly in the service sector. METs can be integrated into products leading to product innovations and can be part of production processes implying process innovations. Of course, one's firm new product may be another's firm new process. Technical development in the products of many engineering firms represent process innovations for their customers. Despite of this fact, the distinction between product and process innovations is a useful one. For the employment effects of product innovations are quite different from those of process innovations.

In principle, the above mentioned devices can be used in almost any product where something is to measure, monitor or control. In particular, the opportunities for including micro-electronic components in engineering products (such as automobiles, computers, consumer electronics, household appliances, etc.) is vast. These components often replace mechanical and/or electro-mechanical components. Typical examples of product innovations range from consumer goods (such as digital watches, electronic calculators, microprocessor controlled household appliances, robotic toys, emergency and security systems) over manufacturing machinery (such as presses, spot welders, sewing machines, fabric cutting machines, machine tool controls), office equipment (such as text and data processors, executive work stations, automatic cash dispensers, electronic mail), vehicle and transport equipment (such as air conditioning units, air navigation equipment, train controls, aero engine controls, automatic pilots) to medical equipment (such as infusion pumps, haematology analyses) and defense equipment (such as guidance systems, radar). Micro-electronic based circuits can replace a large variety of assembled devices utilized throughout the economy, such as for example, measurement, monitoring and control devices using any mix of mechanical, electromechanical, electric, discrete or integrated circuit technology. Product applications, however, are less common than process applications up to now, partly because they tend to be more difficult to develop and need more technical expertise, longer development time and higher development costs.

The micro-electronic revolution is not only creating new goods and services, but also altering how they are produced. In the last fifteen years developments in micro-electronics based technologies have opened prospects for the rapid evolution of equipment used in performing a wide variety of tasks that involve receiving, processing, transmitting and acting upon information. In the service sector process applications cover the use of electronic data processing in production processes accompanied by electronically enhanced equipment and its application. In manufacturing microprocessors gradually penetrated into all aspects of the production process, especially in the engineering industries which have been in the forefront of developments of automation relevant to industrial manufacturing. Applications cover the use of micro-electronics based equipment in the design, fabrication, assembly, handling, quality control and testing or other operations on site necessary to make a product ready for sale. Typical process and production applications include the use of computer-aided design (CAD) equipment, computer-aided manufacturing (CAM) systems and computerized tools and strategies for manufacturing management (such as management information systems (MIS) and computer-aided planning (CAP)). CAD encompasses the use of a microcomputer based interactive graphics system in product design, production analysis and other engineering analysis.

CAM is concerned with the application of programmable automation technologies into the production process. Programmable in this context means that these technologies can easily be switched from one task to another by changing the computerized instructions. The term automation refers to the fact that these technologies perform a significant part of their functions without direct human intervention. They tend to provide flexibility in contrast to special-purpose automated machines because they are reprogrammable. Flexibility means that these technologies are suitable for small, medium and large volume production. Flexibility also allows to incorporate product changes if necessary (see OTA 1984a). Prominent examples of CAM technologies are numerically controlled (NC) machines, computerized numerically controlled (CNC) tools, robots and flexible manufacturing systems which will be briefly described in turn.

Numerically controlled (NC) metal cutting and forming machine tools- developed already in the period 1948-53 under the sponsorship of the US Air Force to help to produce complex parts for aircraft - are devices which cut a piece of metal according to programmed instructions about the desired dimensions of a part and the steps for the machining process. Numerical controls were first offered commercially in 1955. They characteristically consist of a machine tool, which is equipped with motors to guide the cutting process, and a controller which receives numerical control commands (OTA 1984a). This technology effectively achieves a drastic reduction in the setting-up of machine tools which is mainly the result of the ability to program off the machine and the ability to store a suite of tapes or programs from which a number of repeat items or variations of these can easily be made if necessary.

Around 1975 machine tool manufacturers have begun to use microprocessors in the controller of NC machines, i.e. to construct NC tools with integrated computer control (so-called computerized numerically controlled or CNC machines). CNC can be used, like NC, with separate programmers and operators, but allows also the reintegration of operating and programming and facilitates moreover editing and programming on the machine. Both modes of CNC usage can be found in practice. CNC machines are substantially more reliable than ordinary NC machines. Moreover, they offer the operator more complex information about the status of the machining process and provide the ability of linking CNC machine tools with various other machines (see OTA 1984a). NC and CNC machine tools are far more efficient than conventional automation machinery. Of course, their efficiency in actual production work varies, depending on the type of machine and the degree of precision and difficulty of the work assigned to it. As regards the scale of production they are considered to be most efficient for the recurrent production of small and medium-sized batches. The advent of CNC also allowed another development, namely the simultaneously control of a number of NC machines by a single computer (known as direct numerical control or DNC).

Robots, another example of CAM, may be roughly defined as reprogrammable manipulators designed to move workpieces or tools along various paths for the performance of a variety of tasks. Typically an industrial roboter has three main parts: the controller consisting of the hardware and software (involving a microcomputer or microelectronic components) which guides the motion of the robot; the manipulator consisting of a base which is usually bolted to the floor, and the arm itself which can be configured in various ways to move through particular patterns. Robots with point-to-point controls for simple material handling tasks were first introduced commercially in 1959. The first robot with path control capability appeared in 1961. This type of robot was suitable for a number of purposes, such as spray painting, spot welding, arc welding and investment casting (see Ayres 1988).

Among users of robots, the car industry strongly dominates. Other important users include aerospace, electronics, foundries, machinery and miscellaneous light manufacturing (see OECD 1983, OTA 1984a). There is a significant difference between Japan and other countries with respect to the tasks that have been robotised. In Japan robots are concentrated in welding and painting while in the US and in Western Europe their use is spread over a wider range of tasks (including machine (un)loading material handling) (Watanabe 1986). It is worthwhile to note that in Japan where a high level of automation had been attained already prior to the introduction of robots the labour-saving effects have been much more restricted than elsewhere.

The largest potential use of robots is in labour intensive mass production assembling and inspection of light industrial and consumer goods. But there are still many important technical problems to overcome. It is worthwhile to mention that the next generation of robots will be differentiated from previous generations by its sensory ability and capability to react to changes in the surrounding work environment. In particular visual and tactile sensing is necessary for the application to more complex assembly tasks.

Robots, NC and CNC machine tools with automated part-handling can be used in a machining cell or so-called flexible manufacturing system (FMS). FMSs, which have evolved particularly, but not exclusively in the metal forming and mechanical engineering industries, are production units which enable to produce a range of discrete products with a minimum of manual intervention. The first attempts to combine NC machine tools with an automated materials-handling system under computer control appeared around 1967. Applications have focused on mid-volume batch production of moderately complex parts of volumes of 2,000 to 50,000 units/year (Ayres 1988).

In more sophisticated modern flexible manufacturing systems the workpieces are processed at various programmable multipurpose machine tools and other workstations. They typically consist of production equipment work stations (NC, CNC or other machine tools for fabrication, assembling or treatment) linked by a materials handling system to move parts from one workstation to another. It operates as an integrated system under the central control of a computer which maximizes work flow and minimizes in process queue. These systems may involve in built tool changing, component (un)loading, checks for wear and failure, product testing, etc. (Wilson 1987) and allow small quantities of goods to be produced inexpensively and permit rapid change-over to the production of other products.

At the beginning of 1985 there were 46 FMSs in the US, 100 in Japan, 60 in the USSR and 36 in the FRG (see Ayres 1988). The reasons for the scarcity of application include the newness, the high costs and complexity of the systems. The high complexity implies the need for a high degree of technical expertise which manufacturers may not have to date. In addition, there are still serious reliability and versatility problems which need to be solved before a more widespread diffusion might be expected.

It is worthwhile to note that FMS may be further integrated and controlled by a CAD/CAM system into an overall automated factory which is usually called a computer-integrated manufacturing system (CIM). CIMs are still in an early stage of diffusion and consequently their total employment impact is marginal up to now. A major problem usually involved with linking CAD and CAM is that of incompatability of different systems. The need for standards in interfaces and program languages is strong and important. Without standards it is extremely difficult to combine equipment of different vendors and to proceed incrementally towards CIM. Typically CIM involves the integration and coordination of design, manufacturing and management using computer-based system (see Vickery et al. 1987). These systems, however, are still in the process of development and there is some uncertainty regarding their final configuration. Figure 1 illustrates the relations between the systems and their components.

There are several other kinds of CAM equipment. For example, programmable process controllers are used quite extensively in both discrete manufacturing and in continuous-process industries (such as oil, chemical, pulb and paper). They are utilized to control a variety of production processes and to collect information about such processes. NC tools may be considered to be specialized forms for controlling a machine tool.

To sum up, programmable automation technologies offer improvements in material handling, fabrication, finishing, assembling, quality assurance and control by applying computerized techniques to control tools of production, to gather and manipulate information about the manufacturing process, and to design and plan the whole process. Moreover they promise an increase in the degree of control over the whole enterprise. Thus, they improve coordination and increase efficiency and flexibility (in terms of both the range of products and volume of a specific product) and increase productivity and control over the manufacturing process (see OTA 1984a). Some of the most significant productivity gains relate to significant energy, materials, capital and labour savings as well as to a more efficient

Figure 1: Relations between Programmable Automation Technologies and Their Components: Horizontal and Vertical Integration

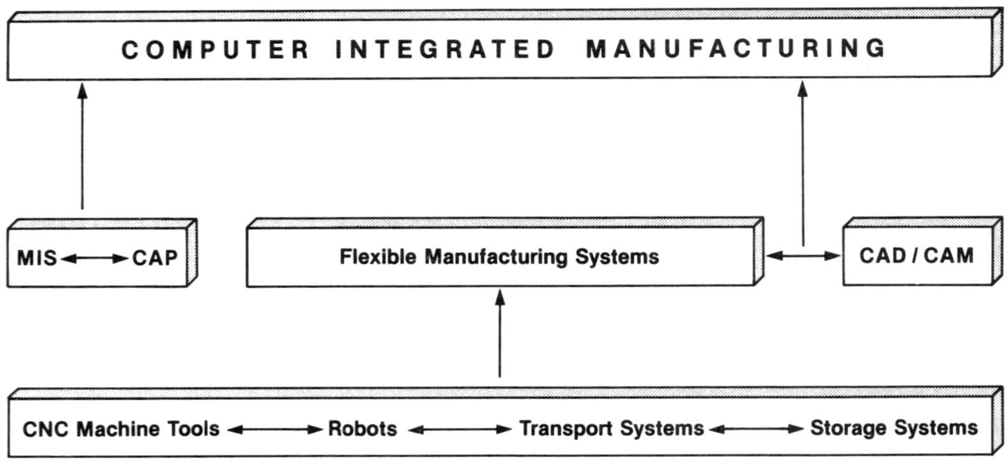

Note: ↔ refers to horizontal integration and ↑ to vertical integration

Source: Adapted from Vickery et al. (1987)

inventory control (see Schonberger 1982). Some of the elements under CAM such as NC machine tools have an important, but a more specific sectoral impact; while many of the elements have only an indirect impact on major manufacturing sectors so that it is difficult to follow through and assess the impact of these elements (see OECD 1983).

Each of these technologies is in a relatively early stage of development and in even earlier stages of applications to date. Only NC and CNC machine tools as well as CAD are somewhat more mature, but even in these cases there are still many unsolved problems. But they continue to develop rapidly, at a rate of growth similar to the computer technologies as a whole. These developments tend to focus on five major directions, namely to increase their power in terms of speed, accuracy, reliability and efficiency; to increase their versatility in terms of the range of problems to which the technologies can be applied; to increase the ease of use in the sense that they require less operator time and training, can perform more complex operations and be adapted to new applications more rapidly; to increase what is commonly called the intelligence of the system; and finally to increase the ease of integration into more comprehensive and coordinated devices (see OTA 1984a, 74).

Even if the rate of use of micro-electronics in products and processes has climbed rapidly since the late 1970s (see Northcott and Rogers 1984; Northcott et al. 1985; Northcott 1986) their potential impacts are still a long way from being fully realised. Despite the potential of micro-electronics applications there are several problems which have slowed the rate of diffusion at the firm level. The technical factors tending to slow the rate of adoption include the complexity of the new equipment and problems with software. The non-technical factors affecting the use include the high costs of development and implementation, lack of capital and know-how, organisational resistance to change, problems of standardization and the availability of appropriate education and training programs. A series of comparable surveys of manufacturing industry shows clearly that the set of problems encountered is similar in different countries and moreover that many countries face skill shortages as one of the principal barriers when applying micro-electronics in both products and processes (see Northcott et al. 1985). This is illustrated in Table 1.

The overall impacts of METs have been masked by the uneven pattern of applications which are concentrated in particular countries and industries. There are marked differences in the rates and patterns of application in both products and production processes between different countries and

Table 1: Main Disadvantages and Problems in the Use of Micro-Electronics Based Technologies: Product and Process Users (1983)

(percentages)

Disadvantages and Problems	Product Users			Process Users		
	France	FRG	UK	France	FRG	UK
Lack of people with expertise	58	65	50	51	54	43
Problems with software	10	38	14	11	27	16
High costs of development	30	34	38	24	23	25
General economic situation	26	25	45	23	19	42
Lack of finance for development	18	23	30	19	17	29
Problems with microprocessors/ components	9	19	9	8	16	14
Problems with sensors	9	19	11	15	14	8
Higher production costs	19	17	18	4	14	7
Opposition from shop-floor/unions	16	17	6	9	13	10
Difficulties of communications with subcontractors or suppliers	6	15	9	15	12	15
Opposition in management	1	6	4	2	4	5
Opposition from other groups in firm	3	3	4	2	3	4

Source: Northcott et al. (1985)

different industries. Some countries such as the USA, the FRG, Sweden and especially Japan have been clear leaders in both the rate of uptake and in the breadth of product applications, whereas other countries such as the Netherlands and France have trailed significantly. Although potential product applications are widespread, metal working/engineering, electrical/ electronics and transportation equipment industries have taken the lead in all the five countries for which detailed information is available (see Table 2). Their products evidently show the greatest potential for applying sensing, processing and transmitting functions in order to enhance performance and extend the range and quality of products.

Table 2 shows that the use of microelectronics in production processes is much more widespread than in the products themselves. There are also quite clear country-specific differences. Manufacturing establishments in Japan and the FRG use microprocessors and their electronic equivalents as part of process and production equipment more frequently than in the Netherlands and France. Also some industry-specific differences do exist. Printing, transportation equipment, electrical/electronics, metalworking/ engineering and chemicals industries are twice as likely to use micro-electronics based process equipment than clothing, some parts of textiles and wood products industries. The use of more advanced programmable automation technologies is restricted to a small set of industries up to now, even if their potential for wider applications is high. For example, industrial robots have been mainly used for material handling, machine (un)loading, spray painting, welding, machining and assembling in general and precision machinery and electrical/electronics industries up to now (see Vickery et al. 1987).

3. General Issues in Measuring Employment Effects

This section intends to discuss some major types of employment effects (see Figure 2) first and then some methodological problems involved in the task of identifying and measuring such effects.

The overall impact of METs on employment at the macro level may be considered as outcome of direct and indirect, labour displacing and job creating effects. Direct effects may be regarded as the net changes in jobs at the point of introduction of the new technology (Brainard and Fullgrabe 1986). If micro-electronics is introduced to improve production processes, the direct employment effects tend to be negative, and if the technology is used to create improved or new products, additional jobs may be created, especially if the product innovations are radical (i.e. not substituting for

Table 2: Use of Micro-Electronics Based Technologies in Selected Countries and Industries (1983)

Percentage of Manufacturing Establishments Using	Metalworking/ Engineering	Electrical/ Electronics	Transportation- Equipment	Print	Chemicals	Food	All Industries
Product Applications							
France	23	28	23	1	1	0	6
FRG	42	42	9	6	7	0	13
Japan (1982)	30	42	10	0	0	0	13
Netherlands	10—30	>30	10—30	n.a.	n.a.	n.a.	5
UK	28	50	14	0	0	0	10
Process Applications							
France	31	33	58	50	43	39	35
FRG	59	54	39	77	52	46	47
Japan (1982)	71	72	73	69	62	44	59
Netherlands	10—30	10—30	10—30	>30	n.a.	10—30	30
UK	45	51	33	62	51	60	43

Source: Vickery et al. (1987)

n.a. not available

Figure 2: The Impact of Innovations on Labour and Employment: A Categorization of Different Effects

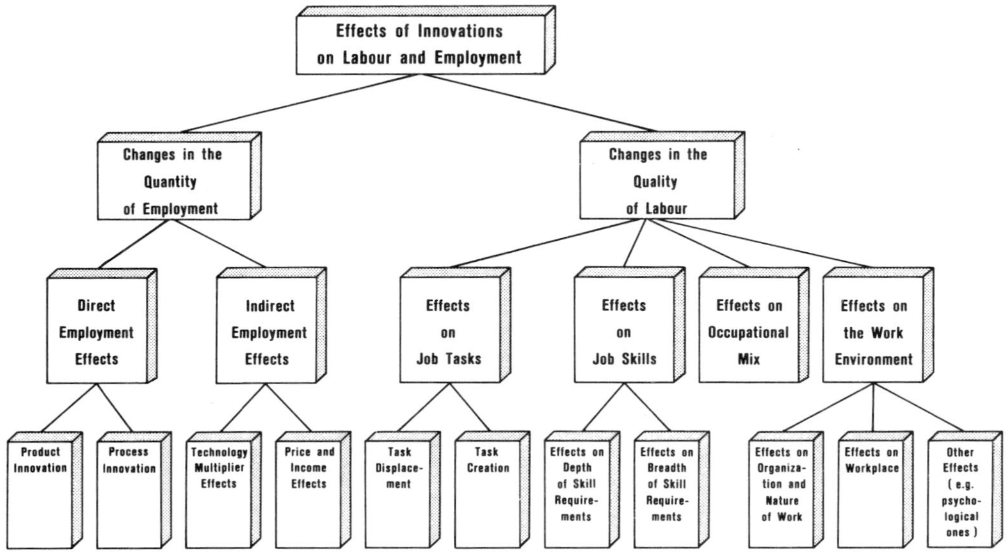

established products and services). The direct employment effects are only one aspect of the full impact of METs. The introduction of a new technology in general and a new process innovation in particular and the effect it has on relative costs of production induces several processes which tend to increase the level of employment at the macro level usually after some time lag. These effects are termed indirect labour or compensating effects (see Brainard and Fullgrabe 1986).

There are various indirect effects which tend to offset some, if not all of the direct employment displacement effects (see Stoneman, Blattner and Pastre 1982). The two major types of such effects are technology multiplier effects on the one side and price and income effects on the other. Technology multiplier effects account for the required increase in gross investment to install the new technology. This increase may generate demand and employment (for a period) in supplier industries and the capital goods sector. There may be also implications for intermediate demands in the input-output sense.

Income and price effects act to yield increased profits or wages or a reduction in prices in the case of process innovations, all of which increase real income which can stimulate demand and consequently may lead to more jobs. Product innovations resulting in improved or new products should lead to better price/quality ratios for goods to the consumer. This may imply increased incomes and higher demand. But there is no guarantee that compensation via increased demand will be realised.

The strength of such indirect effects is of central importance in the impact of METs on employment at the macro level. Strength and the time profile of this type of effects depend inter alia on investment behaviour, success in foreign markets, the characteristics of the new technology, the skill structure of the labour force and especially on the price and wage flexibility which is considered to be a key determinant of whether the impact of the new technology falls on employment or real wages (Stoneman, Blattner and Pastre 1982).

Of course, employment induced by compensating effects need not necessarily to match the labour force released through direct effects in skills, other characteristics and in space. Such mismatches might result in technology induced structural employment. Moreover, it is worth mentioning that labour released directly may not come from the innovating firm or sector, but from non-innovators.

The employment impact of METs on any individual country strongly depends on the diffusion speed and pattern of the technologies throughout the economy which theirselves are conditioned by several factors such as the industrial structure of the country, the technological position of the economy, the availability of labour skill, social attitudes to technological change etc. Employment effects are also strongly influenced by the overall economic and social environment in which the technologies are being introduced. The employment effects vary from country to country, from industry to industry and from company to company due to differences in the levels of automation and organisational efficiency attained prior to the introduction of the new technologies, the labour and management practices, the workers' adaptability, etc.

Direct and indirect effects are especially conditioned by macro economic conditions. Conditions of slow growth and low demand encourage process innovations and delay possible indirect or compensating effects. This seems to be the case in many countries in Europe to date. In contrast rapid economic growth and high demand stimulate product innovations and increasing capital investment in new technology based machinery which can result in shorter time lags of the indirect or compensating effects.

Moreover, employment impacts have to be considered in relation with the country's role in international economic and trading relations. This international dimension of the relationship between technological change and employment is becoming more and more important nowadays. Countries which are experiencing slower technological change in comparison to other countries might forego some or all of the positive price and income effects while countries leading in innovation may obtain extra employment via expanded exports and internalisation of compensating benefits. Moreover, if such countries import new technology capital goods, the technology multiplier effects will be reduced. All this means that differences in the competitiveness between countries induced by technological change can lead- via international trade - to the transfer of jobs from non-innovating to technologically leading countries (Brainard and Fullgrabe 1986).

Regardless of whether METs have a positive or negative effect on aggregate employment, they will change certain job tasks, affect job skills required, the occupational mix and the work environment. The scope of such changes - which may be termed changes in the quality of labour in contrast to the changes in employment levels or changes in the quantity of labour- is quite often neither obvious nor immediate because micro-electronics based technologies are often accompanied by significant transformations of manufacturing organisation. The more extensive such transformations, the broader the set of the personnel affected by the introduction of microelectronics based technologies, and the more difficult it is to attribute employment effects to METs, per se.

The identification and the measurement of both changes in the employment levels and changes in the quality of labour is a complex and difficult task. There are several conceptual and technical problems and difficulties involved in this task. The issues which will be discussed here refer to the problem of isolating employment effects induced by technological change, the choice of an appropriate methodological approach and technical statistical problems.

One of the major analytical problems one faces in analysing the technical change-employment relationship is the problem of isolating employment effects induced by the introduction of METs from effects resulting from all the other changes, taking place at the same time, internal and external to the establishment. A change in employment may be the result of individual factors such as governmental fiscal and monetary policy, world trading conditions, energy prices, exchange rate changes, exogenous shocks and technological change, but may be also affected by the numerous interactions between these factors. As a consequence it is generally agreed that it is not only conceptually difficult, but also infeasible to measure direct and indirect employment effects with great precision.

A second major difficulty is the choice of a methodological approach. In principle there are two major categories of approaches: micro-oriented and macro-oriented ones. Both perspectives have shortcomings. The case study approach relying on the fact that the impact of technological change on employment can be most readily observed at the plant level inherently neglects any compensating effect and is not able to capture the aggregate trends associated with the introduction of a major new technological system. Consequently, the estimates derived generally overstate the negative effects of technological change.

The macro-economic perspective derives employment estimates from models of the interaction among industries based on their requirements for labour and other production inputs and is able to take into account direct and indirect employment effects. But it generally lacks the detailed information provided by the case study approach, even in the case of a disaggregated input-output framework, and is often rather poor in its assessment of new technologies. Traditional neo-classical compensation model approaches generally identify only the effects of changes induced by process innovations within a broadly static framework with demand and substitution elasticities. The introduction of new products and services is by and large ignored in most partial or general equilibrium approaches (see Freeman and Soete 1985). Evidently, there is a need to integrate micro- and macro-approaches on the one side and to explicitly incorporate the essentially dynamic nature of the relationship between technological change and employment on the other.

Finally, there are significant technical problems of quantifying employment effects of METs because of inadequacies in industrial and occupational statistics. The current national and international classification schemes have been only poorly adjusted to reflect the recent changes in production technologies, products and services even if considerable efforts are being made to devise and collect new statistics in some countries such as France and Japan and at the international level by the OECD Information, Computer and Communications Policy Group.

The reliability of inferences about quantitative effects on national economies and industries is, thus, limited because of the scarcity of good data describing shifts in skill requirements, types of jobs or the structure of manufacturing and especially service industries producing and using METs. One consequence of this lag in satisfactory classification and measurement in most countries is that the impact of METs is quite often underestimated, especially with respect to changes in the quality of labour.

In spite of these problems and difficulties there is a large and increasing number of studies in several countries which have attempted to assess and to forecast the impacts of METs on employment and labour. It is the purpose of the next two sections to summarize and integrate some of the major findings. In section 4 attention is being paid to employment effects on aggregate and sectoral employment while section 5 considers changes in the quality of labour due to the introduction of METs. Stress is laid here upon the identification of trends and direction of likely changes rather than their magnitudes.

4. Changes in the Level of Employment: Aggregate and Sectoral View

In view of the large body of literature dealing with the relationship between micro-electronics based technologies and employment at the aggregate or at the sectoral level it would be not practical to refer to all the individual studies which have been studied in detail. Instead a general overview of their conclusions will be presented.

All the studies show more or less severe methodological deficiencies. Most ignore the productivity effects on inputs other than labour. Only few take explicitly the international dimensions of the technological change-employment relationship into account. Relatively little attention is given to the factors influencing the speed and pattern of applications of METs. The overall impact on employment in any particular country, however, crucially depend on the domestic rate of diffusion of the METs. The faster they diffuse, the more rapidly the potential productivity gains will be realised. The slower the new technologies diffuse relative to other countries, the more likely it is that other countries will gain a permanent advantage via the indirect employment effects.

Most studies devote a great deal of effort to identifying the direct effects, especially the negative implications induced by process innovations. There have been only few attempts to assess the importance of compensating effects. Epecially price and income effects are rarely taken into consideration. Some of the few exceptions can be found in an OECD Information Computer Communications Policy report (1981) on the issue of micro-electronics, productivity and employment. These studies attempt to stimulate the possible employment effects using different macro-economic approaches to assess the implication of an acceleration in the pace of technical change where various feedback mechanisms are taken into account. The authors inter alia conclude that faster diffusion of METs has greater negative direct effects, but also greater positive effects on employment as it would be the case if the METs would diffuse more slowly. Usually, however, positive direct and indirect employment effects are strongly underestimated or even neglected. These shortcomings certainly do notarise primarily from a lack of understanding the factors and the dynamic nature of the technological change-employment relationship, but are associated with the methodological and statistical problems already discussed in section 3 (see also Brainard and Fullgrabe 1987).

Most studies have a relatively narrow focus, dealing with the effecs of specific new technologies (such as, for example, robotics) and/or considering specific sectors (such as, for example, engineering). The few studies addressing the issue of current employment effects at the macro level tend to agree that employment and unemployment levels in recent years were not significantly influenced by technological change (see, for example, the OECD (1982) report of the effects of micro-electronics, robotics and jobs based on studies from thirteen countries). Up to now, there is no strong evidence to support the fear of widespread unemployment caused by METs even if there is a weak indication that METs may have a labour saving bias (i.e., a slight tendency to increase the capital/labour ratio ceteris paribus). Shifts in demand patterns, international competition and economic growth patterns were much more important. A comprehensive OECD review of recent evidence on the impacts of technological change - carried out by Brainard and Fullgrabe (1986) - suggests that the increase in jobs seems to have exceeded losses induced by the introduction of METs, mainly due to job growth in business, financial and communications services and to a much lesser extent in high technology industries.

Perhaps more important than the impact in the current period is the potential significance for future patterns. The effect of new technologies on future employment levels is extremely difficult to forecast with confidence because such assessments have to be based upon assumptions on several key factors, such as the rate of diffusion of the technologies, the size and distribution of productivity gains and general economic trends. Nevertheless, there is a wide range of forecasts of future employment effects in several countries. These studies often differ in their forecasts due to several reasons, including the use of different methodologies, different theoretical assumptions, differences in the structure of national economies and in the characteristics of the labour forces, differing extents of use and rates of diffusion of the new technologies (see Brainard and Fullgrabe 1986). Despite of all these differences they can be broadly grouped into three types.

The first type of studies forecast a more or less significant net growth in aggregate employment and assume that any possible loss in direct employment will be more than offset by compensating employment effects in the long run. Only rather few studies take on this optimistic view. Wilson and Whitley's (1982) UK study, for example, belongs to this category. This study finds the possibility of a large net gain in employment from greater applications of micro-electronics based technologies. The results indicate that productivity could be raised by 0.5 per cent per year (over the trend growth rate) during the period of 1985-1990 through increased use of these new technologies. Job loss due to direct employment effects will be significantly offset by the strong compensating effects.

The second type of studies, on the contrary, argues that the macro impact of micro-electronics applications on employment is bound to be negative, at least in the short and medium term periods. This pessimism derives from the labour saving nature of much MET when applied to production processes, coupled with its pervasiveness and the limited prospects for the generation of new products in general and in particular from considerations, such as the possibility of jobless growth; the widespread applicability of

METs, especially in the service sector where compensation effects may not be strong enough to balance direct job losses; and a change from the present pattern of step-wise introduction of new technologies to the introduction of large integrated technology systems which may have a great labour saving potential that may not be compensated by job creation from increased demand. A recent ILO study carried out by Kaplinsky (1987) on the likely future employment impacts of micro-electronic applications predicts that radical shifts will be most likely anticipated if wide micro-electronics based changes will occur.

These pessimistic studies tend to overestimate the labour saving effect of METs due to several reasons. First, they ignore the possibility of different modes of application by smaller firms and underutilization of the new machinery installed. Second, the fact that the very flexibility of METs tends to augment the average amount of work incorporated in each unit of final output, inter alia by encouraging product differentiation and more frequent model changes (the so-called amplifying effect), is neglected (see Watanabe 1986). Third, they are strongly concentrated on process innovations. Finally, they fail to take product innovations and the many compensating effects into account. All this means that they tend to overstate the early negative impacts on aggregate employment and to understate the later positive effects.

The third type of studies finds that job gains and losses are and will be approximately equal in number (i.e, that the employment impact is neutral). The actual and the likely future effects of the new technologies on the levels of employment are likely to be minor, especially in comparison with the effects of factors such as fluctuation in macroeconomic demand and growth. This point of view is supported by an increasing number of recent studies from different countries (see, for example, OTA 1984; Brainard and Fullgrabe 1986). These studies emphasize that the major impacts are reflected less in increases or decreases in levels of total employment rather than in the distribution of labour in terms of sectors of employment, occupations, skills and regions.

Technological change has quite different employment effects in different economic sectors and industries. There seems to be a general consensus of opinion that the introduction of METs has contributed to the overall decline in manufacturing employment in most countries, but the magnitude of the impact was relatively small in comparison with the total losses of jobs in the sector. A recent study by Northcott et al. (1985) based on a series of plant level surveys and weighted to a national basis provides a good overview of the current state as far as the application of METs in both products and processes in the manufacturing industries of France, the FRG and the UK during 1981-83 is concerned. Applications in products were generally associated with increases in jobs or small decreases while larger declines were generally associated with applications in production processes over the two-year period. The overall assessment for users of METs is summarized in Table 3. It should be noted that these estimates do not include indirect effects, particularly those of non-users which may result from a failure to remain competitive. Moreover, they do not include the impact of METs on office employment within manufacturing. The term MET is restricted to the use of micro-electronics in products and its applications for the direct control of the manufacturing process and materials handling.

Even if the figures are subject to such uncertainties and limitations they give a good indication of the broad orders of magnitude, better in any case than any other figures presently available. It is important to note that the net losses were rather small in comparison with the total reductions in manufacturing in each country. Moreover, much of the loss appears to have been absorbed in natural wastage. In the British case the total figure of actual redundancies due to the use of micro-electronics amounts to only 12,000 (equivalent to about 0.6 per cent of total employment in manufacturing at the time of the survey). About half of them are suggested to have been involuntary redundancies.

Table 3: Changes in Manufacturing Employment due to Use of Micro-Electronics Based Technologies in Selected Countries (1981 — 1983)

Country	Gross Decrease (in thousands)	Gross Increase (in thousands)	Net Change (in thousands)
France	—25	13	—12
FRG	—47	17	—30
UK	—54	20	—34

Source: Northcott et al. (1985)

A more recent plant-level survey of manufacturing establishments in the UK for the period of 1983-85 undertaken by Northcott (1986) shows an accelerating loss of manufacturing jobs with the increasing application of METs. The net losses amount to 87,000 (about 2.7 per cent of total UK unemployment) for this more recent time period in comparison with 34,000 during 1981-83. Most of the increase in job losses are attributed to the greater use of more sophisticated CAM systems.

These and similar case studies indicate that job losses are expected to increase substantially as companies introduce more sophisticated METs into production processes. Currently the full impact on employment is being eased temporarily, because the displaced workers are often assigned to other jobs in the firm. This mode of adjustment, however, diminishes prospects for hiring new employees in future (see Brainard and Fullgrabe 1986). But it is not likely that technological change will generate significant unemployment in the near future. It may, however, exacerbate regional unemployment problems, especially in regional economies where metal working industries with mature and declining products are concentrated.

Of course, the introduction of METs has quite different employment effects in different manufacturing industries. Some have been more affected than others. The principal users of METs have been firms in the metal working industries, especially in car industries, electrical and non-electrical machinery. The effects in the watch industry are well known. But also other industries such as printing have begun to be significantly adversely affected. Also firm size may influence the incidence of employment effects. To date large firms dominate as users (see, for example, Northcott and Rogers 1984).

Finally, the service sector deserves some brief notice. Estimates of the impact of METs (such as office automation, advanced telecommunications, automated transactions, expert systems) on employment are not so readily available. Up to now, only little attention has been paid to the relationship between technological change and employment in the service sector, partly because of the more difficult conceptual and statistical problems in pursuing studies in this area, in comparison to manufacturing industries (see Gleave 1986). Nevertheless it is quite clear that the effects strongly depend on the nature of METs, on institutional and regulatory conditions and also on the labour management methods. The impact of technological change on service employment has been most obvious and strongly positive in producer services (such as financial services, computer services, etc.). Employment growth in these services is primarily attributed to high levels of capital investment in information technology which enabled to improve the quality of services to customers, and most recently to develop wholly new types of services as part of the intensifying competition and diversification. The net impact of new technologies on employment in public and distributive services are much more ambiguous, but seems to have been rather small while other factors dominating the trends in job growth (Brainard and Fullgrabe 1986).

In so far as likely future impacts in the service sector are concerned there is considerable variation in the predictions. Some argue that increasing automation in the service sector will change the trend of employment growth, making the sector a net loser of jobs. Others like Brainard and Fullgrabe (1986) expect that a more extensive application of new technologies is likely to create more jobs in services.

Barras (1987) argues in a more differentiated way. He predicts that in the short term (0-5 year horizon) the predominant effect of further investment in information technology may be labour displacing because the major impact of the investment is likely to be manifested in terms of further process innovations in existing services, while in the medium term (5-10 year horizon) employment generating effects might predominate.

5. Changes in the Quality of Labour

Regardless of whether the micro-electronics based technologies are essentially neutral or not in their impact on employment levels they will affect job tasks, job skills and the work environment (compare Figure 2). This section attempts to discuss the likely impact of the introduction of new technologies in this respect and basically draws on case studies in manufacturing industries and in particular on the OTA (1984a) study on programmable automation in the USA.

The introduction of METs alters the tasks to be done in manufacturing. It eliminates tasks and ultimately jobs and/or creates tasks and consequently jobs. Programmable automation technologies perform tasks - such as, for example, welding, materials handling and assembling - previously done by production and related workers. In particular, FMSs and CIMs show a great potential for displacing labour, greater than stand-alone MET applications. Because these systems are still in a rather early stage of development, significant labour displacement by such systems is unlikely to occur in the near future. On the other hand, new tasks and, thus, jobs may be created, especially in the maintenance of the automated equipment, but also in the design of products stimulated by the introduction of CAD systems and computer-aided engineering applications of CAD.

Micro-electronics based technologies affect also the job skills required in manufacturing. The issue of how technological change affects the skills of those who operate the new technologies has received increasing attention in recent years (see, for example, Attewell 1987, Zicklin 1987). Much of this research draws its inspiration from Braverman's (1974) view on the radical deskilling of machinists who operate NC machines. Empirical research about the effects of NC machines on the skills of machinists, however, is rather ambiguous. Some authors suggest that a radical deskilling of machinists has been taking place. Others claim that the operation of NC machines requires more skills than that of conventional machines. Still others report that only the types of skills may be different (see Zicklin 1987). In general programmable automation technologies give rise to a greater need for conceptual skills (e.g., diagnosis, problem solving and programming) and a lesser need for motor and routine mental skills (e.g., machining).

The OTA (1984a) study on programmable automation in the USA argues that METs will alter both the depth and breadth of skill requirements. Skill depth refers to the input necessary to perform an individual task or a cluster of interrelated tasks. Skill breadth refers to the input necessary to perform a set of (non-similar) tasks and, thus, more to jobs or occupations than to single tasks. Skill depth shows two dimensions, namely time to proficiency and judgement. Jobs consisting of tasks requiring little or no time to master and limited judgement tend to be low skilled jobs which have a broad access and a relatively low pay. It is argued that programmable automation technologies tend to lower skill depth in both dimensions for many tasks, including those performed by professionals and craftworkers. The potential effects on skill breadth are less evident than effects on skill depth. In some cases the introduction of automation, especially of more sophisticated technologies, may require a less intimate knowledge of a single process of task, but a broader familiarity with production activities and their interconnections.

New technologies do not only alter skill requirements, but also occupations and their distribution. The overall patterns of change in occupational mix which is expected to arise may be characterized by the following broad trends (OTA 1984a, 144-148):

- in all sectors, a decreasing proportion of lower managerial and supervisory occupations, with remaining personnel more restricted to prepare and transmit information to upper management;

- in manufacturing, a large decline in the proportion of production and related workers engaged in low skill and routine activities, such as materials handling and assembling, but also inspection and quality control;
- in manufacturing, an increasing proportion of occupations installing, operating and repairing automated equipment and providing related support;
- in manufacturing, some deskilling of tasks in some craft occupations (for example fitting and toolmaking) due to transfer of prior operator functions to machines and
- in services, a decreasing proportion of more routine information handling occupants, such as low skilled clerical.

These broad trends identified in several studies will reinforce the long term shift from blue collar to white collar employment as well as the growth in professional, technical and higher management occupations.

Research on the social impacts of METs on the (manufacturing) work environment is modest in scope and support. This issue seems to be more common only in those countries with a stronger tradition in labour unions. The new technologies may affect the work environment (the organization of work, the nature of work, the workplace) of most manufacturing personnel. To date, however, it is certainly too early to predict how these technologies will affect the work environment. Nevertheless there seems to be a general agreement in the literature that there is nothing inherent in micro-electronics based technologies to impose a particular form of work organisation. Thus, it might be expected that they will be adapted to traditional structures of work organisation in the short term. Major changes will arise only in the longer term because of two major reasons. First, micro-electronics based technologies (except NC and CNC machines) are still in a relatively early stage of development and in even earlier stages of applications. Second, managers seem to lack up to now the necessary background to assess the technological options on the other (see OTA 1984a, 204 pp.). The effects on the work environment will be primarily determined by the management's motivation for the introduction of METs and by the labour-management relations. Cooperation between employers and workers in union or non-union settings to determine the design and implementation as well as the pace of change might be a useful approach to minimize potential negative effects such as, for example, automation related stress and boredom.

6. Some Concluding Remarks

It is unlikely that the impact of micro-electronics based technologies will be revolutionary in nature. The impact will be certainly more likely evolutionary. To some degree the extent of gains and losses in employment will depend on the rate at which the new technologies diffuse. For some industries, in particular those facing strong international competition (such as engineering), a fast introduction of new technologies may be beneficial in terms of output and employment. A failure to adopt in these cases will result in greater job losses even in the medium term through lack of international technical change competitiveness. In other cases, for example, in office-based services and banking, a very fast rate of new technology introduction may be to the detriment of employment in the absence of product innovations.

Up to now and most likely also in future the major impact of micro-electronics based technologies is reflected less in decreases or increases in levels of aggregate employment than in changes in the occupational structure and in skill requirements of the labour force. The increasing adoption of micro-electronics devices is likely to increase the mismatch between skills and job opportunities and may lead to severe adjustment problems of technical change, in particular in the absence of comprehensive retraining schemes. Labour training and retraining policies are, thus, of central importance. The efforts undertaken by private firms have to be complemented by public programs in order to minimize the adjustment problems of structural change, especially for young people. The major concern of training and education policies should focus on the adaptability and flexibility of the labour force. For flexibility will be a key feature to retraining especially if one can not predict the future development in the skill composition of labour demand with great precision (see OECD 1987, 61). But it would be a mistake to assume that training and retraining policies alone would be able to solve the present and especially the upcoming labour market adjustment problems.

Governments have also an important role to play to manage technological change and its impact on employment by policy decisions taken at a macro level, for example, in setting the general economic conditions which may stimulate the development of MET applications and investments in them, in identifying market failures in the supply of information, finance and technical expertise as well as in developing and maintaining the supporting infrastructure. Soete and Freeman (1985) claim for a greater emphasis on diffusion of the new technologies as well as on the further development of key technologies. They identify three major sets of technology policies which seem to be particularly relevant to affect the rate of technological change:

- policies aiming at stimulating directly firms to take up radical inventions/innovations;
- policies aiming at promoting widespread diffusion of successful new technologies and lowering the institutional barriers and
- promotion of international transfer of new technologies.

These policies ranging from direct financial support to various forms of indirect risk-taking support to promote the emergence of radical innovations seem to be important to stimulate productivity growth based upon technical change, especially in times of recessions when private firms tend to hesitate to invest in radical, but risky innovations.

References

Attewell, P. (1987), The deskilling controversy, Work and Occupations 14:323-346.

Ayres, R.U. (1988), Future trends in factory automation, Manufacturing Review 1:93-103.

Barras, R. (1987), The Case of Office Based Services, Organisation for Economic Co-operation and Development, Paris.

Bertrand, O. and Noyelle, T. (1986), The Evolution of New Technology, Work and Skills in the Service Sector, Organisation for Economic Co-operation and Development, Paris.

Bhalla, A.A. (1986), Impact of new technologies on employment and social conditions. Paper presented at the Conference on 'Technology Alternatives and Employment Policy Options to the Year 2000', IIASA, Laxenburg, March.

Bjorn-Andersen, N. (1981), The impact of electronic digital technology on traditional job profiles, in: OECD (ed.) Microelectronics, Productivity and Employment, pp. 48-58, Organisation for Economic Co-operation and Development, Paris.

Blackburn, P., Coombs, R. and Green, K. (1985), Technology, Economic Growth and the Labour Process, Macmillan Press, London.

Brainard, R. and Fullgrabe, K. (1986), Technology and jobs, STI Review 1, 9-46.

Braverman, H. (1974), Labor and Monopol Capital: The Degradation of Work in the Twentieth Century, Monthly Review Press, New York.

Brotchie, J.F. (1986), Industrial interdependence via information technology and transport interaction - employment impacts, in: Nijkamp, P. (ed.) Technological Change Employment and Spatial Dynamics, pp. 115-130, Springer, Berlin et al.

Browa, H. (1984), Auswirkungen der technischen Entwicklungen in der Mikroelektronik auf Wirtschaft und Arbeitsmarkt in der Schweiz, Verlag Rüegger, Diessenhofen

Bullinger, D. (1985), Mikroelektronik - Ökotronik: Entwicklungsstand, Perspektiven und voraussehbare Auswirkungen der neuen Technologien, in: Öko-Institut/Projektgruppe Ökologische Forschung (ed.), Arbeiten im Einklang mit der Natur - Bausteine für ein ökologisches Wirtschaften, pp. 257-281, Dreisam, Freiburg.

Canadian Department of Employment and Immigration (1985): Changing skill requirements, in: OECD (ed.), Employment Growth and Structural Change, pp. 83-89, Organisation for Economic Co-operation and Development, Paris.

Carnoy, M. (1985), High technology and international labour markets, International Labour Review 124:643-659.

Crompton, R. and Reid, S. (1982), The deskilling of clerical work, in: Wood, S. (ed.) The Degradation of Work? pp. 163-178, Hutchinson, London.

Cross, M. (1983), Technical change, the supply of new skills, and product innovation, in: Gillespie, A. (ed.), Technological Change and Regional Development, pp. 54-67, Pion, London.

Dostal, W. (1981), The contribution of information and related technologies to productivity growth, in: OECD (ed.), Microelectronics, Productivity and Employment, pp. 38-47, Organisation for Economic Co-operation and Development, Paris.

Dostal, W. (1982), Fünf Jahre Mikroelektronik - Diskussion, Mitteilungen aus der Arbeitsmarkt- und Berufsforschung 15:151-166.

Dostal, W., Kamp, A.-W, Lahner, M. and Seessle, W.P. (1982), Flexible Fertigungssysteme und Arbeitsplatzstrukturen, Mitteilungen aus der Arbeitsmarkt- und Berufsforschung 15:182-191.

Fischer, M.M. and Nijkamp, P. (eds.) (1987), Regional Labour Markets, Analytical Contributions and Cross-National Comparisons, North-Holland, Amsterdam et al.

Freeman, C. and Soete, L. (1985), Information technology and employment: An assessment, Science Policy Research Unit, University of Sussex.

Gershuny, J. and Miles, I. (1983), The New Service Economy: The Transformation of Employment in Industrial Societies, Frances Pinter, London.

Gleave, D. (1986), The impact of innovations on service employment, in: Nijkamp, P. (ed.), Technological Change, Employment and Spatial Dynamics, pp. 177-195, Springer, Berlin et al.

Goddard, J.B., Gillespie, A.E., Robinson, J.F. and Thwaites, A.T. (1985), The impact of new information technology on urban and regional structure in Europe, in: Thwaites, A.T. and Oakey, R.P. (eds.), The Regional Economic Impact of Technological Change, pp. 215-241, Frances Pinter, London.

Group of Experts on Technology, The Business and Industry Advisory Committee (1981), Impact of microelectronics on employment, in: OECD (ed.), Microelectronics, Productivity and Employment, pp. 196-210, Organisation for Economic Co-operation and Development, Paris.

Hanisch, T. (1981), The impact of information technology on employment from the standpoint of labour market theory, in: OECD (ed.), Microelectronics, Productivity and Employment, pp. 63-74, Organisation for Economic Co-operation and Development, Paris.

Institut für Wirtschaftsforschung (1981), Technischer Fortschritt - Auswirkungen auf Wirtschaft und Arbeitsmarkt, Berlin.

Jones, B. (1982), Destruction or redistribution of engineering skills? in: Wood, S. (ed.), The Degradation of Work? pp. 179-200, Hutchinson, London.

Kaplinksky, R. (1987), Microelectronics and employment revisited: A review, ILO, Geneva.

Katsoulacos, Y (1984), Product innovation and employment, European Economic Review 26:83-108.

Kimbel, D. (1981), The macroeconomic relevance of advanced microelectronics and telecommunications technologies, in: OECD (ed.), Microelectronics, Productivity and Employment, pp. 15-21, Organisation for Economic Co-operation and Development, Paris.

Kimbel, D. (1986), Information technologies and output and employment movements. An analysis of some recent quantitative observations and likely future developments. Paper presented at the Conference on 'Technology Alternatives and Employment Policy Options to the Year 2000', IIASA, Laxenburg, March 1986.

Kohler, C. and Schultz-Wild, R. (1983), Flexible manufacturing systems - manpower problems and policies. Paper presented at the 1983 World Congress on the Human Aspects of Automation, Ann Arbor, Michigan, August 1983.

Lakshmanan, T.R. and Chatterjee, L. (1986), Facilitating technical change. The role of women. Paper presented at the European Summer Institute of the RSA, University of Umea, Sweden, June 1986.

Leontief, W. and Duchin, F. (1984), The effects of automation on employment (1963-2000). Abstract and executive summary of final report. Institute for Economic Analysis, New York University.

Maeda, N. (1981), A fact finding study on the impacts of microcomputers on employment, in: OECD (ed.), Microelectronics, Productivity and Employment, pp. 155-180, Organisation for Economic Co-operation and Development, Paris.

Markmann, H. (1985), The role of trade unions in coping with the labour implications of technological change, in: OECD (ed.), Employment Growth and Structural Change, pp. 141-146, Organisation for Economic Co-operation and Development, Paris.

Marsh, P. (1982), The Robot Age, Abacus, London.

Miles, I (1986), Information technology and service activities: Two Scenarios. Paper presented at the Conference on 'Technology Alternatives and Employment Policy Options to the Year 2000', IIASA, Laxenburg, March 1986.

Northcott, J. (1986), Microelectronics in Industry: Promise and Performance, Policy Studies Institute, London.

Northcott, J. and Rogers, P. (1984), Microelectronics in British Industry: The Pattern of Change, Policy Studies Institute, London.

Northcott, J. et al. (1985), Microelectronics in Industry. An International Comparison: Britain, Germany, France, Policy Studies Institute and the Anglo-German Foundation, London.

OECD (ed.) (1981), Microelectronics, Productivity and Employment, ICCP 5, Organisation for Economic Co-operation and Development, Paris.

OECD (1982), Micro-Electronics, Robotics and Jobs, ICCP 7, Organisation for Economic Co-operation and Development, Paris.

OECD (1983), Industrial Robots. Their Role in Manufacturing Industry, Organisation for Economic Co-operation and Development, Paris.

OECD (ed.) (1985), Employment Growth and Structural Change, Organisation for

Economic Co-operation and Development, Paris.

OECD (1987), Information Technology and Economic Prospects, ICCP 12, Organisation for Economic Co-operation and Development, Paris.

Österreichisches Institut für Wirtschaftsforschung (1981), Anwendungen, Verbreitung und Auswirkungen der Mikroelektronik in Österreich, Vienna.

Office of Technology Assessment (1984a), Computerized Manufacturing Automation: Employment, Education and the Workplace, OTA-CIT-235, U.S. Government Printing Office, Washington, D.C.

Office of Technology Assessment (1984b), Technology, Innovation and Regional Economic Development, OTA-STI-238, U.S. Government Printing Office, Washington, D.C.

Patel, P. and Soete, L. (1987), Information Technologies and the Rate and Direction of Technical Change, Organisation for Economic Co-operation and Development, Paris.

Petit, P. (1987), Automation of Services: The Case of the Banking Sector, Organisation for Economic Co-operation and Development, Paris.

Pietchinis, S. (1983), Computer Technology and Employment - Retrospect and Prospect, MacMillan Press, London.

Rees, J., Briggs, R. and Hicks, D. (1985), New technology in the United States' machinery industry: Trends and implications, in: Thwaites, A.T. and Oakey, R.P. (eds.), The Regional Economic Impact of Technological Change, pp. 164-194, Frances Pinter, London.

Rothwell, R. and Zegveld, W. (1981), Industrial Innovation and Public Policy: Preparing for the 1980s and the 1990s, Frances Pinter, London.

Rumberger, R.W. (1981), The changing skill requirements of jobs in the U.S. economy, Industrial and Labor Relations Review 34:578-590.

Rush, H. and Hoffman, K. (1987), Microelectronics in the Clothing Industry, Organisation for Economic Co-operation and Development, Paris.

Sorge, A., Hartman, G., Nicholas, I. and Warner, M. (1981), Microelectronics in the Workplace: Unity and Diversity of Work under CNC in Great Britain and West Germany, International Institute of Management, Berlin.

Schonberger, R.J. (1982), Japanese Manufacturing Techniques, Free Press MacMillan, New York.

Soete, L. (1986), Employment policies with regard to the new technologies: The case for technology diffusion policies. Paper presented at the Conference on 'Technology Alternatives and Employment Policy Options to the Year 2000', IIASA, Laxenburg, March 1986.

Soete, L. (1987), The impact of technological innovation on international trade patterns: The evidence reconsidered, Research Policy 16, 101-130.

Soete, L. and Dosi, G. (1983), Technology and Employment in the Electronics Industry, Frances Pinter, London.

Soete, L. and Freeman, C. (1985), New technologies, investment and employment growthi, in: OECD (ed.), Employment Growth and Structural Change, pp. 52-82, Organisation for Economic Co-operation and Development, Paris.

Solomon, J.S. (1987), Union responses to technological change: Protecting the past or looking to the future? Labor Studies Journal 12:51-64.

Spenner, K. (1979), Temporal changes in work content, American Sociological Review 44:965-975.

Stoneman, P. (1987), An analytical framework for analysing the impact of information technologies on economic perspectives, in: OECD (ed.), Information Technology and Economic Prospects, pp. 70-93. Organisation for Economic Co-operation and Development, Paris.

Stoneman, P., Blattner, N. and Pastre, O. (1982), Major findings and policy responses to the impact of information technologies on productivity and employment, in: OECD (ed.), Micro-electronics, Robotics and Jobs, pp. 17-131, Organisation for Economic Co-operation and Development, Paris.

Stooß, F. and Troll, L. (1982), Die Verbreitung 'programmgesteuerter Arbeitsmittel', Mitteilungen aus der Arbeitsmarkt-und Berufsforschung 15:167-181.

Thwaites, A.T. (1983), The employment implications of technological change in a regional context, in: Gillespie, A. (ed.), Technological Change and Regional Development, pp. 36-53. Pion, London.

Vickery, G., Blair, L., Estime, M.-F and Hayakawa, H. (1987), Diffusing new technologies: Micro-electronics, STI Review 2:41-80.

Watanabe, S. (1986), Labour-saving versus labour-amplifying effects of micro-electronics, International Labour Review 125:243-259.

Weiss, M.A. (1985), High-technology industries and the future of employment, in: Hall, P. and Markusen, A. (eds.), Silicon Landscapes, pp. 80-93, Allen and Unwin, Boston, London and Sydney:

Werneke, D. (1983), Microelectronics and office jobs: The impact of the chip on women's employment, ILO, Geneva.

Whitley, J.D. and Wilson, R.A. (1987), Quantifying the impact of information technology on employment using a macroeconometric model of the United Kingdom economy, in: OECD (ed.), Information Technology and Economic Prospects, pp. 176-220, Organisation of Economic Co-operation and Development, Paris.

Wilson, R.A. (1987), IT in the Engineering Industry, Organisation for Economic Co-operation and Development, Paris.

Wilson, R. and Whitley, J. (1982): Quantifying the employment effects of microelectronics, Futures 14:486-495.

Wood, S. (ed.) (1982), The Degradation of Work? Skill, Deskilling and the Labour Process, Hutchinson, London.

Zicklin, G. (1987), Numerical control machining and the issue of deskilling, Work and Occupations 14:452-466.

4 Innovation, new information technologies, and the structure of urban regions

R. H. FUNCK AND J. S. KOWALSKI

1. Introduction

We live in a time of discontinuities. The rate of technological change and diffusion of innovations in all areas of human activities has accelerated to an awesome extent. New communication technologies and data processing techniques seem to have invalidated most of the elements of and procedures in reality to which we are accustomed.

In economics, in regional studies and related fields we re-discover the classics of innovation research (Schäffle, Schumpeter), we also tend to emphasize dynamics, motion, disequilibrating forces generated by change, rivalry, competition, by the endless pursuit by all human beings of new ideas, of technological solutions, of social rearrangements, instead of focusing our attention on the theoretically possible, but in reality irrelevant equilibrium, steady states. Synergetics, the theory of chaos and related theoretical concepts incorporating the non-linearity of the real world as well as the disruptive, 'catastrophic' quality of many societal changes provide tools for the conceptualization -if not always operationalization- of such phenomena (see Weidlich, Haag, 1983).

The acceleration of technological change, and especially the new ways of connecting people and institutions in space, as well as of moving them from one location to another - all this must and will bring economic-spatial consequences, which, as to their range and scope, are still very vague. It seems, however, certain that new technologies, particularly in the field of communication, and innovation-rich activities will, to a considerable extent, determine a region's competitiveness and future opportunities for development. Areas where these activities are relatively abundant will very probably leave their less intensively participating competitors behind in the game in which the stakes are high indeed: growth, wealth, living standards and political power.

In the 1970s and in the early 1980s important studies (Van den Berg et al., 1979; Klaassen et al., 1984) established the concept that large

European cities had entered the disurbanization phase, and that not the hinterlands of these large cities but the small and medium-sized cities (SMCs) would absorb the outflow of people from large agglomerations. This process was postulated to have resulted from a combination of existing agglomeration diseconomies in large cities, and of agglomeration economies in SMCs. In a number of our own studies the thesis of global deconcentration (resulting from agglomeration diseconomies) and local concentration (i.e., concentration of activities leading to higher densities in SMCs as a result of agglomeration economies) could be formulated (Funck, Blum, 1981, 1986; Kowalski, 1983). Empirical research by other authors also seemed to support this theoretical statement (cf., Hall, 1982).

How will the advent of new technologies and the growing role of research and innovation-oriented activities influence this trend? Will they reinstate the advantages of large cities, as some scholars maintain, or will they lead to further spatial deconcentration to the advantage of hinterlands and of smaller centres by enabling people to communicate and to move in space with more ease than in the past (for a review of arguments of both viewpoints see Funck, 1986)?

As in many other mature economies, in the last years the economy of the Federal Republic of Germany experiences serious structural changes connected with the rapid development and spread of new technologies leading to creation of new economic activities, reshaping the character of the old ones and influencing the structure of the labour force.

The above mentioned influences are likely to reflect (Lakshmanan, Chatterji, 1986, p. 15):

- the cumulative effect of a cluster of new technologies, and
- the continuing structural evolution of the sectoral composition of the economy leading to the growing dominance of the service sector and to the changes within this dominant sector (see Table 1 on the development of employment by sectors in the Federal Republic of Germany).

Development and diffusion of new information technologies are enhanced by various factors which are regionally unequally available. Some of those factors are (Funck, 1986):

- regional or urban economic structure as described by sectoral control relationships;
- the quality of regional or urban transportation and communication infrastructure;
- existence of the Schumpeterian dynamic, risk-willing entrepreneurs;
- provision with venture capital;

Table 1 Sectoral structure of employment in the Federal Republic of Germany

Employment by sectors in %	1950	1960	1970	1983	1984	1985
Agriculture Forestry and Mining	28,2	16,5	10,6	7,5	7,4	7,4
Manufacturing and Construction	39,1	45,1	46,8	39,7	39,4	39,1
Services	32,7	38,4	42,6	52,8	53,2	53,5

Source: Institut der Deutschen Wirtschaft, Zahlen zur wirtschaftlichen Entwicklung, 1986, p. 6.

- accessibility of technological information;
- presence and activity of innovation-supporting institutions in the public and private sectors.

Similarly, it can be said that regional or urban deficits in the supply of relevant information and of other above enumerated factors constitute spatially variable detriments or barriers for the development of new technologies.

2. Spatial Aspects of New Information Technologies

The possible spatial and employment impacts of the introduction of the new information technologies came to the attention of spatial planners in the Federal Republic of Germany only very recently. In a forecast of the future spatial organization of the country prepared by PROGNOS and Bundesforschungsanstalt für Landeskunde und Raumordnung in 1983 several scenarios of future employment patterns include different assumptions concerning the growth of productivity of labour, but no spatial consequences of new communication technologies are discussed.

A distinct shift of viewpoints in considering the role of technological information in urban and regional development has occurred in recent years. While, during the 1970s, the economic-technological content of information, the degree of access to it and the spatial patterns of information flow have been predominant as objects of research, in the early 1980s the advent of new communication technologies has moved to the centre of analytical interest (Funck, Kowalski, 1984).

Under the first approach, spatial access to innovation has been recognized as the most important factor for enabling or facilitating technology-based activities as a prerequisite for regional development especially in backward, lagging regions. The peak of this type of approach to the role of information in regional development may be epitomized by the formulation of the 'innovation-oriented regional policy' and related proposals (Ewers, Wettmann, 1980; Kowalski, 1982).

In contrast to this concept of indirect effects of information -through innovation in production techniques- on regional development, the second approach visualizes a more direct impact of new communication technologies on locational requirements of production and on mobility patterns. At the same time, fears and anxieties concerning the negative impacts of these technologies come to the fore. The ease with which the 'friction of distance' may be overcome with the help of these technologies, is held to constitute a threat for jobs and for the traditional role of cities, without leading by itself to a diminishing of the differences in the development levels between core and peripheral areas.

At present, a real danger exists that the spread of new information technologies will be directed predominantly to already developed areas whereas the peripheral regions will stay largely devoid of information transfer networks (see Hoberg, 1983; Schulz-Trieglaff, 1982).

Referring to the long-term effects of these technologies on employment as a whole, no consensus has, so far, been achieved in the literature. Some studies maintain that the overall impact will be neutral or negligible, since the disappearance of workplaces will be compensated by the creation of others (see Dostal, 1982). Other authors are more pessimistic in predicting considerable loss in the employment level (e.g., Henckel et al., 1984), still other reports recognize that the results of forecasts are extremely sensitive to the underlying assumptions, so that very little in the way of hard facts can be achieved (OECD, 1981). Generally, the overall effects of new technologies on society and its functioning can be estimated only in a value-loaded way.

It seems obvious that the advent of new communication technologies will lead at least to serious structural changes in the labour market. Many routine jobs will vanish and certainly others will be created. But with respect to the possible overall decrease of the number of workplaces some fears seem exaggerated, at least in a long-term perspective, considering the present demographic tendencies in industrialized societies (Funck, Kowalski, 1987).

As mentioned above, new technologies also influence locational behaviour. Generally they increase the degree of freedom of locational choice of firms and individual households. At the same time they may widen the attractiveness gap between core and periphery should the latter for some reason remain outside the umbrella of new information networks. Also, regarding some types of service activities (banking, insurance, etc.), new information technologies through permitting largely increased scales of firms, and as a result of an increase in the degree of complexity of operation may lead to a stronger reliance on locations within the large agglomerations.

In a recent study of the spatial effects of new communication technologies on urban and regional development a team of German authors (Henckel et al., 1984) investigated these issues in a rather speculative manner. They conducted questionnaire analyses and interviews with experts in information technology-dependent enterprises. As the authors state, their conclusions must be treated with caution due to the incomplete data and the 'non-scientific' methodology. They may be summarized as follows (op. cit., pp. 161-162):

(1) Economic factors and increased acceptance levels will lead to a relatively fast diffusion of new information technologies.

(2) Information technology does not in itself cause spatial changes. However, it reinforces existing migratory trends.

(3) Information technology is a necessary but not a sufficient condition for the acceleration of spatial deconcentration.

(4) Spatial effects of information technology are not uni-directional. Both centripetal and centrifugal effects are discernible.

(5) Since out-migration from metropolitan agglomerations to smaller centres is, at present, the dominant tendency it will be reinforced by information technologies. However, some countertendencies may also

occur, like the concentrating effects of rising energy costs (see Kowalski, 1983).

(6) No equalization of development levels between agglomerated areas and the countryside should be expected. In the foreseeable future the spatial effects of new communication technologies will be a strengthening of the advantages of metropolitan areas and local centres as compared to the periphery.

(7) The North-South shift of population and economic activities which prevails in the Federal Republic of Germany will be reinforced.

(8) New work-places will be created predominantly in new firms. These will tend to locate outside the city centres and thus form one of the factors of metropolitan local deconcentration.

(9) The potential of creation of tele-work-places is limited. The influence of this type of work on traffic flows and land use patterns will be very limited if not negligible.

With respect to the impact of new communication technologies on traffic flows various tendencies have to be regarded: a slight decrease of flows as a consequence of tele-work; a decrease of flows of commuter traffic as a result of the local concentration tendencies, a decrease of flows as a result of consumer use of information technologies (tele-banking, tele-shopping etc.); increased flows in consequence of increasing leisure time activities.

The conclusion can be stated that the new information technologies will exert diverse effects on spatial structure and development. They may have deconcentrating or concentrating impacts with respect to physical distribution of activities in space, with the deconcentration tendency prevailing in all probability. On the other hand they may lead to decentralization or centralization of power and control throughout space with the probability of centralization higher than vice versa (see on those points Spehl, 1985 and Funck, 1986). Left to themselves the new technologies will most probably reinforce existing trends in development, i.e., lead to increased disparity between the strong and weak regions. Their effects in the future will depend on the distribution in space of new communications infrastructure. It must be considered that the explosive spread of these technologies, while in principle making it easier to provide information facilities and to enable the processing and transfer of information in practically every location in peripheral areas, will also lead to tensions concerning the viability of the centres of large cities and increase the relative backwardness of the periphery (Castells, 1985).

Thus, the introduction of new communication media does not diminish the relevance of regional policy measures including the provision of innovation-relevant information and conditions. On the contrary, the growing importance of information technologies in the reshaping of society will enhance the potential role of regional policy in the future.

3. Empirical Investigation[1]

The development of new information technologies and their spread into commercial applications depend on the innovative behaviour of entrepreneurs and other economic agents. Hence an attempt has been made to empirically analyze the differences in 'innovative propensity' in various urban and non-urban areas in the Federal Republic of Germany. We assumed that this 'innovative propensity' will be responsible, among other factors, for the future development of new technologies, and consequently for the spatial impacts of these technologies. We were interested in the question whether there are any visible differences in this innovative behaviour between cities and rural counties, between cities of various sizes and between cities and rural counties located in expanding, fast growing Southern Germany (the States of Bavaria and Baden-Württemberg) and Northern Germany, which is often said to lag behind. We also wanted to gain some insight into the particular situation of selected Old Industrial Regions, which face almost insurmountable problems due mainly to the structure of their economies (see Funck, 1985; Funck, Kowalski, 1987).

Even a very tentative answer to those questions can provide valuable insights into the possible future fate of various urban centres and regions in the high-technology era.

Data base, methodology and results

In our investigation we relied on data from the study of innovation densities conducted at the Fraunhofer Institute of Systems Analysis and Innovation Research in Karlsruhe (ISI, 1984) recomputed, extended and adapted to more spatially desaggregated units (Blum, 1985; Blum, Koblo, 1986).

Innovation density in a given region is defined as the ratio of the number of enterprises which applied for grants for employment of R & D personnel in the framework of regional innovation support programmes in the

---

[1] This section is partly based on Funck, Kowalski, Koblo, 1987.

Federal Republic of Germany to the total number of enterprises in this region. It may be argued, that this ratio is at best a very rough indicator of the innovation activities in the region, expressing to some extent the bureaucratic skills of managements of the enterprises and not their innovativeness, but at present it is the only indicator available for all regions in the Federal Republic of Germany.

The second indicator relates to patent applications by private persons and firms. This is not an ideal indicator either: not all inventions become innovations in the Schumpeterian sense, also the granting of a patent rather than applying for it should be taken into account. Unfortunately, however, data for patent granting are not available in adequate spatial disaggregation.

Additionally we look at the total spending on research personnel in the firms which took part in the above mentioned government support programmes. This provides some clues with respect to the scale of research efforts being conducted in various counties and cities.

We investigated 326 counties (Kreise) in Germany, of which 91 are so-called city-counties (kreisfreie Städte), meaning a city constituting a county in the administrative sense. Furthermore, in some cases we divided the group of city-counties into large (over 200 thousand inhabitants), medium (100 thousand to 200 thousand) and small cities (less than 100 thousand).

First of all we have estimated the impacts of various regional variables on innovation density for the whole Federal Republic of Germany by means of cross-regional regression analysis in order to gain insights into the general determinants of innovation activities in Germany. Then we investigated in greater detail spatial differences in innovation density for various types of regions.

The regressions for the whole of the Federal Republic of Germany have been conducted with the use of the L-1.1 Programme (Liem et al., 1985). In comparison with other methods, which assign a linear or logarithmic functional form to the data, L-1.1 has the great advantage that it allows the data to determine the functional form. It also considers statistical irregularities in the data like heteroscedasticity, dependencies in the data or autocorrelations.

To find a best fitting functional form for the underlying data several functions are estimated with the programme. In every estimation one or a few parameters are excluded. At first a very restricted linear or loglinear function is estimated, then a more flexible functional form is developed.

The estimations are compared with each other through a likelihood value which is computed by the programme. The difference in the likelihood value of two estimations provides a measure of improvement in the functional form from one estimation to the next. Finally a best fitting functional form for the data is derived.

The effects of the exogenous variables on the endogenous ones are measured by the elasticities of the exogenous variables. In non-linear functions the often used regression coefficient is inappropriate for this purpose.

We call the effect of an exogenous variable 'stable', if the elasticity has the same sign in every estimation. If the elasticity changes in any estimation or if the value of the t-statistic is very low, we call it 'unstable'. The results are provided in Table 2.

In the table we have no absolute values for the elasticities because in non-linear transformations which have been used in the L-1.1 programme, we cannot take the absolute values to quantify the effects. For this reason the table contains only '+' and '-' signs to indicate the direction of the influence.

One should notice that the 'best', i.e., the strongest impacts were obtained for variables which may be called 'classical', like population, fixed assets, income and unemployment. The results for infrastructure variables and variables expressing administrative status of a region did not provide any clear indications on the direction or strength of the influence.

In the 'classical' variables one can observe the positive impact of the population density (which might be interpreted as agglomeration factor). Interestingly the impact of fixed assets per capita is strongly positive in the non-city counties and negative in the city-counties. Possibly this could be explained by the high levels of fixed assets values in heavy industry and mining, which are not as innovative as other sectors and which are concentrated in cities, especially in many cities in the Northern part of the Federal Republic of Germany. The fact that income per capita levels in the adjacent counties exert positive influence seems plausible.

We are somewhat at a loss trying to explain the impacts of income per head and unemployment levels on the innovation density. In the case of non-city counties it seems that low incomes coincide with less innovativeness, but if employment is high it leads to more innovative efforts.

In the cities higher incomes go hand in hand with increased innovation density, but higher unemployment levels decrease it. One would have to study economic structure and organizational features of firms in particular cities

Table 2 Determination of innovation densities in the Federal Republic of Germany 1980

	All regions	Non-city counties	Cities
Population density[1)]		++	+
Fixed assets per head		++	-
Income of adjacent[2)] counties or cities per capita	+	++	+
Income per capita[2)]	-	--	+
Unemployment[3)]		++	-
Interstate highway density 4)		-	+
State highway density[4)]		-	-
Other roads density[4)]	-		
Density of road net-[5)] work nodes		+	-
Number of beds in hotels per 1,000 inhabitants		+	
High level center[6)]			
Site of regional authorities 7)	-		-
Capital of State or of the Republic	+	-	+

Legend:

++ (--) high and stable positive (negative) effect
+ (-) low positive (negative) effect

1) Number of inhabitants per km^2

2) In 1000 DM

3) Number of unemployed persons per 1000 inhabitants

4) Length in km per 1 km^2

5) Number of nodes per 1 km^2

6) Oberzentrum

7) Bezirksregierung

in depth in order to be able to provide definite interpretations of those results. This was impossible within the scope of this paper.

We shall subsequently describe and analyze innovation density levels, patent applications, and the spending on research and development personnel for various types of regions and for selected cities in the Federal Republic of Germany.

Table 3 and Figure 1 summarize the results for innovation densities; Tables 4 and 5 provide additional information on patent applications and R & D spending.

From Figure 1 it can be seen that no clear-cut statement considering innovation densities of the 11 West-German states, depending on their location and economic prosperity, is possible. High densities are recorded in Baden-Württemberg as expected, but also in Bremen, with its stagnating ship-building industry and high unemployment. On the other hand, as can be seen from Table 3, if we compare Northern Germany with Southern Germany as a whole, the advantage of the South becomes visible. This advantage is especially pronounced in the case of non-city counties. The low density levels for the city-counties in the South should not come as a surprise. The share of relatively small cities in the urban system there is higher than in the North. As can be seen in Table 3, small cities as a group record the lowest density levels, compared with medium and large cities, but also with non-city counties. The advantageous position of the large cities should be noted.

Table 4 shows indicators for the whole 91 city sample. With respect to patent applications medium sized cities perform better than the large ones, which, in turn, outperform the small cities.

Small and medium-sized cities record higher spending for R & D personnel than their employment and GDP shares. Thus, on the whole it seems that the small and medium-sized cities do not perform too badly, but also not too well, in the innovation landscape of the Federal Republic of Germany.

Taken together those results indicate that large cities and being in the South coincides strongly, with innovation activities. Being in the non-city region is no disadvantage, in comparison with small and medium cities, especially if the non-city region in question happens to lie in the vicinity of large cities, possibly in the South.

This last statement is also supported by the evidence given in the bottom part of Table 3 where we provide innovation densities for a selected new industrial region and its hinterland (Stuttgart) and for two peripheral

Table 3 Innovation densities in the Federal Republic of Germany *) 1980

Federal Republic of Germany average	16,627
small cities	11,331
medium cities	15,143
all cities	15,870
non-city counties	16,627
cities in the North	16,296
cities in the South in the North	14,947
non-city counties in the South	18,500
Northern Germany	15,961
Southern Germany	17,718
Saarland	16,984
Ruhrgebiet cities	13,557
Ruhrgebiet other counties	16,847
Stuttgart-city	23,515
Stuttgart-hinterland	25,652

*) Computations based on data from ISI, 1984

FIGURE 1

Innovation densities in different regions in the Federal Republic of Germany *

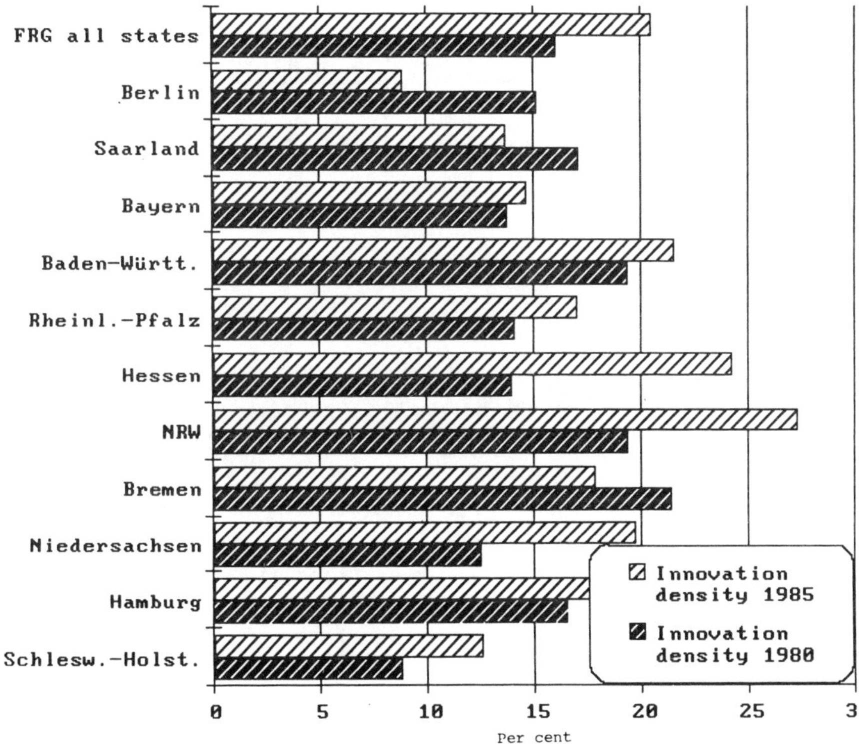

*) Computations based on data from ISI

Table 4 Patent applications and R & D spending in the Federal Republic of Germany (1981)

	FRG	All cities	Small cities	Medium cities	Large cities
Patent applications per 100 thd. inhabitants	62.59	122.32	81.76	168.64	113.92
Spending on research staff as % of such spending in all cities	-	100	17.01	34.31	48.68

Table 5 Innovation densities and patent applications in county-cities of Baden-Württemberg (numbers in brackets apply to hinterlands of the given cities)

	Number of patent applications, 1982	Innovation density, 1981
Stuttgart	2,288	23.51
Mannheim	585	12.95 (28.70)
Karlsruhe	184	5.73 (34.67)
Freiburg	219	14.02
Heidelberg	239	20.0
Heilbronn	228	15.87
Pforzheim	106	14.44
Ulm	116	11.11
Baden-Baden	29	8.93

Old Industrial Regions (Ruhr and Saarland): Stuttgart shows a very highvalue but its hinterland has an even higher level. The Ruhr-area records the expected low density, but its non-city counties do not deviate from the Federal Republic of Germany average. The same is true in the case of Saarland.

We shall briefly turn our attention to that part of the FRG in which technical change, introduction of high-tech related activities and new communication technologies are pursued fastest within the country, namely the State of Baden-Württemberg. A very recent study, conducted on behalf of the state government, on the economic situation of rural areas (Struktursituation, 1987) provides a very similar picture as the one gained from the analysis of the whole of the Federal Republic of Germany: the rural areas and smaller urban centres performed economically quite well in the last decade, in comparison with the larger urban areas and their hinterlands (see Figure 2). But the patent applications are concentrated very strongly in the largest urban centres of the region (Table 5). Also, similar to the situation for the country as a whole, the hinterlands of the large cities record the highest levels of innovation densities.

4. Concluding Remarks

On the whole the empirical evidence presented in this study indicates that the distribution of innovation activities which are as a rule connected with the information based new technologies, seems to favour the economically stronger highly developed areas of large cities and their neighbourhoods as opposed to the lagging peripheral regions and old industrial areas. The long-term spatial effects of new information technologies are not certain: to a large extent they will depend on the efficiency of future regional and urban policies and on technology policy pursued by the federal and state governments. Left to itself the development of new technologies will reinforce rather than cure, various regional 'diseases' of which the North-South disparity and the problems of Old Industrial Regions are the prime examples.

In the short run no policy is possible which would improve the disadvantageous position of regions which do not participate in new technological revolution at an equal rate with the rest of the country. Only a well conceived, concerted and persistent application of a package of measures, aimed at an increase in the locational attractivity of those areas, mainly by improvement of the quality and supply of specialized labour force, provision of research and education centres and specialized financial

Figure 2 Relative economic situation of types of regions in Baden–Württemberg

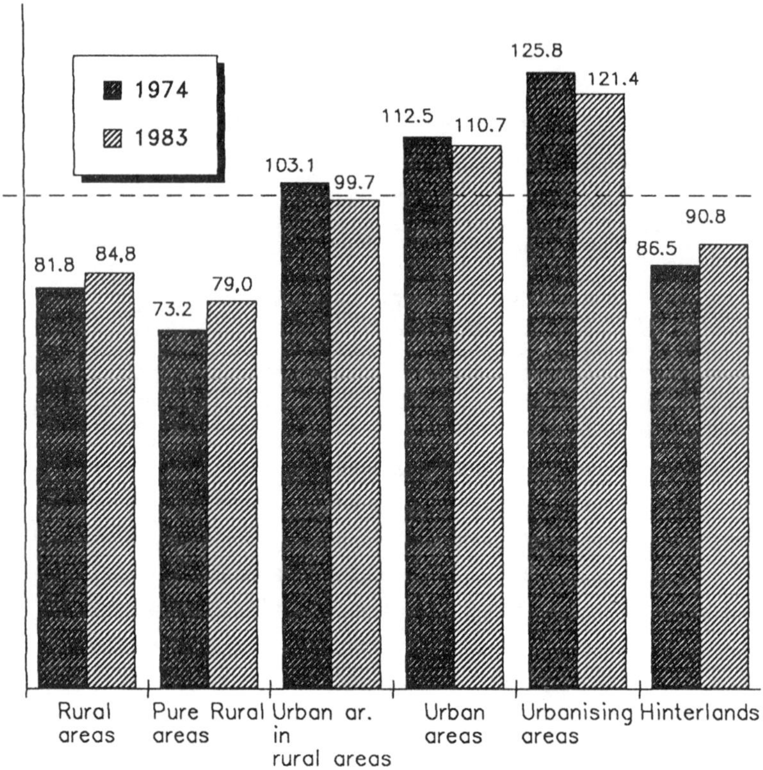

Baden–Württemberg = 100

Source: Struktursituation 1987 p. 79

and other services (Amberger et al., 1987) can help to avoid the growing polarization between the strong and weak regions as a consequence of the development of new information technologies.

References

Amberger, J., Böttcher, H., Funck, R. and Keller, M. (1987), Innovationspotential im Wirtschaftsraum Karlsruhe, Lörrach, Karlsruhe.

Van den Berg, L., Drewett, R., Klaassen, L., Rossi, R. and Vijverberg, C. (1979), Urban Europe, CURB, Erasmus University, Rotterdam.

Blum, U. (1985), Karlsruher Regionaldatenbank, Institut für Wirtschaftspolitik und Wirtschaftsforschung, Universität Karlsruhe, available from the author.

Blum, U. and Koblo, R. (1986), Regionale Determinanten des Innovationsprozesses, Institut für Wirtschaftspolitik und Wirtschaftsforschung, Universität Karlsruhe, Discussion Paper 8/1986.

Castells, M. (ed.) (1985), High Technology, Space and Society, Urban Affairs Annual Review, Vol. 28, (Sage Publications), Beverly Hills, London, New Delhi.

Dostal, W. (1982), Fünf Jahre Mikroelektronik - Diskussion, Mitteilungen aus Arbeitsmarkt- und Berufsforschung, H. 2, pp. 12-31.

Ewers, H.-J. and Wettmann, R.W. (1980), Innovation oriented regional policies, Regional Studies, vol. 14, pp. 161-180.

Funck, R.H. (1985), Die Obersteiermark hat eine gute Chance - vergleichbare Regionen in der Bundesrepublik Deutschland, paper presented at a conference Neue Kraft für traditionelle Industrieregionen, Graz, 2nd October, Mimeo, Karlsruhe.

Funck, R.H. (1988), Technologiepolitik und räumliche Struktur, in: Funck, R.H. (ed.), Technologiepolitik und Regionalentwicklung, Heidenheimer Schriften zur Regionalwissenschat August Lösch in memoriam, H. 9, pp. 10-23.

Funck, R.H. and Blum, U. (1981), Urban Policy Analysis, in: Nijkamp, P. and Rietveld, P. (eds.), Cities in Transition: Problems and Policies, Sijthoff and Noordhoff, pp. 243-255.

Funck, R.H. and Blum, U. (1986), Processes of Urban Development, in: Paelinck, J. (ed.), Human Behaviour in Geographical Space, Gower, Aldershot, Hants.

Funck, R.H. and Kowalski, J.S. (1984), The role of information in regional and urban development, in: IARUS (International Association for Urban and Regional Statistics), 11th Meeting, Copenhagen, pp. 53-89.

Funck, R.H. and Kowalski, J.S. (1987), Innovation and Urban Change, in: Brotchie, J., Hall, P. and Newton, P. (eds.), The Spatial Impact of Technological Change, Croom Helm, London, New York, Sydney, pp. 229-239.

Funck, R.H, Kowalski, J. and Koblo, R. (1987), New Technology, Innovative Activities and the German City System, in: Opitz, O. and Rauhut, B. (eds.), Ökonomie und Mathematik, Festschrift für Rudolf Henn, Springer Verlag, Berlin, Heidelberg, New York, pp. 533-543.

Hall, P. (1982), Urban Europe, Gower, Aldershot, Hants.

Henckel, D., Nopper, E. and Rauch, N. (1984), Informationstechnologie und Stadtentwicklung, Kohlhammer, Stuttgart, Berlin, Köln, Mainz.

Hoberg, R. (1983), Raumwirksamkeit neuer Kommunikationstechniken - innovations- un diffusionsorientierte Untersuchungen am Beispiel des Landes Baden-Württemberg, Raumforschung und Raumordnung, November/Dezember, pp. 211-221.

ISI (Fraunhofer-Gesellschaft, Institut für Systemtechnik und Innovationsforschung), Erfassung regionaler Innovationsdefizite, in: Schriftenreihe des Bundesministers für Raumordnung, Bauwesen und Städtebau, Heft Nr. 06.054, 1984.

Institut der Deutschen Wirtschaft (1986), Zahlen zur Wirtschaftlichen Entwicklung, Köln.

Klaassen, L., Molle, W.T.M. and Paelinck, J. (eds.) (1984), Dynamics of Urban Development, Gower, Aldershot, Hants.

Kowalski, J.S. (1982), Information Development Poles: the West German Context, Research Memorandum no. 8219, Department of Economics, University of Amsterdam, The Netherlands.

Kowalski, J. (1983), Energy Costs and the Changing Spatial Structure of Cities, Institut für Wirtschaftspolitik und Wirtschaftsforschung, Universität Karlsruhe, Discussion Paper 6/83.

Lakshmanan, T.R. and Chatterji, L. (1986), Technical change and metropolitan adjustments, Regional Science and Urban Economics, 16, pp. 7-30.

Liem, T.C., Dagenais, M.G., Gaudry, M.J.I. and Koblo, R. (1986), L-1.1: Ein Programm für Box-Cox-Transformationen in Regressionsmodellen mit heteroskedastischen und autoregressiven Residuen, Institut für Wirtschaftspolitik und Wirtschaftsforschung, Universität Karlsruhe, Discussion Paper 4.

OECD (1981), Microelectronics, Productivity and Employment, Paris.

Prognos und Bundesforschungsanstalt für Landeskunde und Raumordnung (1983), Kurzfassung der Raumordnungsprognose 1995, Basel, Bonn.

Schulz-Trieglaff, M. (1982), Können neue Medien einen Beitrag zum Ausgleich von Entwicklungsunterschieden einzelner Regionen leisten?, Informationen zur Raumentwicklung, H. 3, pp. 17-34.

Spehl, H. (1985), Räumliche Wirkungen der Telematik. Stand der Diskussion und Programm des Arbeitskreises der Akademie für Raumforschung und Landesplanung, Raumforschung und Raumordnung, November/Dezember, pp. 254-270.

Struktursituation (1987), Die Struktursituation des ländlichen Raums in Baden-Württemberg 1974-1984, Ministerium für Wirtschaft, Mittelstand und Technologie des Landes Baden-Württemberg, Stuttgart.

Weidlich, W. and Haag, G (1983), Concepts and Models of a Quantitative Sociology: The Dynamics of Interacting Populations, Springer, Berlin, Heidelberg, New York.

5 The programmable automation trajectory in time and space

F. ARCANGELI AND R. CAMAGNI

Prologue (*)

The aim of this contribution is to clarify a number of unsettled issues concerning the new industrial location forces that are at work in the world wide massive redeployment of manufacturing. There are a number of economic, social and technological reasons for this phenomenon that we have become familiar with during the past two decades.

First, the current locational forces depend on the social and competitive mechanisms which simultaneously regulate the pervasive diffusion of Programmable Automation across industries and regions. Consequently, after the mid 1970s the acceleration of the penetration of microelectronics has been paralleled by an increasing interregional competition in the world economy[1].

Second, a major drive towards industrial agglomeration, even for manufacturing 'mature' products, has overcome the previous trends of geographical dispersion of routine activities. We have witnessed an inversion of the patterns described in the neoclassical stylization of the 'product life-cycle' by Hirsch, Vernon and Hymer[2].

The theoretical framework adopted in the present contribution is, on the contrary, the one under development in the heterodox stream of evolutionary economics.

A vast amount of recent empirical findings point towards the existence of new forces of industrial agglomeration. A striking fact that is often mentioned is the move of world industrialisation from the Atlantic to the Pacific ring, with the inversion of their relative 60 to 40% shares in the last 20 years (Burenstam-Linder, 1985).

Our aim is to clarify the change of the location model behind this shift: conclusions will not be so pessimistic for the 'old staple' Atlantic basin, provided it will be fast moving beyond second generation, micro-electronics-intensive programmable automation, towards next generation, A.I.-intensive flexible automation[3].

The paper is divided into three sections. In the first one we shall give a definition of the single subsystems of industrial automation processes, and analyse them through a 'map of industrial automation' charting each subsystem according to two main dimensions: hardware vs. software and standalone products vs. systems. On a third dimension of the map, the value of the international market for each subsystem is estimated and charted.

In the second section, the time trajectory of these systems is analysed, both at the theoretical and at the empirical level, with particular emphasis on the forthcoming transition from programmable to flexible automation, in the use of a number of process innovations such as robots, Flexible Manufacturing Systems and Computer Integrated Manufacturing (FMS and CIM, respectively).

The main hypothesis in this section is that organisational factors are the most decisive ones affecting diffusion speed, much more than purely technical ones; inertia to organisational change is therefore a major factor of retardation for techniques that have already become cost-effective.

In the third section, the characteristics of the spatial diffusion process of these technologies is analysed, through a review and comment of some major empirical findings in both the US and Europe. From these studies the evidence emerges of the strong rejuvenation potential of these technologies for 'mature industries' and old industrial regions. The increasing and cumulative organisational implications of technical change, in the evolution of the factory automation trajectory, is likely to reinforce the spatial concentration towards rejuvenated old industrial areas and a few selected 'first-comers' among new high-tech R & D poles.

The paper also includes a number of conclusions about the consequences of structural change upon regional policies.

1. A Map of Factory Automation

Factory automation may be defined as the set of systems and components, incorporating information technologies and in particular computer systems, specialised electronics, telecommunications and software, addressed to the performance of the following broad functions within the factory: design, production management and manufacturing.

This definition is consistent with the contemporary Computer Integrated Manufacturing philosophy, which includes all the previously mentioned functional areas; moreover, it is wide enough to include the specialized electronic instruments such as sensors, actuators and terminals.

An accepted definition and assessment of factory automation which is clear is still lacking; this is mainly due to the complexity of the technology itself, at the crossroads between different but convergent technologies, and to its continuous and fast evolution.

Therefore a quantitative assessment of the size of the market is highly hampered by this condition, and the single reliable evaluations of some specialized submarkets are never consistent with the aggregate evaluations of the global market.

A recent inquiry has tried to bring some transparency and order in this picture, not really adding a new definition to the existing ones but rather following a bottom-up procedure of comparison and evaluation of the single submarkets, and their presentation within a coherent framework (Camagni, 1988).

The proposed general framework is based on the well-known 'Map of the information business', developed at Harvard, charting the single sectors and products on a Cartesian plane along two dimensions: the product/service and the form/substance ones.

In building the Map of Factory Automation, the second dimension (on the x axis) was held, as it represents the first principle governing the structure of the sector, the principle of integration, between hard and soft components, between mechanics, electronics, and software. However, on the second dimension of the Map, a different dichotomy was employed, the product-system one, expressing the second genetic principle of factory automation, i.e., networking.

As Figure 1.1 shows, the left hand column of the Map encompasses mainly hard elements, going from supervisory systems to single, stand alone mechatronics products. The right hand column encompasses soft elements ging from high order integration systems of the three, usually still distinct, functions of design, production management and manufacturing, down to specific applications of Artificial Intelligence and expert systems to factory automation. The central column encompasses 'integrated' products and systems such as Computer-aided design and engineering (CAD/CAE) and automation testing equipment.

CIM is located logically in the central upper part of the map, among the systems but evenly integrating hard and soft components.

On the third dimension of the map, the value of the market is charted. In Table 1, an assessment of the world and of the Italian market is presented, excluding 'in-house' developments.

In box no. 2, only the electronics parts and control systems of the different machines are included. In box no. 3, encompassing the mechanical

Figure 1.1 A Map of Factory Automation

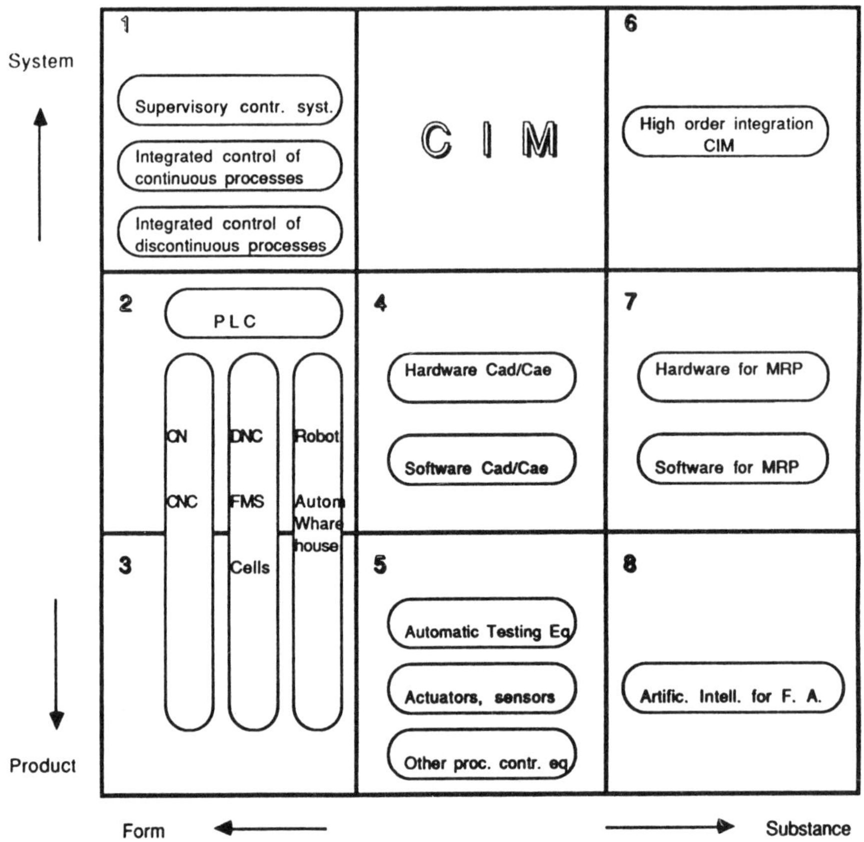

1 = Supervisory control systems
2 = PLC and Controllers
3 = Mechatronics
4 = CAD/CAE
5 = Actuators, sensors, testing eq.
6 = High order integration (CIM)
7 = Production management (MRP)
8 = Artificial intelligence for factoı automation

Table 1 THE MARKET OF FACTORY AUTOMATION
Italy (1986) and world (1985)

	WORLD MARKET '85		ITALIAN MARKET '86	
	bill $	%	bill lire	%
Supervisory control systems	7,0	17,5	410	17,4
PLC and controllers	4.9	12,3	295	12,8
Mechatronics	2,9	7,25	179	7,6
CAD/CAE	4,6	11,5	260	11,0
Actuators, sensors and testing eqms.	13,0	32,5	825	34,9
High order integration (CIM)	0,6	1,5	20	0,8
Production management (MRP)	6,6	16,5	374	15,8
Artificial intelligence for F.A.	0,4	1,0	-	-
TOTAL	40,0	100	2363	100

parts of mechatronics systems, it was decided to exclude the mechanical parts of stand alone NC and CNC machine tools and automatic warehouses (whose controls are included in box 2). Hence, box 3 includes just the mechanical parts of robots, DNC (Direct Numerical Control) systems, flexible manufacturing cells and systems (FMC, FMS).

2. The Time Trajectory of Programmable Automation

The aim of this section is to reconsider briefly the recent evolution of some main techniques of programmable automation in manufacturing, as well as the gradual emergence of new generations of more flexible techniques.

Programmable automation emerged inside a technological paradigm of manufacturing, that was previously dominated by dedicated machinery or 'hard' automation, within Taylorist or Fordist production systems (Arcangeli, 1989, ch. 5).

The first generations of programmable machinery were introduced in the 1950s, first of all in discrete piece manufacturing, with the early Numerical Control applications to machine tools; at the end of the decade, the introduction of mainframe computers to continuous process control systems followed.

The scarce efficiency and high cost of these techniques was only overcome by the positive impact of the microelectronics revolution in the 1970s, that is as a spillover effect of a broad technological trajectory in electronics and information systems.

Some anticipations of this acceleration in the pace of technical change had already been produced endogenously, i.e., inisde the automation and process engineering world. Namely, with the introduction in the 1960s of a flexible and radically new antropomorphic machine: the robot. This was followed at the end of the decade, by the substitution of Programmable Logic Controllers (PLCs) for electromechanical switches (relays) in the automotive industry.

But the major start of second generation, microelectronics-based programmable automation was not achieved until the mid 1970s, when Honeywell introduced its first Distributed Control System (DCS) in continuous processes, and Fanuc entered the market with its first Computerized Numerical Control (CNC).

Since then, microprocessor units (MPUs) and microcomputers have been diffusing more and more at shopfloor level, allowing for an increasing flexibility in the use of programmable automation, a much wider variety of applications and lower costs.

A major novelty was the parallel technical change in software that followed the fast pace of change in hardware: this has lead to the elimination of the previous barriers in terms of technical expertise required to use programmable machinery (due to the increasing user-friendliness of Basic-like programming languages for the machine controllers).

Parallel developments were the applications of computer-based Production planning and control systems, such as the MRP-I, MRP-II and Just-in-Time techniques; the diffusion of CAD induced by microcomputers and work-stations in the 1980s, and the beginning of CAD-CAM linkages in some leading (electronics and automotive) industries.

These evolutions have been generating the need for more efficiency, such as the one allowed by a higher level of integration of heterogenous devices and programming environments in multi-layer networks: a direction now called Computer Integrated Manufacturing (CIM).

During the current decade, CIM is the subject of a large R & D activity by automation users and producers, with the aim to develop cost-effective solutions; the major step into this direction has been the agreement about a common standard of communications within factories (MAP: Manufacturing Automation Protocol). But its uncertain start has not yet produced the expected standardisation and cost decrease effects.

Figure 2.1 gives a schematic view of the evolution across different generations of programmable automation, leading in the near future to CIM.

In fact, R & D, standard-setting agreements and new commercial developments in CIM are expected to give rise in the next decade to a new, third generation of 'flexible' control and supervisory systems. They will make an intensive use of advanced software products (Artificial Intelligence) and processes (CASE). This will allow cost-effective solutions to the integration of the spheres of design, manufacturing, logistics and management.

Nonetheless, as we argue in the following, the next decade is not likely to see the Factory of the Future yet. The progress towards cost-effectiveness will create just one of the two conditions for Flexible Automation to diffuse. Concerning organisational change and social adaptation, it is likely to be a very slow social process, accompanied by power conflicts and open resistance at all levels of the professional community.

FIGURE 2.1 The time trajectory of programmable and flexible automation. (1960-2000)

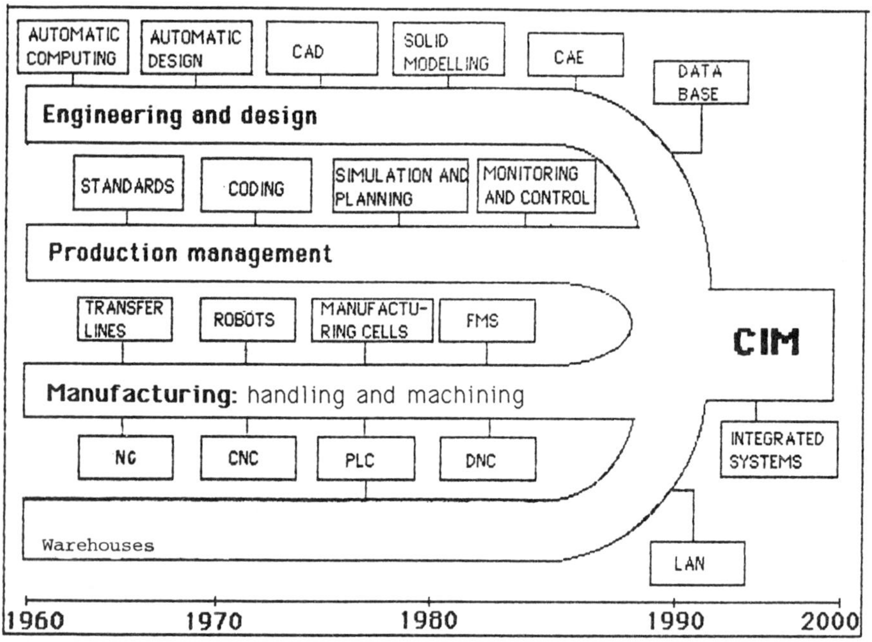

Source: Reseau, in Camagni 1984.

Programmable automation in machining and processing

CNC devices in machine tools eliminate the burden, scarce efficiency and chance of mistakes in managing a large library of tapes carrying the programs relative to different programming languages, machines and operations. A dedicated and user-friendly microcomputer is installed in the machine, that is directly commanded and programmed by the operator.

DCS systems reduce the costs of early systems of direct process control via a mainframe computer; this centralisation, as well as the one typical of traditional panelboard rooms is eliminated: a microcomputer-based operator's station directly controls a section of the process, operating through a number of programmable or dedicated controllers.

Both radical innovations are based on the use of MPUs, allowing for cheaper and more flexible controls; hence, both of them greatly enhance the potential for diffusion of digital control systems towards new industries and smaller scale operations, in the two areas of pertinence of continuous (bulk or fluid) processes and discrete manufacturing (in large scale or small batches), respectively.

DCS (Distributed Control Systems) were first commercially introduced by Honeywell in 1975 with its TDC 2000 system: although quickly followed by imitation from the other main competitors, Honeywell took the lead and is still number 1 worldwide in digital process control systems (although Philips and Siemens are narrowing the gap in the European market, and a major Japanese supplier has emerged: Yokogawa).

CNC (Computer Numerical Controls) in machine tools were first introduced in 1976 with Systems 5 and 7 by Fanuc (43% is owned by Fujitsu; see Sciberras and Payne, 1985, ch. 3 for the reactions by main rivals such as Siemens, GE and Allen-Bradley).

Fanuc is by far number 1 in CNC supply and its position is still unchallenged. In 1986 many sources and interviews with suppliers actually confirmed an estimate of 40 thousand CNC units sold by Fanuc, equal to 50% of the world market. Siemens followed with about 8,000 units sold of the Sinumerik series (source: authors).

Such an amazing concentration rate for a rather simple and commoditised device is due to a number of factors. First of all, the lead factor for the first entrant was so large as to allow the company to cumulate a significant gap in terms of learning and scale economies. This affected both costs, and the capability of matching the machine tool manufacturers demands. This gap was created and maintained both with the sales network, and through flexible specialisation in the production of CNCs (Sciberras and

Payne, 1985, p. 41). Watanabe (1983, p. 38) suggests that the scale economies achieved in marketing, software and post-sale servicing was a key to success for Fanuc.

Hence both radical innovations, with first innovators still ahead of rivals ten years after, confirm the role of cumulative factors that favour an innovative strategy, vs. an imitative one, in order to maximise Schumpeterian profits; their size was proportional to the fast diffusion rates of these cheaper and more efficient alternatives to first generation electronics-based machine and process controls.

Figure 2.2 illustrates recent diffusion rates of CNCs in selected countries. CNC investments as a share of total expenditures on machine tools rose in OECD countries from about 20-25% in 1978, to more than 50% in 1984 (except in the US); new industrialising countries (NICs) lag behind, but in a few cases (like Brazil) they are catching up very quickly.

Programmable automation in handling and integrated systems

Robots evolve across three generations: limited or fixed sequence (pick-and-place) devices, servo-controlled robots (with servo-mechanisms and a digital memory), and finally artificial intelligence based ones. With increasing flexibility, new applicable fields have been opened, ranging from the first simple workpiece and tools handling uses in welding and machine feeding, to the increasing applications in electronics and automotive assembly.

Data on their diffusion by country (Flamm, 1986; Edquist and Jacobsson, 1988) show that it was only when cheaper and more flexible versions reached the market at the end of the 1970s, that diffusion took off, first in Japan and a few years later in Western countries. 1982-84 were the first boom years of robotic investments, with the number of world installations increasing from 24 to 70 thousand in just three years. Diffusion in NICs is still negligible, except in Singapore, owing to labour scarcity and public incentives to adoption in this country.

Flexible manufacturing cells and systems (FMC, FMS) mingle programmable control devices (such as PLCs and CNCs), with automated handling devices, robots and system supervisory computer control. They are still the most immature technique amongst the ones described here (apart from the still experimental CIM networks)[4], lacking standardised and transferrable approaches from one system to another.

Nonetheless, prototypes have already started to be developed and installed, and some leading users and producers have accumulated relevant

Fig. 2.2 ADOPTION RATES OF NCMTs 1978-85
(% share in total machine tools inv. in monetary terms)

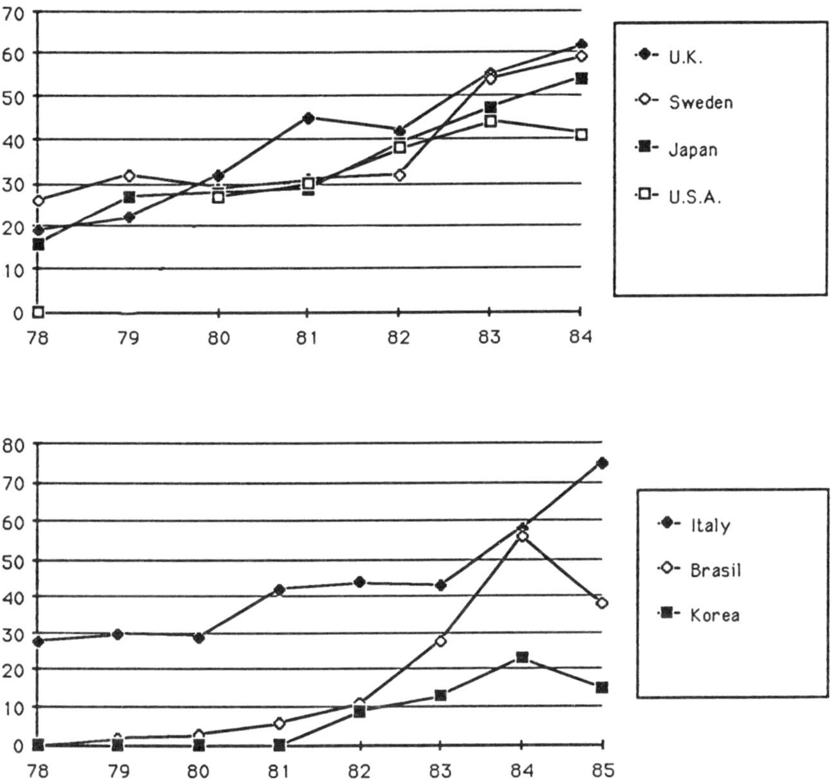

Sources: Edquist and Jacobsson (1988), UCIMO (1987), American Machinist (various years), and authors' estimates.

experience and learning in this field. Moreover, around the mid 1980s some standardisation emerged in smaller systems (simple cells or 'modules' based on a machining centre endowed with a robot and other peripherals).

This lead to an acceleration of diffusion, with estimated world installations doubling between 1983 and 1985. The take-off of the diffusion curve is expected during the second half of the decade,with a similar boom as the one already experienced by robots after the 1981-82 general crisis.

Figure 2.3 shows diffusion patterns in a few selected countries, Japan leading again at the end of the previous decade, and the socialist countries catching up along with Western ones in the 1980s.

Table 2 illustrates the global picture of diffusion of the different techniques of programmable automation in major OECD countries and in selected NICs as well. In order to compare data cross-country, we have adopted the measure of diffusion proposed by Edquist and Jacobsson (1988): units installed per employees in the engineering industry.

The Japanese leadership in adoption does not extend to a software-intensive technique like CAD (Computer aided design); this is the only technique where the US lead. Western Europe has already caught up with Japan in most techniques, except in robots.

Finally, a number of empirical generalisations about the cross-industry patterns of diffusion and firm size will be introduced.

With the evolution of programmable automation towards robotic cells and systems, away from a simple shift in the technique of machine control, on a standalone configuration (from manual to dedicated devices such as servo-mechanisms, and then from computer-based controls towards microelectronic devices), diffusion patterns modify as well.

But there is no clear cut tendency towards any homogeneous diffusion trend, for example in a downward direction along the firm size distribution (as one would predict on the basis of neoclassical threshold models of diffusion).

One has to consider significant differences across technologies in these diffusion patterns (Thwaites and Oakey, 1985; Acs et al., 1988). Particularly, robotic diffusion seems to be much more hierarchical and selective, at least in the US (not in Japan), compared to CNCs. The latter have actually represented a unique window opportunity to small batch operations, and mechanical engineering jobshops, to enter the micro-electronic trajectory.

Nonetheless, a number of general patterns appear to emerge as the typical ones of programmable automation.

Figure 2.3 Growth of FMS in the Federal Republic of Germany, Japan the USSR and the United States

(Units)

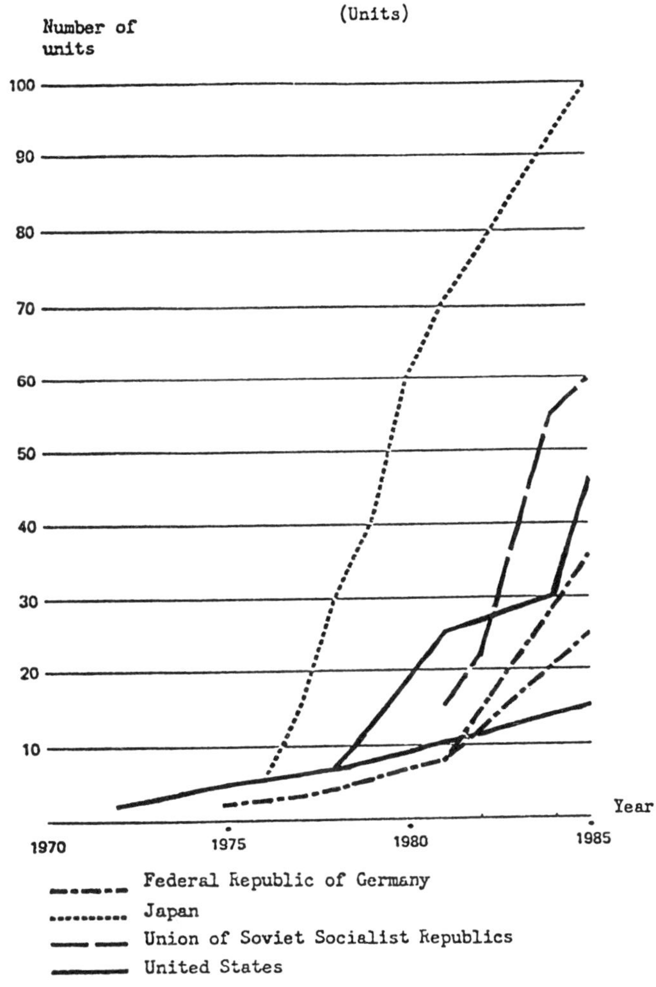

Note: The two curves for the United States apply to a wider and a stricter definition, respectively, of FMS. For the Federal Republic of Germany, an interval estimate is given for the number of FMS installed at the end of 1984.

Source: ECE, 1986 (Figure III.1)

Table 2 **The stock and density of programmable automation techniques in selected OECD and developing countries**

		NCMTs	CAD	ROBOTS	FMS
JAPAN	stock	118,157 (1984)	7,300 (1984)	21,000 (1984)	60 (1985)
	density	22,399	1,384	3,981	11
SWEDEN	stock	6,010 (1984)	1,900 (1984)	2,400 (1984)	16 41985)
	density	22,177	7,011	8,856	59
ITALY	stock	23,119 (1984)	3,000 (1985)	2,585 (1984)	13 (1985)
	density	16,752	2,174	1,873	9
USA	stock	103,308 (1983)	59,400 (1984)	13,000 (1984)	67 (1985)
	density	11,728	6,743	1,475	8
FRG	stock	46,435 (1984)	11,000 (1983)	6,600 (1984)	20 (1985)
	density	11,376	2,694	1,617	5
UK	stock	32,566 (1984)	9,000 (1983)	2,253 (1984)	35 (1985)
	density	10,505	2,903	724	11
KOREA	stock	2,680 (1985)	743 (1986)	55 (1985)	3 (1985)
	density	5,176	1,437	106	6
SINGAPORE	stock	700 (1985)	290 (1985)	313 (1984)	0- 3(1975)
	density	4,526	1,875	2,024	0-19
YUGOSLAVIA	stock	1,232 (1983)	100 (1983)	32 (1983)	N.A.
	density	1,720	140	45	N.A.
BRASIL	stock	1,711 (1985)	340 (1985)	35 (1985)	0 (1986)
	density	1,033	205	21	0
INDIA	stock	1,178 (1985)	770 (1987)	N.A.	0-2 (1984)
	density	807	527	N.A.	0-1
ARGENTINA	stock	500 (1985)	74 (1983)	N.A.	N.A.
	density	N.A.	N.A.	N.A.	N.A.

Note: DENSITY is number of techniques by million employees in the engineering industry (SIC 381-5)

SOURCE: Italy from different studies; FMS in OECD countries from IIASA; robots in OECD countries from CERIS, Torino; other data from Edquist and Jacobsson (1988).

At the aggregate level, the adoption of programmable automation processes has followed a previously unexpected direction: small and medium size firms were supposed to be the natural receivers of these technologies, at least in discrete manufacturing, on the basis of the easier application to small batch production runs of the much more flexible and 'convertible' machines.

On the contrary, up to now diffusion has mainly regarded large firms, and very often the ones operating in mass productions like automobile and electrical appliances. Such firms have increasingly introduced versatility (i.e., working on different pieces, in random sequence, with the same equipment) in dedicated machining and assembly operations[5] (Figure 2.4).

The persistence of scale economies at the plant level, even under the new form of economies of joint production or 'scope economies'; the risk of losing elasticity in dimensioning the volume of production over time; the necessity and convenience of integrating robots in complex networks of computer-controlled production systems; the difficulties and costs of programming such integrated cells and systems; and the importance of the organisational and strategic implications of these new system technologies: all these as well as other similar reasons may hamper a faster diffusion of programmable automation among SMEs (Camagni, 1984 and 1985; Colombo and Mariotti, 1985; Lafaille, 1985).

On the one hand, cheaper microelecronics hardware and complementary user-friendly software have accelerated the diffusion of simple devices in standalone machines also to SMEs, that were completely cut off from computer based manufacturing until one decade ago.

But, on the other hand, the shift of the trajectory towards system integration and increasing SW sophistication creates barriers to further adoptions by SMEs, the ones that would most need an increasing flexibility in their small batch processes.

The transition from programmable to flexible automation

One has to consider a second, qualitative dimension of diffusion, together with the quantitative one measured by adoption rates. The efficiency of the investment in a specific programmable automation technique, highly depends upon the current position and the past experience of the user firm along its adoption path.

A simple 'substitution' philosophy, mainly pursued through the adoption of standalone robots and CNCs, may reach a region of decreasing 'dynamic marginal returns' after a while (Von Tunzelmann, 1986).

FIGURE 2.4 FLEXIBLE AUTOMATION TRAJECTORIES

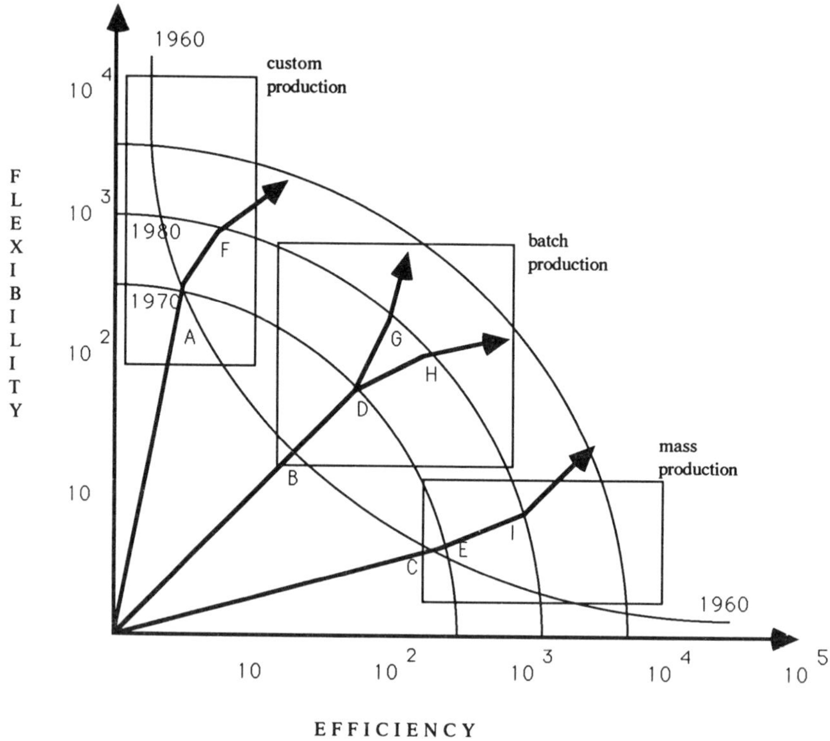

A = standalone manual machine tools
B = semi-automatic machine tools
C = dedicated automatic machines
D = Standalone robots, NC, MC
E = transfer lines with CNC
F = semi-automatic or automatic machine tools
G = manufacturing cells
H = flexible manufacturing systems
I = flexible transfer lines, robogate, FMS

Source: Camagni, 1989

This is the stage when it is worth adopting a different 'production integration' approach, exploiting synergies among the different components (controls and actuators, workpiece transfer and machining, parts machining and assembly).

Finally, when a layer of flexible production cells has diffused enough, it is time for the firm to organise a control network: we enter a third stage of strategic or systemic integration among different functions of the firm, or CIM (Camagni, 1987b and 1989).

As programmable automation diffuses, it reaches smaller enterprises and it increases intra-firm penetration in larger ones. From the very start large users face a problem: they are spending large sums of money on devices that do not talk to (comprehend) each other. Besides this, gains in productivity are limited by the scope of automation by separated islands. The CIM or flexible automation trajectory aims to overcome these limits of the first wave of penetration of microelectronics into automation.

The basic mechanism of substitution of programmable for dedicated automation is illustrated in Figure 2.5, which refers to the assembly phase. With increasing costs of labour and cheaper cost of electronics (components and hardware), the scope for application of programmable machinery is increasing, in small to medium batches of production.

The incentive to move from simple programmable to programmable and flexible automation is less straightforward.

First of all, the motivation of saving labour has lost its relevance, after the first phases of introduction of robotics. For example, in the case of Fiat, the second wave of robotization does not affect employment levels (sharply reduced in the first wave), except for the automation of maintenance.

Even the higher efficiency of direct labour has lost its centrality. The objectives of adopting firms are those of increasing connectivity and compactness of automation systems, and harmonizing them with the global organisation of design, production and marketing within the firm.

Another motivation for a departure from early programmable automation is the following one. Manufacturing firms, both in new and in mature industries, tend to become system integrators, supplying only the core components, assemblying final products and delivering them with complementary services to clients; hence increasing productivity in parts manufacturing has a very limited impact upon total productivity (Ayres, 1987, pp. 20-21).

Direct manufacturing labour has often decreased below 20% of total

FIGURE 2.5

DIFFUSION OF PROGRAMMABLE VS. DEDICATED ASSEMBLY AS A FUNCTION OF RELATIVE COST TRENDS

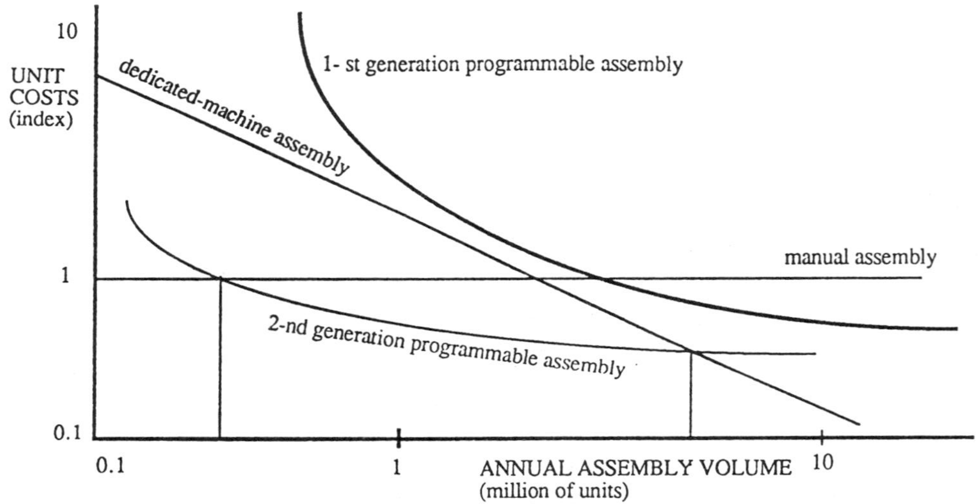

SOURCE:P.M. LYNCH (1976, fig. 6.2, p.125), with Authors' modifications

manufacturing costs, while logistics, assembly and quality control account for above 50%. Hence product and process redesign using CAD-CAM systems and other innovations leading to CIM become appealing.

The beginning of this further transition shows that technology push factors are inadequate to play a leading role from now on. In fact, the technology is already there, at least for hardware components; what is lacking is the capability of user firms to integrate new systems in their growth strategy, to reshape their traditional organisation inherited from the previous mode of production, to anticipate technical change with appropriate organisational change.

Demand pull factors will dominate the automation trajectory from now on; this will delay the take off of CIM for a while. However, the incubation phase of flexible automation within large user corporations will give them new opportunities for entering the new CIM markets, on the basis of learning and lead factors.

With the transition of large adopters across the three stages named above (substitution, production integration or FMS, and CIM), technology determinants lose ground in favour of organisational and strategic ones as major determinants of diffusion patterns.

The type of changes that the organisation adopting complex automation systems has to face, are twofold. On the one hand, there is a **micro-organisational** level, which concerns the immediate measures that the company takes to respond to the novelty introduced by technical change (labour process tasks redefinition, new jobs profiles, formation of work teams).

On the other hand, at the **macro-organisational** level, one has to consider the more general innovations in the structure of the firm, aiming to optimise the use of New Information Technologies (restructuring the decision and planning process, co-ordinating the DP and automation policies within a CIM strategy, creating new functions or Divisions, and revising existing functions).

If increasing professional job contents, extending control tasks and polyvalence are the key words at the micro level, at the macro level we have a number of converging theories and proposals. Their common aim is introducing 'intrapreneurship' in the large firm; but solutions differ about the optimal strategy to favour creativity, innovativeness and promptness in decisions and business strategies.

Solutions range from matrix structures, to ad-hoc and informal configurations; reaggregation of the functional units into smaller, product/ technology specific groups; 'project teams', and, finally, the return to the generalist (i.e., the 'global manager') that brings to mind the small entrepreneur.

Field research (Camagni, 1987a,b) found that the majority of firms introduced complex organisational changes: on the one hand, by introducing new co-ordinating functions, in a barycentric position with respect to the existing ones; on the other hand, moving to complex matrix structures (only in large firms).

As production processes increase in complexity, organisations adapt to this with more integration tools, ranging from ad-hoc meetings to committees, from additional directive roles to new bodies. The production management function is at the centre of criticism and reform, but strategic marketing is even more emphasised by interviewed firms. The search for more integration between production, marketing and engineering is seen as more necessary than in the past.

At the micro level, findings by one of the authors confirm similar trends to the ones detected in other surveys. The trend is towards a decrease of unskilled and semi-skilled labour, with a sharper decrease for women in this category. Professional figures such as the machine tool controller, product quality controllers and generic or polyvalent maintenance workers tend to disappear; on the contrary, new jobs require much less 'craftmanship' and more capacity of the 'decisional' type, that is higher level control functions (as the lower ones tend to be absorbed into automated feedback loops).

3. The Spatial Diffusion Process

This section explores a number of spatial implications of the restructuring of industry that have come along with the rise of a microelectronics-based industrial complex producing industrial automation goods and services, and with the start of a pervasive diffusion of these new capital goods during this decade.

The first two parts will summarise some empirical findings on the spatial impact of this restructuring, and the third one will explore some possible interpretations of these findings.

The last part of the section will introduce some speculation about the locational patterns induced by the structural change occurring in the forthcoming transition from programmable to flexible automation.

Polarisation effects of programmable automation

The producer industries of programmable automation exhibit a high concentration rate and regional specialisation pattern, both in developed and in developing countries. Major areas of concentration are old industrial regions, where the producers of dedicated automation were already present, as well as the main plants of large users.

The causal mechanism of this polarisation and scarce diffusion into high-growth and new industrialising regions is twofold: technology linkages on one side, and the location of major users of programmable automation on the other side.

First of all, the automation industries are vertically and horizontally (i.e., across broad families) linked into an industrial complex. Backward and forward linkages exist inside the member industries of the complex, with basic Information Technology sectors and with leading user industries as well. These linkages consist of labour mobility in new professions (software), commodity and technology exchanges, inter-firm co-operation and cross-industry diversification by leading firms in each sector.

So some spatial polarisation should be expected in order to take full advantage of the agglomeration economies and optimise the cost-effectiveness of managerial, financial, human and R & D resources channelled within this web of inter-industry linkages.

Second, the technical quality and commercial availability of the new techniques has been evolving on the basis of learning-by-using and innovative interactions between leading users and producers; this is true even in basic control devices, as the case of PLCs developed by a team of engineers at GM shows. This innovative interaction is becoming more and more relevant with the increasing complexity of automation systems: from robots to robotic cells, from standalone programmable machines to FMS (Ettlie, 1985).

Even the market leaders in general automation commodities, that are located far away from important markets, must join alliances or joint-ventures with leading users or specialists that are true 'insiders' in these outlet industrial districts or regions.

On a wider, world scale, this mechanism induces increasing alliances of Japanese automation leaders with U.S. firms, the most striking example being the joint-venture between Fanuc and General Motors in robotics (other examples in the same industry are the acquisition of Unimation by Westinghouse, and those of Trallfa and ESAB by the Asea group, now merging with Brown Bovery; finally, in-house developments by all major car producers).

But the same phenomenon can be observed at the interregional scale too, in major industrial countries: that is, joint-ventures and new specialist advanced automation suppliers do not locate randomly.

They actually cluster in regions characterised by those sections of electromechanical and Fordist industrial complexes that are faster adopters of advanced automation; this is particularly true for the supply of FMS and related services and devices (where the demand from the electronics complex is negligible: see Gros-Pietro and Gaibisso, 1988).

On the contrary, both specialised automation tools for micro-electronics and CIM-related innovations are much more software-intensive and less related to the engineering industrial complexes. But this does not imply that they actually locate randomly or in scattered patterns.

In fact, they are very sensitive first of all to the location of the most scarce resource, that is the labour pools for programmers and analysts; and, second, to the actual locations where microelectronics and computer system firms clustered in their postwar emergence and growth (e.g., Milan and Munich in Europe, California and Boston in the US).

It is interesting to observe that there is a very clear separation , with scarce intersections between the regions specialising in Information Technologies at large (such as the ones just mentioned), and the old, Fordist industrial regions now emerging in the field of Programmable Automation.

As for the latter, we refer namely to the following ones: the new Torino-Milano-Bologna industrial triangle in Italy (as opposed to the old one, which had Genova as a third pole instead of Bologna); the city of Stuttgart and the South-West of Germany (Baden-Wurttemberg is actually the 'automation' region as opposed to the basic electronics one, Munich and Bayern: Electronics, Sept. 1988, pp. 8-10); the North-East corridor in the States, and particularly the 'Automation Alley' stretching from Detroit to Ann Arbor, Michigan (The Economist, April 11, 1987, p. 15); Toyota city and other concentrations in Japan.

First the location of automation producers tends to polarise and, second, it is attracted by leading user areas. Now the 'diffusion paradox' that we have commented on in the previous Section enters: leading user industries are, on a quantitative scale, the old industries that restructure their products and processes; only in the first phases of early applications of new techniques, electronics industries have played, together with defence and aerospace, a major role, while the bulk of engineering industries have always been late-comers, with the main exception of Japanese machine tool makers pioneering in FMS adoption[6].

But once, with cheaper microelectronics and cost-effective second generation solutions for factory automation, the bulk of engineering industries invest in Programmable Automation, they become the crucial market outlet. This phenomenon started at the end of the 1970s.

Hence only research and software-intensive activities tend to locate in 'Silicon Valleys', while the bulk of production, and even more final assembly and system integration of automation is located in automotive industrial districts and engineering industrial regions.

We shall consider two examples in the U.S., in order to characterise these alternative and divergent locational tendencies: CAD supply and the adoption of CNCs.

The origins of the CAD technology and of its leading firms are strictly linked with military programmes and the aerospace industry, where the technology was first generated. Hence one would expect a locational pattern similar to microelectronics or the computer industry, given this similar origin[8].

IBM is the first world vendor, after having increased its share of the US-based turnkey CAD systems sales from 12% in 1980 to 21% in 1985; the two leading SW packages it has sold worldwide are both produced by aircraft firms (Dassault and Lockheed). Number 2, with a 16% market share in 1985, is Intergraph: located in Alabama, near NASA Headquarters, the firm took full advantage from this location in its start-up phase (manpower and technical flows with NASA-R & D activity; see Kaplinsky, 1982).

Therefore professional labour markets in software, location of R & D activities and new, aerospace or microelectronics-based industrial areas have a potential attraction for a SW-intensive industry like CAD; the technological frontier of CAD-CAM linkages, a major step towards CIM (see Figure 2.1), is still located in the microelectronics industries, and precisely in Integrated Circuits and PC boards design and fabrication, consequently this locational force is likely to persist.

Adoption patterns of CNC by user industry show a close similarity in each major industrial nation (Edquist and Jacobsson, 1988; Arcangeli, 1989). Hence the regional patterns of adoption studied for the cases of Italy and the U.S. may have a more general representativity.

In the U.S., the adoption of CNCs by plants located in the four large regional divisions, shows the highest level in the Mid-West area (47%), and the lowest one in the South (28%). Another interesting feature is that old plants adopt more than new ones: this finding applies not only to CNC but to every programmable manufacturing technique; moreover, it holds both at the national scale and within each main region.

It is exactly for CNC adoption within the Mid-West region that the lead of old plants is by far the widest. This is the region where the plants opened before 1960 show a 60% diffusion rate, while only 30% of the younger ones have moved to the new technique.

This set of converging findings has been interpreted by Rees et al. (1985) according to the following arguments: as one would expect from the spatial product cycle theory, new branch-plants (both in old and in new industrial regions) are less innovative than old ones; old industrial regions, with skilled labour, show a direction towards faster equipment renewal (a kind of Ricardo effect stimulating technical change); finally, industrial restructuring has been more intensive in old plants and regions, in order to face market crises with production rationalisation: moreover, head-plants need more flexibility in production than new specialised ones where recently installed dedicated automation can still be in use for a longer period.

We shall add another line of arguments as well in order to explain these findings. As we know from the Italian experience, small specialist firms that tend to cluster in old industrial areas have a large activity of continuous innovation, very often through retrofitting of CNC to old machinery; this is particularly true for machine producers themselves, who have more expertise and can retrofit machines or restructure old plants with better cost-effectiveness than user firms.

Finally, the dematuration of Fordist industries (electrical engineering and vehicles) stimulates the revitalisation of older regions and the attraction of advanced automation suppliers, that is a cumulative process of fast technical change in products and processes.

Why interregional diffusion in the 'Third Italy' got exhausted

In the case of CNC, the results in Italy confirm only partially the US ones. In fact, 83% of CNC units are concentrated in the North and 35% in just one region, Lombardy: but this reflects the locational pattern of the engineering industry and of its equipment. Lombardy has just an average diffusion rate (in terms of stock of machines), while Southern Italy is above average because of the younger industrial park (note that a different adoption measure is used here; this may affect the results); Piedmont, the second industrial region, has even lower than average penetration rates, probably as an effect of the lower CNC adoption in the automotive industry.

The new industrialising regions of the North, Emilia and Veneto, deserve particular attention. The region of Venice has been an earlier adopter of NC and CNC. The highest rate of increase in CNC in later years is in Emilia, where a sample survey has shown that adoption motivations are different from the growth-oriented ones that prevail in Lombardy and in Mezzogiorno. For small jobshops doing custom operations for larger firms, that characterise the engineering industry in this region, CNC adoption is not motivated by the target of increasing market shares, but by achieving more flexibility and lower unit costs, hence bargaining higher profit margins from clients (Ciciotti, 1982).

Nonetheless, when we move to more complex and demanding techniques, the polarisation effects in favour of old industrial regions emerge again, as in the US case, in Italy too.

From the start robotics showed a concentration in diffusion, even more than in production, on a regional scale (Camagni, 1984 and 1985). The evolution toward system integration and CIM shows a cumulative advantage of leading adopters in the Lombardy region (Camagni, 1987a, b) and a very poor performance of the industry located in Mezzogiorno, with the exception of CAD-CAM applications within the aerospace complex near Naples (Cainarca et al., 1987). But differentiations are likely to emerge within the North as well: the CIM trajectory is likely to be more selective than first adoptions of microelectronics.

This implies that the actual gap between Lombardy on the technological frontier of 'production integration' (the second stage of adoption described in the previous section), and Veneto, where the majority of firms are smaller and still developing their first stage or 'substitution' approach, is likely to persist (Camagni and Gambarotto, 1988).

We shall now summarise the conclusions that can be drawn from a detailed review of the empirical evidence available in Italy (see also Pegolo, 1989 for a thorough discussion of inter- and intra-regional diffusion of PA in Italy).

To a certain extent Italy confirms the US findings. However, the Italian case points to the active role in diffusion played by regions of recent industrialisation, based on local entrepreneurship (Emilia, Veneto) and not on branch plants.

This finding confirms the analysis developed by Piore and Sabel (1984), two scholars who have paid attention to the new industrialising regions of the 'Third Italy' (as it has been called after the analysis by Arnaldo Bagnasco).

The jump onto the PA bandwagon of the Third Italy, which is

characterised by a diffuse and multicentric urban system, reinforces the non-metropolitan patterns of industrial location, by modernising the diffuse industrial capacity.

On the contrary, there is no general evidence that strong interregional equilibrating trends are favoured by PA diffusion; the opposite might hold: both the US and the Italian case show that it deepens the 'dependent' character of the more backward and heterogeneous regions which have attracted branch plants.

In sum, this is what has recently happened in Italy: the Third Italy was a fast follower in the programmable automation trajectory, while the gap between the whole North (old and new industrial regions) and the South has increased more.

Nonetheless, one should carefully rethink a very popular view on these phenomena of stuctural change that has been focusing the Schumpeterian small entrepreneur and its flexible specialisation (namely, the one proposed by Sabel and Bagnasco, in a number of contributions).

The analysis outlined in this contribution, but developed with more depth in other studies (Cortese, 1987; Arcangeli, 1989; Pegolo, 1989) points to an alternative interpretation. During the declining and stagnation phases of the postwar Fordist long wave, some semi-peripheral regions in Italy had a unique window opportunity to enter the set of core or central reigons in the European industrial space, and they have caught the most of it (particularly in Veneto and Emilia, as symbolised by the rise of the new emerging industrial groups Benetton and Ferruzzi).

The basis of this upgrading has been established in their success as 'semi-peripheral' regions in the postwar long cycle, particularly because of the new spatial division following the 1957 EEC agreement, allowing a vast and increasing market in many mature industries for the 'New Industrialising Regions'.

As for the upgrading in the tail of the past long cycle, it was favoured by the combination of: increased cost competition in the EEC markets, easy access to niche markets (also by independent, not so innovative neither flexible small firms), absolute appropriability of standardised and unprotected product and process technologies; and, finally, favourable conditions in local labour markets and labour force reproduction costs.

This exceptional and provisional 'small is beautiful' paradise at the end of the Fordist wave in Italy has been a matter of diversion from the real issue at stake, which in fact has always been the restructuring of the big corporations and their return to profitability on international standards. After the 1981-82 crisis, the return to profitability has actually favoured the largest firms and the industrial groups in Italy, so that the national structure was prepared to enter the phase of industrial concentration at the European scale that is now occurring.

Also process innovations contribute now to reverse the advantage that small firms and late-industrialising regions enjoyed in the late 1960s and most of the 1970s.

The new challenge now facing the Third Italy is the move towards CIM, that might be much more selective than the past trajectory, and reproduce some of the disadvantages already experienced by the receiver regions of branch plants. In fact, the convenience to introduce CIM only above a high intra-firm adoption threshold of programmable automation systems will favour only large firms as CIM adopters for a long time ahead.

Finally, microelectronics adoptions in products as well as in processes, have been studied from the regional viewpoint: in the US this is the only case where the advantage of metro compared to non-metro locations is statistically significant (Rees et al., 1985, Table 7.8, p. 184); in the UK, the gap in favour of the Midlands and the South-East, vs. development areas, is much stronger than for process innovations (Gibbs and Edwards, 1985).

This is far from unexpected, and constitutes a further element for arguing against the persistence of the 'Third Italy' model of catching-up. In fact, the close link between product and process innovations, and the systemic drive of both trajectories, militate in favour of big corporations, contrary to the Piore-Sabel (1984) hypothesis of a continuous evolution from traditional craft flexibility to CIM.

As far as microelectronics applications to products are concerned, selective technological barriers are much stronger than for pervasive process innovations (computers, PLCs).

The survey on European diffusion of microelectronics in products and processes shows that in 1983 product applications were much less; they were more selective across industries, with electronics outstanding as a user industry, followed by a number of engineering applications and scarcely any other user industry.

Finally, product applications of microelectronics were very rare in plants with less than 500 employees; process innovations were much more even, reaching about one third of smaller plants as well; hence, 2/3 of plants from 20 to 50 employees ignored microelectronics use in Europe in

1983, and the majority of plants up to 100 employees did the same (see PSI et al., 1985). On the contrary, almost no plant abvoe 1,000 employees ignored the use of microprocessors, either in products or in processes.

A tentative explanation of the spatial patterns of Programmable Automation

In concluding these arguments suggested by the empirical evidence, we propose the following interpretation of the locational forces at work.

The spatial impact of programmable automation on adopting industries appears to counterbalance the previous trend to an increasing share of foot-loose industries and productions, that characterised the tail of the postwar growth epoch, and was limited only by the polarisation forces prevailing in the early phases of the product cycle.

Now, systemic links between R & D, product and process innovations, together with increasing R & D expenditures also in traditional sectors, and shorter product life cycles modify the picture.

The results are: a strong polarisation of R & D activities; a rejuvenation of old industrial regions; and the regionalisation of the new industrial complexes that characterise the new techno-economic paradigm (as it happened before, with the Factory-Towns of the steel and car industry, and the coastal petrochemical and integrated steel complexes).

R & D and SW intensive activities are attracted by resource locations in such a way as in a Weberian locational model.

R & D and science-based activities are mainly attracted by new industrial poles where microelectronics is advancing faster. Besides chips, the other pervasive resource is the labour force for SW, which tends to be attracted by microelectronics-based new complexes, but it is less scarce in far away, less developed areas.

Cheaper labour pools are increasingly looked for by electronics firms, until a higher level of automation in SW engineering will become viable: e.g., Bristol in the UK, Spain and Mezzogiorno (namely: the city of Bari) in Europe, India and other Asian underdeveloped countries for Japan (Gregory, 1986, ch. 19).

As chips become increasingly a software product too, their design and conception phase might gain some degrees of freedom in its location, as SW does, unless cumulative factors will keep it within 'Silicon Valleys'; in fact, technology trends might, in the long run, move R & D on hardware towards 'Artificial Intelligence Valleys'.

This creates a first layer of locations of basic technologies, exhibiting new locational patterns, but following a model not so different from the previous one explained by the product cycle approach (where the location model in the early phase is actually of the Weber type).

But this is not the entire picture. Again, Weber might be of help. Let us imagine a locational triangle, where the basis is formed by two resource pools: microelectronics (HW) and analyst-programmers (SW). Resource-oriented industries are attracted towards this base, but system integration activities will need to be closer to end users (the upper vertex of the triangle).

The fact that some major user concentrations are the old industrial regions, whose mature industries have been rejuvenated by the pervasive diffusion of the microelectronics paradigm, explains the **conservative and consolidation** spatial effects of Programmable Automation.

The crucial fact is that the technology chain reaches a point where microelectronics, HW and SW, blends with electric, mechanical engineering and process technologies in vertically integrated user sectors.

Rejuvenating old industrial regions appear to be the best candidates as attractors of these inter-industry technology flows shaping new industrial complexes, under the initial stimulus of leading user industries, reshaping their environment, the external network of suppliers and inducing a major shift of the associated machinery industry from dedicated to programmable automation.

The cases of West Germany and the UK differ from this model of geographical stability for specific reasons. In fact, from the theoretical viewpoint the spatial rooting of a new industrial complex is not necessarily linked to the previous one. Schumpeterian processes of creative destruction do not apply only to firms, but to their linkages and networks too: a new web of producers-producers and producers-users relationships emerges with the new complex (Lundvall, 1986).

What one should expect and predict from this theoretical approach, is a selective process among old industrial regions, and particularly between the ones that are more or less fast and efficient in the Penelope's task of remaking their co-operative networks, the 'nervous systems' of their Marshallian industrial districts.

Hence the case of the above named European countries follows this rule: with the move from the Rhine valley to Bayern of the industrial centre in one case, due to slowness in restructuring of the old steel-based complex; and with the decline of the Midlands in the UK, due to the scarce innovativeness of British management compared with the management of multinationals that locate their branch plants in the new growth areas (M4 and the peripheral regions: see Morgan and Sayer, 1988).

The institutional aspects of this move to a deeper adoption of the new paradigm (Perez, 1985) brings to the fore again the classical dichotomy between the economic and the planning or administrative region: industrial districts cannot solve all the new infrastructure problems by themselves, unless they are supported by a social consensus and an implementation of diffusion policies at the regional planning level.

Finally, institutional and infrastructure policies, as well as the peculiarities of social and institutional change on a broad regional scale (encompassing a number of specialised industrial poles or 'districts'), are likely to exhibit inertial factors also across techno-economic trajectories.

For example, although it is still premature to predict the future spatial patterns of the new, incubating paradigm of biotechnologies, there is some reason to expect it to be somewhat similar to the new geography of microelectronics and automation[9].

Functional integration and the spatial impact of CIM

We shall observe the historical evolution of regional systems and how they develop: by combination of various segments of technological cycles, by learning from import substitution and gradually evolving toward industry mix differentiation and export orientation (Jacobs, 1984). We can note a mix of elements of continuity and bifurcation: what Aydalot (1986) had labelled the model of 'rupture/filiation'. As the trajectory of a technology goes through phases of stability and instability, so does the trajectory of an urban or regional system.

The component of continuity is represented by the local learning processes consolidating into externalities, through spinoffs from large firms, service networks, public infrastructures and cross-fertilisations (Camagni and Malfi, 1986).

When discontinuity appears, in the early phase of a new paradigm, oligopolistic competition in the new leading industries tends to estabish a new economic order soon, while bargaining about the distribution of the surplus from the more efficient basic technology tends to converge toward a new social regime.

A similar process occurs in space: elements of a re-ordering of spatial relations after a new innovation wave (Pred, 1969) can be found in the conditions for appropriability of Schumpeterian profits, in the new relations between production and markets, and in the structure of the firm affecting the location of its activities.

We have seen in the previous parts of this study that programmable automation, as a part of a restructuring of industry in the 'tail' of the past long cycle of development, tends to confirm the spatial structure of industry, reinforcing only some, more vital established industrial areas (both old/central and new/peripheral ones), and selecting only the most efficient ones among the new growth regions.

It is interesting to observe that the spatial innovations coming with programmable automation (PA) can hardly be defined. In a sense, this selective but not so innovative spatial impact of PA is consistent with its likely historical role: a bridge in between two long waves: the past Fordist and the next one. For the same reason, one would expect that many factors might modify this conservative scenario, with the start of a new trajectory of flexible automation (FA).

Intuitively, one would imagine that a number of new industrial poles and regions would emerge, while selection would become much sharper among the old ones, with just a few of them capturing new growth waves. Hence, the top hierarchy of urban systems would be affected, starting from the Braudelian question: 'Where will the centre of the world-economy locate after New York? Along the Milan-Stuttgart, or along the Tokyo-Osaka axis?' Moreover, window opportunities, as the ones captured by the 'Third Italy' in the past, might appear also for the Italian and the World South, that is for 'hard' periphery, as opposed to the Northern semi-periphery favoured in the recent past: the Third Italy and the Far East (see on this issue Perez, 1989).

In the following paragraph we limit ourselves to the more cautious task of focusing on the new forces that are likely to reshape world and continental economic spaces. We will argue here that:

(1) new appropriability conditions are modifying the locational pattern described within the traditional product cycle models;
(2) user-producer interactions tend to confirm existing industrial locations, given actual diffusion patterns; more generally, co-operative networks have similar effects;
(3) functionally integrated plants (blending development, engineering and production, extending just-in-time and quality control to the network of suppliers) and new flexible firms will also deviate from the predictions of product cycle models.

First, appropriability conditions for the profits from innovation exhibit these trends:

- the product cycles are shorter and lose the 'tail', i.e., the phase when production tends to diffuse from centres to peripheries;

- the low and uncertain rentability of high R & D costs in shorter payoff periods, induces strategies of lead time, where more integrated structures of firms and plants are required, as well as closeness of management, research and production to leading market areas;
- product differentiation strategies are also pursued by large, multiplant firms, in order to split high R & D and marketing costs into a whole product range; when successful, they restrict the markets for small, single plant firms.

In extreme synthesis, these factors tend to redistribute integrated firm structures (from Headquarters to assembly of final products) in the main advanced market areas: Japan, US and Europe (Ohmae, 1985); R & D and components manufacturing are not necessarily attracted by this market pull, and can continue to be located according to a resource orientation.

Second, producer-user interactions reinforce these trends: first, the main adopting firms are located in old industrial areas as we have stressed before; second, technology trends tend to reinforce this pattern due to cumulative effects.

As for concentration, on the one hand vertically integrated system solutions, application-specific components (like ASICs in microelectronics) and application-specific SW should favour small specialist firms.

But on the other hand, large firms adopting flexible automation processes and flexible organisational structures tend to diversify into these new vertically integrated product markets. The major phenomenon is the increase of strategic alliances among firms, in order to implement this wide penetration strategy into diversified markets and segments. The analysis of these alliances in the case of factory automation confirms that they tend to reinforce large oligopolistic firms, much more than they open new markets to automation specialists and leading user firms (Arcangeli, 1989).

This group of factors tends to confirm the role of large firms, that articulate their technology and organisation, within the firm and in networks of alliances, in order to be 'closer' simultaneously to a number of user industries and areas.

Third, large firms tend to adopt the flexibility characteristics that had favoured small innovative firms, as well as they try to produce internally the sinergies of Marshallian industrial districts (Arcangeli and Camagni, 1987). Functionally integrated new units of multiplant firms will not act and locate according to the pattern of branch plants by multinationals that have been stylised within a product cycle framework. A new locational pattern, much more similar to the traditional concentrated one, is likely to emerge from the diffusion of new organisational structures and procedures such as Just-in-Time (Sayer, 1985, Estall, 1985, Zysman, 1987).

All these trends combined will favour industrial as well as spatial concentration in the first phases of development of flexible automation. Only in the long run, imitation and standardisation of the new techniques will allow industry deconcentration and spatial diffusion in new and more peripheral growth poles.

But it is not known where this concentration will occur. This is why our analysis does not imply any deterministic approach to the issue: as any transition, the one starting now from PA to FA can reshape urban and regional established hierarchies. Hence, every country and region that performed well in the first one, will try to diversify into the second one, as well as new emerging areas that were late-comers in microelectronics. They will try to leapfrog into Artificial Intelligence and CIM.

This will be the unique window-opportunity for the Brazils, the Koreas and the Mezzogiornos of the various Continents. These most mature NICs and regions might save decades of efforts, by achieving in a chaotic environment, what might be impossible to get in a more stable and hierarchical one.

The opportunity is about the same as the one seized by Japan with microelectronics and PA, that is: first, avoid the enormous R & D expenditures; second, be fast in reverse engineering and entering new markets as imitators; third, enter the world markets gradually, as soon as you have accumulated competitive assets within national user-producer industrial complexes (Lundvall, 1986).

Conclusion

The paper has considered some of the more relevant dimensions of analysis, in the actual transition of factory automation from second-generation programmable devices (based on microelectronics) to third generation ones (based on new developments in software and communications, such as CASE, AI, new open architectures and communication standards).

In the first section we have drawn a map of industrial automation, classifying its component families along two dimensions: hardware vs. software intensity, and standalone products vs. integrated systems.

The evolution in time shows that software intensity is growing for many reasons: first the cost decrease of HW within each subsystem, second the technological and demand evolution towards more SW-intensive subsystems (such as Cad-Cam and CIM), and, third, the drive towards integrated and complex systems (FMS and CIM).

The most important consequence is that the diffusion of methodologies and tools for the automation of software production (SW engineering) is going to become the crucial factor for cost-effectiveness of advanced automation (Ayres, 1987; Gaio, 1988; Arcangeli, 1989).

In the second section we have emphasised the learning process, by producers and users as well, that has characterised the evolution of the first prototypes of flexible automation systems. What emerges from this evolution, is a new type of filière or industrial complex, where different industries converge to produce and implement flexible automation. The degree of convergence of innovative industries creates locational forces favouring the spatial concentration of these industries, for many reasons. Analysing the locational impact of second generation programmable automation, and the first trends in flexible automation, we come to the following conclusions in the third section of this contribution.

First, the more complex are the technological systems and the more different technological families are converging, the more innovative industries are creating webs of co-operation between firms, diversification across industries, joint developments of new processes by automation suppliers and users. This is likely to create industry hierarchies and polarisation effects for the core industries of the new trajectory.

Second, what appears at a macro level as a locational shift of fast growth industries, both across and within world regions, is the outcome of precise micro-mechanisms, of obsolescence of inter-industry old links, as well as of inter-firm co-operation webs in industrial districts, on the one hand. And emergence of new districts, growth poles, and innovative co-operations producer-producer and user-producer, on the other hand (Lundvall, 1986).

Third, the capability shown by some, selected old industrial regions, to re-industrialise, and the failure of other ones (particularly in Europe) to do the same, can be explained within this framework as well. The main variable we suggest is the capability of local firms and policy-makers to modernise economic and social networks, so that they can compete with the new emerging growth poles and regions. That is, a Penelope's task of undoing and reshaping the inter-firm co-operative networks in industrial districts and regions.

This analysis constitutes the basis for redefining regional policies suitable for a period of fast structural change. A major source of inspiration might be a revisitation of the classical literature on growth-poles, because in fact a much similar class of developments is taking place now, as in the immediate postwar period, with new industries and infrastructures assuming the lead of a new long swing of investments.

We suggest here a provisional list of issues for these policies. Structural change increases the North/South technological gap, but also creates new growth opportunities for some sections of the 'South'.

A SW-intensive growth model is the best chance from now on, because of the observed trends in automation systems. It requires preconditions in terms of investments in human capital, first layers of industrialisation under a Fordist regime, and technical capabilities to master Information Technologies (IT) and their applications.

Regional policies accompanying such a catch-up strategy might:

1) accelerate penetration in SMEs with diffusion agencies (Brown, 1981); this is important, since the Brazilian case shows that policies for technological independence from the North have a trade-off in terms of lower opportunities for diffusion in SMEs;
2) target specific user industries and applications, in order to concentrate catching-up efforts in a sequential strategy (like Japan and other Far Eastern NICs did);
3) support the constitution of a local network of specialist suppliers of components, services and software;
4) favour spin-offs and externalities from large users and innovative centres in Information Technologies (Universities, research labs, big corporations, and public expenditure centres);
5) modernise metropolitan areas and CBDs with a view to their incubators' functions (avoiding nonetheless the illusion of 'science cities', that is the unsuccessful imitation of earlier and irrepetible patterns of the North).

In fact, two distinct social processes have been observed at the regional level in the North: first, a 'Silicon Valley' type, attracting science-based activities, has emphasised interpersonal communications in a community of engineers, specialising in a radically new basic field (semi-conductors and their applications to information systems; see Saxenian, 1989).

And an 'Automation Alley', second-generation type, attracting application-specific technology developments; this is focusing upon inter-firm co-operation, in an industrial environment accumulating learning about interfacing basic technologies in end-user systems.

New window-opportunities for the South (Perez, 1989) might arise from a third generation of IT-based growth poles. They will retain both science-based features (but this time in SW, not in HW), and end-user orientation. Their new, typical co-operative networks are likely to be among specialist firms that control distinct poles of complex technology systems (for example, AI and automation specialists).

Notes

(*) This paper is the outcome of a joint reflection by the two authors+ nevertheless Camagni is responsible for Section 1, Arcangeli for Section 3, while Section 2 was jointly written.

1 Microelectronics is the new basic technology trajectory in Information Systems, that develops the use of semiconductors and integrated circuits towards an increasing miniaturization of computers: the first innovation in the field was the microprocessor (MPU) introduced by Intel in 1971-72.

2 See the classical references on the product cycle model: Vernon (1966) and Hirsch (1967). It influenced radical geography as well, through the contribution of Hymer (1972). Many recent analyses at the micro level confirm an inversion towards industrial agglomeration also in mature industries: see e.g., Estall, 1985; Syaer, 1985 and Zysman, 1987.

3 Let us anticipate that three broad families encompass the trajectory of computer-based automation: a first generation of digital devices controlled by mainframe or mini systems, a second one making use of microelectronics in sensors, control and actuators (see a detailed review in Arcangeli, 1989, ch. 5), and, finally, a forthcoming generation of 'intelligent' distributed devices, integrated in CIM networks (see note to follow) and coordinated by AI-based software systems.

4 CIM (Computer Integrated Manufacturing) is still a technological paradigm to be developed, subject to a number of R & D efforts. As Figure 2.1 shows, it is based on the convergence and integration of automation systems in design, production, logistics and management. Seminal contributions that enlighten this new and emerging paradigm are: Hollingum, 1987 and Arcangeli, 1989.

5 In terms of Figure 2.4 one might say that sub-trajectory C-E-I has been the one being adopted and developed faster. See evidence for faster adoption in large scale productions and large plants in: Basic, 1985; Ceris, 1985; Siri-Ucimu, 1985; Camagni, 1986; Gros Pietro and Rolfo, 1989.

6 Flexible Manufacturing Systems are in fact a major exception, because electronics and aerospace are not major FMS users, as they were with NC and CAD (see data in Gros Pietro and Gaibisso, 1988, and comments on the industrial impact of this new trajectory in Arcangeli, 1989).

7 Boston and the M24 axis can be conceived of as an intermediate case in between the two extremes of microelectronics basic locations, and end-user oriented locations for system integration activities. One could imagine a Weberian triangle, with semiconductors as the new raw materials, and old industrial areas as the outlet market: Boston would be a location within the triangle, closer to the end user vertex.

8 See Dosi (1984) and Saxenian (1989), among others, on the role of public demand at the origin of the semiconductors industry and its concentration in Silicon Valley.
A vast literature exists on the changing locations of microelectronics and computer industries. It cannot be summarised there, because it it impossible to isolate spatial patterns from the industry growth mechanisms.
Let us just exemplify: the strong inducements to the worldwide diffusion of these two leading industries were heterogeneous from one growth phase to another. Initially, 'product-cycle' locational policies by multinationals

selected some NICs and 'free enterprise zones' for assembly plants.
But in the last decade, on the contrary, IT industries have been growing in the South mainly because of technology accumulation policies (reverse engineering and learning by doing) pursued by the Government, as well as for the more favourable competitive environment. Gaio (1988) emphasises the importance of the open system architectures prevailing in Personal Computers, as a factor offering new technological and market opportunities to NICs. See also Delapierre and Zimmermann (1986) on this subject.

9 See Orsenigo (1989) for a review of the evidence on the early patterns of the biotechnology paradigm; the author argues that the new institutional environment created by the microelectronics revolution, paved the way for the incubation of biotech as well.
A map of the emerging industry in the US (The Economist, April 30, 1988, Survey: Biotechnology, p. 3). shows two patterns.
First, the majority of firms are attracted by research centres. These are concentrated both in New England and in California.
Second, the only area of 'research-independent' location is the North-East corridor (Pennsylvania-Illinois, plus Michigan) where the automation complex is concentrated as well ('Automation Alley', The Economist, April 11, 1987, p. 15).

Abbreviations

AI	Artificial Intelligence
ASIC	Application Specific Integrated Circuit
CAD	Computer aided design
CAE	Computer aided engineering
CAM	Computer aided manufacturing
CASE	Computer aided software engineering
CBD	Central Business District
CIM	Computer Integrated Manufacturing
CNC	Computerized Numerical Control
DCS	Distributed Control Systems
FMS	Flexible Manufacturing Systems
FA	Flexible Automation
HW	Hardware
IT	Information Technology
LDCs	Less Developed Countries
MAP	Manufacturing Automation Protocol
MPU	Micro Processor Unit
MRP-I	Materials Requirement Planning
MRP-II	Manufacturing Resources Planning
NC	Numerical Control
NCMT	Numerical Control Machine Tool
NICs	New Industrializing Countries
PA	Programmable Automation
PLC	Programmable Logic Controller
SW	Software
SMEs	Small and Medium Enterprises

References

Acs, Z.J., Audretsch, D.B. and Carlsson, B. (1988), Flexible Technology and Firm Size, Case Western Reserve Univ. and International Institute of Management, Berlin.

Arcangeli, F (1989), The Dynamics of Diffusion and the Emergence of CIM in Italy, Ph.D. thesis, Brighton, SPRU, Un. of Sussex.

Arcangeli, F. and Camagni, R. (1987), Sistemi tecnologici e diffusione territoriale. Paper delivered at the Padua seminar on Technical Change and International Competitiveness, May 28-9; forthcoming, in: Benedetti, E. (ed.), Mutazioni Tecnologiche e Condizionamenti Internazionali, F. Angeli, Milano.

Arcangeli, F., Dosi, G. and Moggi, M. (1987), Patterns of Diffusion of Electronics Technologies. Paper delivered at the Conference on Programmable Automation and New Work Modes, Paris, April 2-4.

Aydalot, P. (1986), Milieux Innovateurs en Europe, Gremi, Paris.

Ayres, R.U. (1986), The Man-Machine Interface, Technology Forecasting and Social Change, 29, pp. 99-118.

Ayres, R.U. (1987), Future Trends in Factory Automation, Laxenburg, IIASA, WP-87-22, February.

BASIC, Bureau d'Analyse des Strategies en Informatique et Communications (1985), Lettre 2000, Dossier Robotique, no. 4.

Brown, L.A. (1981), Innovation Diffusion: A New Perspective, Methuen, London.

Burenstam-Linder, S. (1985), Pacific protagonist: implications of the rising role of the Pacific, American Economic Review, May.

Cainarca, G.C., Colombo, M.G. and Mariotti, S. (1987), Etat et perspectives de la diffusion des systemes integres d'automation dans l'industrie metalmecanique italienne, paper delivered at the conference Automatisation programmable et conditions d'usage du travail, Paris, April 2-4.

Camagni, R. (1984), Il Robot Italiano, Milano, Il Sole/24 ore.

Camagni, R. (1985), La diffusione territoriale dell'innovazione: il caso della robotica industriale, in: Balloni, V. (ed.), Esperienze di Ristrutturazione Industriale, Il Mulino, Bologna.

Camagni, R. (1986), Robotique industrielle et revitalisation du Nord-Ouest italien, in: Federwisch, J. and Zoller, H. (eds.), Technologie Nouvelle et Ruptures Regionales, Economica, Paris.

Camagni, R. (1987a), Il Robot lombardo, Milano, Il Sole/24 ore.

Camagni, R. (1987b), Diffusion of flexible automation processes in the Lombardy region: evidence and policy issues, paper delivered at the Conference on Automatisation programmable et conditions d'usage du travail, Paris, April, 2-4.

Camagni, R. (1988), L'automazione Industriale, Milano, Il Sole/24 ore.

Camagni, R. (1989), The flexible automation trajectory: the Italian case, in: Arcangeli, F., David, P. and Dosi, G. (eds.), The Diffusion of New Technologies, vol. 1, Oxford University Press, Oxford.

Camagni, R. and Malfi, L. (eds.) (1986), Innovazione e Servizi nelle Economie Regionali Mature, Angeli, Milano.

Camagni, R. and Gambarotto, F. (1988), L'adozione dei processi d'automazione nel Veneto, Oltre il Ponte, 23, pp 3-36.

CERIS (1985), Sistemi flessibili di lavorazione: un'analisi internazionale, Bollettino Ceris, no. 17.

Ciciotti, E. (1982), Innovazione e Sviluppo Regionale: il Ruolo dell'Impresa pubblica, CSEA, Milano.

Colombo, M.G. and Mariotti, S. (1985), Note economiche sull'automazione flessibile, Economia e Politica Industriale, no. 48.

Cortese, R. (1987), Innovazione tecnologica e servizi reali nei sistemi di piccola impresa: il caso di Montebelluna, tesi di laurea, Venice, Corso di laurea in Urbanistica, IUAV.

Delapierre, M. and Zimmermann, J.B. (1986), L'informatique dus Nord au Sud, La Documentation Française, Paris.

Dosi, G. (1984), Technical Change and Industrial Transformation, Macmillan, London.

ECE (1986), Recent Trends in Flexible Manufacturing, Economic Commission for Europe, United Nations, Geneva.

Edquist, J. and Jacobsson, S. (1988), Flexible Automation: The Global Diffusion of New Technology in the Engineering Industry, Basil Blackwell, Oxford.

Estall, R.C. (1985), Stock control in manufacturing: the just-in-time system and its locational impliactions, Area, 17, 2, pp. 129-133.

Ettlie, J.E. (1985), The Implementation of Programmable Manufacturing Innovations, Industrial Technology Institute, Ann Arbor, Michigan.

Flamm, K. (1986), International Differences in Robot Use, Brookings Institution, Washington.

Fransman, M. (ed.) (1986), Machinery and Economic Development, Macmillan, London.

Gaio, F.J. (1988), The Development of Computer Software Technological Capabilities in Developing Countries, a case study of Brazil, Ph.D. thesis, Brighton, SPRU, University of Sussex.

Gibbs, D.C. and Edwards, A. (1985), The diffusion of new production innovations in British industry, in: Thwaites, A.T. and Oakes, R.P. (eds.), The Regional Impact of Technological Change, Pinter, London.

Gregory, G. (1986), Japanese Electronics Technology: Enterprise and Innovation, Wiley, New York.

Gros Pietro, G.M. and Gaibisso, A.M. (1988), Diffusione internazionale dei sistemi flessibili di produzione, pp. 145 ff, Bollettino CERIS, no. 23, Torino, CERIS-CNR.

Gros Pietro, G.M. and Rolfo, S. (1989), Flexible Automation and Firm Size: Some Empirical Evidence on the Italian Case, Technovation.

Hirsch, S. (1967), Location of Industry and International Competitiveness, Clarendon Press, Oxford.

Hollingum, J. (1987), Implementing an Information Strategy in Manufacture, IFS, Kempston.

Hymer, S. (1972), The Multinational Corporation and the Law of Uneven Development, in: Bhagwati, J.N. (ed.), Economics and World Order, pp. 113-140, Macmillan, London.

Jacobs, J. (1984), Cities and the Wealth of Nations, Random House, New York.

Kaplinsky, R. (1982), Computer-aided Design, Pinter, London.

Lafaille, A. (1984), Strategie robotique des Etats Unis, in Mes-Afri, Automatisation et mutations economiques et sociales, Paris.

Lundvall, B.A. (1986), Technological Revolutions and the International Division of Labour, Aalborg, Aalborg University.

Lynch, P.M. (1976), Economic-Technologic Modeling and Design Criteria for Programmable Assembly Machines, Ph.D. thesis, MIT, Boston, Mass.

Morgan, K. and Sayer, A. (1988), Microcircuits of Capital, Polity Press, London.

Ohmae, K. (1985), Triad Power: The Coming Shape of Global Competition, Free Press, New York.

Orsenigo, L. (1989), The Emergence of Biotechnology, Pinter, London.

Pegolo, G.L. (1989), La diffusione Inter- ed Intra-Regionale dell'Automazione Programmabile, Ph.D. thesis, Venice, DAEST-IUAV.

Perez, C. (1985), Microelectronics, long waves and world structural change: nex perspectives for developing countries, World Development, 13 (3), pp. 441-63.

Perez, C. (1989), Nuove tecnologie e sviluppo economico, in: Benedetti, E. (ed.), Mutazioni Tecnologiche e Condizionamenti Internazionali, F. Angeli, Milano.

Piore, M.S. and Sabel, C. (1984), The Second Industrial Divide, Basic Books, New York.

Pred, A. (1969), Behaviour and Location, Part 2, Lund University.

PSI, VDI and BIPE (1985), Microelectronics in Industry, Policy Studies

Institute, London.

Rees, J., Briggs, R. and Hicks, D. (1985), New technology in the US machinery industry: trends and implications, in: Thwaites, A.T. and Oakey, R.P. (eds.), The Regional Impact of Technological Change, pp. 164-194, Pinter, London.

Saxenian, A.L. (1989), A High Technology Industrial District: Silicon Valley in the American Context, in: Perulli, P. (ed.), Città della Scienza e della Tecnologia, no. 6/7, pp. 29-53, Quaderni Istituto Gramsei Veneto, Venice.

Sayer, A. (1985), New developments in manufacturing and their spatial implications: from FMS to Just-in-Time. Paper delivered at the Conference on Spatial Structures and the Social Process, Lesvos, August, 26-30.

Sciberras, E. and Payne, B.D. (1985), Machine Tools Industry, Technical Change and International Competitiveness, Longman, Harlow.

SIRI - UCIMU (1985), La robotica in Italia: un settore innovativo per la realtà poduttiva del domani, Milano.

Tauile, J.R. (1987), Automacao e competitividade; uma avaliacao das tendencias no Brasil, paper no. 111, IEI, UFRJ, Rio de Janeiro.

Thwaites, A.T. and Oakey, R.P. (eds.) (1985), The Regional Impact of Technological Change, Pinter, London.

UCIMU (1987), Macchine per Produrre Macchine, Il Sole/24 ore, Milano.

Vernon, R. (1966), International Investment and International Trade in the Product Cycle, Quarterly Journal of Economics, May.

Von Tunzelmann, G.N. (1986), Diffusion, Competition and Diminishing Returns, British Energy Technologies, Past and Future, Brighton, SPRU, DRC Discussion Paper no. 38.

Watanabe, S. (1983), Market structure, industrial organisation and technological development: the case of the Japanese electronics-based NC-machine tool industry, ILO, Geneva.

Zysman, J. (1987), From Tokyo to Prato: will American industry respond to the emerging production revolution, paper delivered at the Conference on Automatisation programmable et Conditions d'usage du Travail, Paris, April, 2-4.

6 Patents as indicators of innovative performances at the regional level[1]

A. BOITANI AND E. CICIOTTI

1. Introduction

Although innovative activity, since Schumpeter (1934, 1939) is widely considered crucial for an explanation of economic growth, of relative competitiveness of industries, countries and regions[2], innovation is not an easily measurable phenomenon[3]. Researchers are often forced to rely on some indirect measure such as patent statistics or R & D expenditures, which are (or should be) readily available data[4]. In using such indirect measures of innovative activity one ought to be honest at the outset about their limitations and the likely distortions they may imply.

In Sections 2, 3, and 4 we shall briefly discuss possibilities and problems of patents and R & D expenditures as indicators of technological advance in general and in regional analysis in particular, while in Sections 5, 6, and 7 we shall present a few results obtained by using patent statistics in order to measure the innovative potential of Italian regions.

2. Patents as Indicators of Innovative Activity

It is sometimes maintained that patents are a proper measure of inventive output as distinct from innovation (Freeman, 1982, pp. 53-54). Most of the patents filed by individual inventors never lead to the

---

[1] The authors are grateful to Geoffey Harcourt for his useful comments and to Alessandro Aronica and Alessandro Sattanino for their help with the building of the data base.

[2] There is also a growing body of literature on the relation between innovations and long cycles. See for example Freeman (1982); Freeman and Soete (1982); Freeman (1984) and the reference quoted therein. On the role of technological advance in regional development see the survey by Malecki (1983).

[3] See Archibugi (1987) and Archibugi et al. (1987).

[4] See, among many others, Mansfield (1964) and (1968); Schmookler (1966); Freeman (1982) and (1984); Pavitt (1982) and (1984); Griliches (1984); Griliches and Pakes (1986); Scherer (1983); Archibugi (1987).

introduction of a new product or process and some of the patenting activity of firms is merely part of a defensive strategy aimed at protecting some basic invention, or at preventing competitors from improving a new product or process. In addition, many case studies show that there may be a considerable time lag between patent application and the actual development and marketing of a new product or adoption of a new process **(sleeping patents)**, a time lag that patents statistics cannot capture[5].

However, one could reasonably assume that patents are indicators at least of **innovation potential**, if not of actual innovation. Patents are one of the means of appropriating the benefits of innovative activity, not only of protecting an intellectual achievement (Boehm, 1968; Dossier Brevetti, 1986). Actually, the very reason for the existence of a patent system is to stimulate the innovative activity regarded as a positive externality, by establishing a property right which ensures a temporary monopoly profit to the innovator (Arrow, 1962; Nordhaus, 1969; Taylor and Silberston, 1973; Firestone, 1974).

In any case it seems convenient to disregard patents filed by individual inventors. Though some of these patents may represent 'crucial' inventions they are usually of little economic relevance because they never become actual innovations[6].

When using patent data to make comparisons between the innovation potential of industrial sectors one must take into account that different industries may differ in terms of patent intensity (measured by the number of patents per patenting firm). Chemical firms, for instance, on average show a rather high patent intensity, as imitation is relatively easy in this sector and as firms tend to build up a protective barrier around any foreseeable improvement of a new product or process, which could otherwise be exploited by a rival without costly investments. On the other hand, industrial sectors where the development of new products and processes require huge investments (e.g., aerospace) or where secrecy is legally enforced (e.g., nuclear products) have a lower patent intensity, though they may be no less innovative than the chemical sector (Pavitt, 1982 and 1985).

---

[5] On the strategic aspects of patenting activity see Gilbert and Newbery (1982) and Dasgupta (1987), (1988).

[6] On this subject different opinions are held by Schmookler (1966); Jawkes, Sawers and Stillerman (1969), who stress the role of individual inventors in the process of technological advance, and by the 'neo-schumpeterians', who emphasise the role of large corporations (see Freeman, 1982 and Pavitt, 1982).

Moreover the patent intensity can vary according to the innovation concerned: a firm may have a higher convenience in patenting product innovations than process innovations as these are more protected by secrecy. As a consequence sectors where product innovations prevail over process innovations may become highly patent intensive though their innovative potential is not that high (Antonelli, 1986).

Not all of the patents have the same technological standard or economic significance. In other words different patents reflect a widely different **quality** of innovative activity. This may be due to low requirements of 'innovativeness' imposed by the Patent Office or to lack of technological expertise of the examiners. But even the tightest examination could not prevent the existence of quality differences among patents within a large data base[7]. A possible way to overcome such a problem would be to check the number of patents which have been extended to one or more foreign countries, as extension signal at least the economic relevance of patents beyond the national borders and the willingness of firms to bear additional costs. Unfortunately data on extensions requested by Italian patenting firms are not recorded by the Italian Central Patent Office. However, by disregarding the patents filed by individual inventors the number of worthless patents in the data set should be reduced.

3. R & D Versus Patents as an Indicator of Innovation

R & D expenditures or the number of persons employed in R & D have often been preferred to patents as indicators of innovation potential. It is possible to argue, however, that R & D measures are no less biased than patent measures. Firstly, R & D activities, as well as patenting, do not necessarily lead to actual innovation; they are at most measures of innovation potential. In fact, R & D is often used by firms as a strategic variable in their competitive struggle, and most of the R & D can be purely duplicative. Secondly, it is well known that many small firms do not have specialised R & D departments and that shopfloor research can lead to improvements (both in processes and products) which account for a large proportion of a firm's growth in productivity and market performance (Freeman, 1982). R & D data can therefore overestimate the innovative activity of large corporations with respect to small firms. As a consequence also inter-industry comparisons are likely to be biased, in the sense of

---

[7] An attempt to assess the 'quality' of a small sample of Italian patents can be found in CER (1988), ch. 6.

overestimating innovative activities in highly concentrated sectors.

Even more problematic is the use of R & D expenditures combined with patents in a kind of 'knowledge production function', where R & D is considered as an input and patents as an intermediate output. This approach may be quite appealing to an economist, but the empirical evidence seems to be rather puzzling.

Scherer (1965) tested a knowledge production function for a sample of American firms and found that the 'propensity to patent' (i.e., the number of patents per billion dollars of internally financed R & D expenditure) declines as the size of firms increases and that the proportion of patents to R & D spending has been falling since the late 1950s. In order to explain these results Scherer suggested the existence of diminishing returns in research activity and of declining productivity of R & D over time. Similar results have been recently found by Bound, Cummins, et al., (1984) and a similar explanation has been put forward[8].

This explanation is open to several objections. First of all, it seems implausible that large firms which benefit from incresing returns in production, marketing and financial management have decreasing returns only in their innovative activity. Besides, one may wonder why large corporations appear to have put ever larger sums of money (in real terms) in R & D activities which were proving to be less and less productive over time. One may also wonder why in the past twenty years spending in R & D has been soaring in large firms and growing at a lower rate in small firms (Soete, 1979). One should probably admit some kind of irrational behaviour, but more palatable explanations are not in short supply. Soete (1978, 1979) suggests that the declining ratio of patents to R & D spending as the size of firms increases is better explained in terms of differences in strategies between large and small firms. That is, larger firms tend to use greater R & D spending rather than patents as a protection barrier and do not patents many of the results of their R & D activity. Such behaviour might explain the time trend of patents to R & D ratios in the U.S. as the outcome of increasing industrial concentration.

Pavitt (1982) advances the view that the very nature of industrial R &

---

[8] It is interesting to note that both Scherer and Bound, Cummins, et al., found that, beyond a threshold level the size of firms does not proportionately affect R&D activity (see also Cohen, Levin and Mowery, 1987). Therefore the crude 'Schumpeterian hypothesis' does not seem to hold empirically, while it is unsatisfactory from a theoretical viewpoint (see Dasgupta and Stiglitz, 1980; Kamien and Schwartz, 1983; Dasgupta, 1987 and 1988).

D changes as firms get bigger 'with more emphasis on testing, application engineering and systems and production know-how, all of which are more difficult to patent' (p. 36). Soete's and Pavitt's explanations can be strenghtened by recalling the above mentioned propensity of smaller firms to a less formalised research activity out of specialised R & D departments.

Our opinion is that both patenting and R & D activity are imperfect measures of innovation potential, for there are innovations which are neither patented nor are the outcome of formal R & D, while there are R & D expenditures as well as patents that never lead to innovations. Thus actual innovations, R & D and patenting can be thought of as three partially intersecting sets of innovative activities, as depicted in Figure 1.

Though patents and R & D overlap considerably they are not perfect substitutes, nor can they be regarded as totally alternative indicators. Prudence and completeness suggest that one ought to make careful use of both measures, taking advantage of the additional information about different features of the innovative process that a comparison of the two indicators is able to provide.

4. The Use of Patents in Regional Analysis

As Oakey, Thwaites and Nash (1982) aptly write 'if the rate of technological advance can vary between nations, industries and enterprises, then it seems reasonable to suggest that it can also vary between regions, if for no other reason than because of spatial variations in industrial structure' (p. 1073).

Patent statistics can be a very useful indicator of innovation potential at the regional as well as at the national level, provided one is aware of their biases, which we shall summarise in the present section while giving account of their advantages over other indicators.

a) The regional innovation potential measured by patent statistics can be partially distorted by the behaviour of multi-plant firms. The innovative activity of these firms is often distributed over several plants, located in more than one region, but they usually have just one patents department attached to their headquarters. Hence patent data tend to overestimate the innovation potential of a region where many headquarters of multi-plant enterprises are centred and to change the balance of sectorial innovation potentials between regions.

The case of Alfa Romeo is a useful example. Although an appropriate share of Alfa Romeo's innovative activity can be safely attributed to its Milan plants and research department, almost all of its patents are filed by the Alfa Romeo headquarters in Naples. As a consequece Campania's innovative potential in general and in the car industry in particular is inflated, whilst Lombardia's is underestimated by patent data.

b) If patenting is more frequently used to protect product innovations than process innovations, patent statistics tend to overestimte the technological innovativeness of regions where product innovations constitute a larger share of their innovative activity. Such a bias could turn out to be valuable if it is true that "product innovation is more likely to be a reliable indicator of a region's innovation potential ... because in-house product innovations are mainly dependent on the internal resources of the establishment concerned and the aids to innovation available in the local environment. Conversely, provided capital is available, new process machinery can be readily purchased from outside the local area or abroad with little difficulty" (Oakey, Thwaites and Nash, 1982, p. 1075)[9]. In a sense patent statistics, with their product innovation bias, could be regarded as a better indicator than R & D expenditure in signalling a region's 'indigenous innovation potential'.

c) When making comparisons of innovation potential between regions, attention must be paid to some of the already mentioned biases of patent statistics. For example, the potential technological advance of a region can be overestimated by patent statistics if in such a region there is a heavy concentration of firms belonging to sectors with a high patent intensity, for example the chemical and the mechanical sectors. For this reason in the following sections we shall try, whenever possible, to isolate the truly regional factors from the structural factors affecting the patenting performance of different regions.

Given the declining propensity to patent as the size of firms increases it is possible to say that patent statistics will rate the innovation potential of regions where small or medium sized firms prevail rather higher than R & D expenditures would do. The simultaneous presence in a region of small firms belonging to intensily patenting sectors can give rise to very different pictures coming out of patent statistics and R & D expenditures respectively.

On the other hand, the application for a patent is a fairly administrative, time consuming and sometimes bureaucratic procedure. Know-how of the way to use patent offices is rather limited in peripheral areas,

---

[9] For the empirical evidence supporting this view see Oakey, Thwaites and Nash (1982), _infra_.

especially as far as small and medium sized firms are concerned. This may imply that patents data underestimate the innovative potential of peripheral areas with a large number of small firms. Among the actual and potential advantages of patent statistics over other indicators for regional comparisons of technological advance the following ones should be mentioned.

a) The Central Patent Office attaches to each granted patent a code according to a technology classification system[10]. It is therefore possible to build up a cross tabulation which shows the flow of potential innovations from industrial sectors to technology branches and allows one to measure the degree of 'innovative fertilisation capability' of different industrial sectors, or its reverse, that is the sectorial degree of innovative specialisation. A similar approach can be replicated at the regional level in order to make a comparative analysis of innovative patterns of the same sector in different regions. To allow for regional differences in firm size within the same sector we can examine the degree of specialisation of enterprises belonging to the same dimensional class (in terms of employees or sales) and to the same sector in various regions. The information gained through this kind of detailed analysis can give useful insights for both explanatory and policy purposes.

b) Data on patents licensing (unfortunately not available for Italy) can give rise to a sort of inter-regional balance of technology transfers, in terms of patents 'purchased' and 'sold' by each region to any other one. Such a balance can help to shed light on some features of the regional pattern of technological advance that no other indicator is able to capture.

c) By looking at patenting firms it is possible to calculate an index for each sector in each region which measures the stability of the set of 'innovative' firms over time. Again, this kind of information on the regional innovative process cannot be captured by R & D expenditures or other indicators of innovation potential.

The data base

In order to test possibilities and limitations of patent statistics as indicators of innovation potential at the regional level we use the data set recently built up by Centro Europa Ricerche (CER) on behalf of Ente

---

[10] The Italian Central Patent Office makes use of the World Industrial Property Organisation (WIPO) classification system which is also used by the European Patent Office (EPO) in München. It is a 6 digits classification with more than 600 classes and subclasses. In what follows we employ a 2 digits classification which aggregates the 6 digits one in just 32 branches.

Nazionale Energie Alternative (ENEA). Such a data set records patent applications by Italian firms from 1970 to 1985 and patents granted in the period 1970 to 1979. It does not contain patent applications to the European Patent Office (EPO) which Italy joined in 1979. This is an unfortunate shortcoming of the data set, because many large firms, since 1979-80, have chosen to file their patent applications at the EPO, which ensures protection in all EEC countries and in the US, Canada and Japan, which are the most important markets for Italian firms. Our data are bound to underestimate the patenting performance of larger firms in recent years and to distort the regional patenting activity as it is known that a greater number of firms from northern regions than from southern regions file their patents at the EPO.

Owing to the limited portion of the data base which was available at the moment of writing, the subsequent discussion is incomplete and tentative. Nevertheless it is our aim to present to the reader some ideas of what may be done with patent statistics in regional studies.

5. Regional Distribution of Patents in Italy

At a first glance, Table 1 shows a clear cut concentration of patent applications in only few Italian regions: three regions, namely Piemonte, Lombardia and Emilia account for about 70% of total applications in the period covering 1970 to 1986. On the contrary, eight regions (Trentino, Umbria, Campania, Puglia, Basilicata, Calabria, Sicilia and Sardegna) account for less than 1% each.

Veneto and Emilia have respectively more than 11% and more than 15% of patenting firms while havring just about 8% and 12% of patent applications. The result is a low patent intensity. On the other hand, Lombardia and Piemonte definitely have higher shares of patent applications than shares of patenting firms and, therefore, have high patent intensity.

Apart from Piemonte and Lombardia, Friuli, Lazio Sicilia and Sardegna have a patent intensity higher than the national average, while Emilia and Veneto have a patent intensity definitely lower than the national average. Sicilia and Sardegna actually show the highest ratio of patent applications to patenting firms, due to the very low number of highly patenting firms.

As we mentioned above, patents allow for the calculation of an index of stability of the set of 'innovative' firms over time. Such an index is defined by:

$$\mu = 1 - \frac{PF_{Tot} - PF_t^{max}}{\Sigma_t PF_t - PF_t^{max}}$$

where PF_{Tot} stands for the number of patenting firms in a given region over a given number of years, PF_t^{max} is the maximum number of patenting firms in a single year and $\Sigma_t PF_t$ is the sum of patenting firms over the years. A value of μ equal to one means that the same firms are patenting every year, while μ equal to zero implies that the set of patenting firms changes every year. The higher μ the more stable is the set of 'innovative' firms in a given region.

The index μ for the period covering 1970 to 1985 (see Table 1) shows that regions like Piemonte, Lombardia and Emilia, which have a high share of patenting firms also have an index of stability close to the national value. In these regions, the innovative activity of firms tends to be a fairly 'continuous' and regular process over time. Veneto, though having a high percentage of patenting firms, has a lower value of the stability index. This implies a dynamic and, at the same time, less continuous innovative activity of Veneto's firms. Finally, the high stability of the set of patenting firms in Lazio can probably be explained by their relatively small number.

In order to take into account regional differences in economic structure, patent applications have been compared with economic variables such as industrial value added, industrial investments and industrial employees. Table 2 shows this exercise for three different time periods: 1970 to 1974; 1975 to 1979 and 1980 to 1985.

Apart from Piemonte, Lombardia and Emilia which maintain their primacy with respect to the rest of Italy (the location quotients are higher than one for all the economic variables considered and for the whole period from 1970 to 1980), there are very few regions that have an innovative performance (in terms of shares of patent applications) consistent with their economic role (in terms of the variables considered). Friuli is the sole region which has a pattern similar to that fo the three leading regions. Veneto increases substantially its share of patent applications over time: its location quotient turns out to be greater than one (relatively to industrial value added) only in the period covering 1980 to 1985. Finally, Lazio has an over-representation of patent applications only with respect to industrial employment.

It is interesting to note the difference between northern and southern regions when location quotients are computed with respect to industrial investment. Southern Italy has a share of industrial investments consistently higher than its share of patent applications. This confirms regional differences in the 'qualitative' features of industrial investments, generally not shown by data on fixed investments, as well as the impact of the Italian regional policy which diverted to the South massive investments in mature and less innovative sectors.

The location quotient of patent activity greater than one (with respect to industrial value added) does not necessarily imply a low productivity in the innovation process. The over-concentration of patent activity in few regions can be the result of the already mentioned statistical bias of patent data due to the location of patents departments of multiplant firms. However, it should be stressed that the innovative activity largely benefits from the agglomeration economies of the highly industrialised and technological advanced core areas (Antonelli, 1987 and Epifani, Giannagostino et al., 1987). Patents developed in core regions can then be exploited in other areas directly (by multiplant firms) or indirectly (by licensing or other forms of technological transfers). Finally, structural factors have a strong influence on the regional distribution of patents.

6. Sectoral Specialisation and Patent Activity at the Regional Level

Regional disparities in sectoral distribution of patents, referring to a sample of firms representing an average of 28% of total patenting firms and more than 50% of patent applications, are presented in Table 3.

Only Northern regions, with the exception of Lazio, have more than 20 sectors (out of the 32 considered) with patenting firms, whereas all Southern regions show a very small number of patenting sectors.

In order to account for the effect of the regional industrial mix on patenting activity, patent applications have been standardised with respect to industrial employment by sector and by region.

As depicted in Table 4, firstly there are regions specialised in sectors with a great number of patent applications at the national level ('national patenting sectors'). Generally speaking, highly industrialised regions have a higher than average innovative performance in those sectors in which they are traditionally specialised (e.g., Piemonte in sectors 33, 34, 35, and 36; Liguria in sectors 31 and 48; Lombardia in sectors 25, 31, 32, 34, 36, and 48; Veneto in sectors 31 and 36). In contrast to this, among less developed regions, only Campania in sector 34 and 35 and Sicilia in sector 34 show a good innovative performance in national patenting sectors.

Secondly, Table 5 shows some specific regional specialisation pattern in sectors that, though not very relevant at the national level in terms of patent applications, account for a large share of some regions' patenting activity ('local patenting sectors'). Apart from Liguria, specialised in sectors 22 and 24, this pattern refers to peripheral and less industrialised regions like Trentino (in sectors 22, 45, and 46), Veneto (in sector 45), Friuli (in sector 46) in the North East; Marche (in sectors 46 and 83), Umbria in sectors 41 and 45) and Lazio (in sector 94) in Central Italy and Campania (in sector 50) and Abruzzi (in sector 24) in the South.

Finally, there are sector with a small number of patent applications both at the national and the regional level but highly concentrated in few regions (Table 6); sector 13, which is concentrated only in Lombardia (66% of national patent applications) and Lazio (33%); sector 14, that is mainly located in Lombardia (86%); sectors 16, 21, 91, and 98 which are concentrated in Lazio (83%, 80%, 90% and 100% respectively); sector 26, concentrated in Basilicata (50%).

The existence of a 'regional factor' in patenting activity, is highlighted by analysing the number of patent applications per patenting firms (patent intensity) broken down by sectors and regions (see starred regions of Tables 4, 5, and 6). In such a way it is possible to compare the actual performance of patenting firms in the same sectors across different regions, thus isolating regional factors like firm size, entrepreneurial and managerial attitudes, technological environment, and human capital characteristics. As far as firm size is concerned it is worth mentioning that the data set available provides patent applications broken down by sector, by region and by size class of firms (both in terms of employment and sales). Although these data have not been utilised yet for computing problems, the disaggregation according to firm size would contribute to a better understanding of regional differences in innovative performance.

The analysis of index numbers of patent applications per patenting enterprise confirms the regional disparities in sectorial performance of innovating firms. On the one hand, highly industrialised regions show greater patent intensity in those sectors in which they are specialised and in which they have a high location quotient with respect to industrial employment (e.g., Piemonte in sectors 33, 35, 36; Lombardia in sectors 25, 31, 32, 34, 48; Liguria in sectors 31 and 48 and so on). On the other hand, regions with a large number of small and medium sized firms, like Veneto and Emilia, tend to have lower patent intensity even in those sectors in which they have a strong specialisation and high location quotient. Southern regions show, with few exceptions, a low patent intensity. In this case data are probably biased by the small number of patenting firms per sector, which results in an artificially high (or low) patent intensity index.

A further aspect of regional disparities in patenting activity is depicted in Table 7, where the index of innovative fertilisation capability by sector has been calculated for Piemonte, Lombardia, Veneto, Emilia and Lazio.

If P_{ij} stands for the number of patents granted in technology branch j to firms in the industrial sector i we can calculate $\pi = p_{ij}/\Sigma_j P_{ij}$ and $\beta = \Sigma_i P_{ij}/\Sigma_i \Sigma_j P_{ij}$. The index is then given by $\delta = \Sigma_j \; |\pi_{ij} - \beta_j|/200$, where $\pi_{ij} \geq 0$ for at least one of the technology branches (j) for each i. The lower is the value of δ, the higher is the degree of innovative fertilisation capability of the sector (or the lower is the degree of innovative specialisation).

The index shows wide differences for the same sector in the innovative fertilisation capabilities across the regions examined. Generally speaking, the degree of innovative specialisation of a sector tends to be lower in regions where the sector is characterised by a great number of patents and patenting firms: thus, Lombardia has a low δ in sectors 25 and 32; Emilia in sectors 32 and 33; Piemonte in sectors 34, 35, 36, and 48; Veneto in sector 31. Moreover, in these sectors the value of δ is generally lower than the national average. The behaviour of Lazio is quite different: it has a degree of innovative specialisation higher than the national average in all sectors considered. New features of the regional imbalance in innovatve performance emerge, when the capability of industrial sectors to fertilise technology branches is taken into account. The sectorial specialisation of regions (Table 4) does not necessarily coincide with a corresponding high degree of fertilisation capability (e.g., sectors 31, 34, 36, and 48 in Lombardia; sector 48 in Piemonte).

7. Regional Differentials in Innovations, R & D and Patenting Activities: a Comparison

As we mentioned above, both patents and R & D activity are imperfect measures of innovtion potential; the complementary use of patents (patent application and patenting firms), R & D (expenditure and employment) and innovative firms (by process and/or product innovations) in describing the innovation process is depicted in Figure 1, in which regions are represented according to the prevailing role (in terms of L.Q.) of one (or more) of such indicators. At the intersection of all the three sets (representing patents, R & D and innovation respectively) we find Lombardia, the sole region with

L.Q. greater than 1 for: patent applications and patenting firms; R & D expenditures and employment; product and/or process innovating firms.

Two regions, Piemonte and Lazio, have a strong specialisation in R & D and in patenting activities (more in terms of patent applications than in terms of patenting firms), whereas their share of innovative firms is lower than their corresponding share of industrial employment. These are regions with a strong technological stand and where innovation activity is likely to be more institutionalised, when considering size and nature of the industrial firms (i.e., multinationals in Piemonte and state holdings in Lazio). It is worth reminding that Piemonte has three times as many patent applications as Lazio.

In Emilia, innovative firms and patenting activity prevail over R & D activity. In this region, although specialised R & D activity is not very relevant, because of the prominent role of shopfloor research, there is a strong coincidence between innovation and technology as shown by the high proportion of patent applications, patenting firms and firms that have introduced product and/or process innovation.

In Trentino, Veneto, Toscana and Marche, innovation prevails over patenting activity (although Trentino, Veneto and Marche have a L.Q. greater than 1 for patenting firms) and, even more, over R & D activity. In these regions sectorial specialisation and size of firms tend to emphasise innovations which are based neither on patenting activity nor on formalised R & D.

Liguria is the sole Italian region where R & D prevails with respect to both patenting and innovation activities. On the opposite side, we find Friuli, with a location quotient greater than 1 only for patent applications and patenting firms and a rather low specialisation with respect to innovative firms and, especially, to R & D activity. It is worth noting that the location quotient of innovative firms is 0.90 in Friuli versus 0.44 in Liguria. These findings seem to confirm that patent statistics are a better indicator of indigenous innovation potential than R & D statistics, probably because of their bias toward product innovations.

8. Summary and Conclusions

The preliminary analysis carried out by using patent statistics as an indicator of innovative performance reveals strong regional differences in the Italian context.

- There are very few regions that have an innovative performance (in terms of their share of patent applications) consistent with their economic role in terms of various relevant variables (industrial value added, industrial employment and industrial investment). Lombardia, Piemonte, Friuli, Emilia and, recently, Veneto form the north west-north east strong-hold of the Italian technological advance (Lazio is an exception with very peculiar characteristics).

- There are strong regional differences in terms of sectorial differentiation of patenting firms. Only few regions have a large number of patenting sectors, thus showing a qualitative peculiarity that may increase the technological gap between innovative and non-innovative regions.
- Standardisation by industrial employment and by sector reveals a clear cut 'regional factor' in innovative performance, with strong difference between highly industrialised regions (which exhibit very high location quotients in those sectors in which they are traditionally specialised) and southern regions (interesting exceptions are Campania and Sicilia).
- The analysis of patent intensity by sector and by regions confirms the existence of the regional factors, by strengthening the findings mentioned above and drawing attention to the need for a micro-economic analysis at the firm level (or at least for a disaggregation of patent data by size of firms).
- The same industrial sectors have different technological fertilisation capability in different regions. In order to explain such results and to have useful insights into the regional innovative performance a full-fledged cross tabulation of industrial sectors and technological branches by regions should be used.
- Finally, the analysis confirms our opinion that both patenting and R & D activities are imperfect measures of innovation potential and that they should be jointly used for a better understanding of the regional innovation process, although probably patent statistics more than R & D data can provide the kind of detailed information needed for such investigations.

Table 1. **PATENT APPLICATIONS BY REGIONS 1970-86**

Regions	Patent applications n.	%	Patenting firms n.	%	Pat. appL/ pat.firm n.	Index	Stability Index µ
Piemonte-V.A.	7794	16.05	1709	11.67	4.56	1.38	0.67
Liguria	1030	2.12	256	2.43	4.02	1.21	n.a.
Lombardia	20118	41.43	5368	36.66	3.75	1.13	0.65
Trentino	482	0.99	212	1.45	2.27	0.69	n.a.
Veneto	3687	7.59	1612	11.01	2.29	0.69	0.61
Fiuli-V.Giulia	1725	3.55	499	3.41	3.46	1.04	n.a.
Emilia-R.	6022	12.40	2212	15.11	2.72	0.82	0.66
Marche	641	1.32	349	2.38	1.84	0.55	n.a.
Toscana	2307	4.75	913	6.24	2.53	0.76	n.a.
Umbria	298	0.61	144	0.98	2.07	0.62	n.a.
Lazio	2798	5.76	760	3.68	3.68	1.11	0.75
Campania	445	0.92	152	1.04	2.93	0.88	n.a.
Abruzzo-M.	184	0.38	84	0.57	2.19	0.66	n.a.
Puglia	139	0.29	83	0.57	1.67	0.51	n.a.
Basilicata	22	0.05	11	0.08	2.00	0.60	n.a.
Calabria	23	0.05	20	0.14	1.15	0.35	n.a.
Sicilia	664	1.37	121	0.83	5.49	1.65	n.a.
Sardegna	178	0.37	38	0.26	4.68	1.41	n.a.
Italia	48557	100.00	14643	100.00	3.32	1.00	0.65

Source: data base Enea -Cer

Table 2 **PATENT APPLICATIONS BY REGIONS STANDARDISED BY INDUSTRIAL VALUE ADDED, INDUSTRIAL INVESTMENTS AND INDUSTRIAL EMPLOYMENT(location quotients)**

Regions	**1970-74**			**1975-79**			**1980-85**		
	Value Added	**Investments**	**Employment**	**Value Added**	**Investments**	**Employment**	**Value Added**	**Investments**	**Employment**
Piemonte-V.A.	1.20	1.30	1.03	1.23	1.29	1.04	1.61	1.32	1.00
Liguria	0.61	0.83	0.69	0.66	0.67	0.68	0.52	0.78	0.68
Lombardia	1.57	2.04	1.43	1.50	1.65	1.38	1.92	1.48	1.31
Trentino	0.74	1.13	0.88	0.50	0.59	0.64	0.65	1.07	0.88
Veneto	0.89	1.15	0.79	0.92	1.11	0.82	1.04	1.09	0.93
Friuli-V.Giuli	1.63	1.60	1.36	1.32	1.15	1.13	1.72	1.34	1.45
Emilia-R.	1.25	1.78	1.26	1.43	1.47	1.54	1.46	1.22	1.43
Marche	0.66	0.89	0.61	0.75	0.88	0.73	0.52	0.73	0.55
Toscana	0.55	0.82	0.58	0.61	0.68	0.70	0.72	0.77	0.78
Umbria	0.53	0.58	0.56	0.42	0.45	0.42	0.42	0.51	0.43
Lazio	0.91	0.97	1.13	0.97	1.09	1.08	0.68	0.78	1.15
Campania	0.23	0.19	0.23	0.16	0.16	0.16	0.17	0.21	0.22
Abruzzo-M.	0.27	0.26	0.36	0.20	0.30	0.25	0.20	0.31	0.29
Puglia	0.06	0.01	0.08	0.11	0.06	0.14	0.05	0.06	0.08
Basilicata	0.06	0.04	0.11	0.04	0.05	0.09	0.07	0.07	0.13
Calabria	0.03	0.07	0.07	0.06	0.08	0.10	0.02	0.03	0.08
Sicilia	0.16	0.17	0.27	0.28	0.35	0.43	0.34	0.50	0.85
Sardegna	0.05	0.01	0.06	0.35	0.08	0.45	0.18	0.22	0.31
Italia	1.00	1.00	1.00	1.00	1.00	1.00	1.00	1.00	1.00

Source: data base Enea-Cer

Table 3 **REGIONS RANKED ACCORDING TO THE NUMBER OF PATENTING SECTORS**

Regions	**Patenting sectors**	
	n°	**%** **(out of 32)**
Lombardia	29	90.60
Lazio	27	84.40
Piemonte,Veneto	23	71.80
Liguria	22	68.70
Emilia	21	65.60
Toscana	19	59.40
Friuli	17	53.10
Marche	16	50.00
Trentino	14	43.70
Umbria	11	34.38
Abruzzi,Campania	10	31.20
Sicilia	9	28.20
Puglia	7	21.90
Sardegna	5	15.60
Basilicata	2	6.30
Calabria	1	3.10

Table 4 **REGIONAL SPECIALISATION (Index of patent applications per industrial employee by sector >1) IN NATIONAL PATENTING SECTORS (e.g. sectors with a great number of patent applications at national level)**

Sectors	North West	North East	Cent. Italy	South
25-Chemical industry	Lombardia*		Lazio	Campania
31-Manufacture of metal articles	Liguria* Lombardia	Trentino Veneto* Friuli	Umbria	
32-Mechanical engineering	Lombardia*	Friuli* Emilia	Toscana	Campania
33-Manuf. office mach. and data proc. mach.	Piemonte*	Emilia*		
34-Electrical engineering	Piemonte Lombardia*	Friuli* Emilia		Campania Sicilia*
35-Manufacture of motor vehicles and acc.	Piemonte*	Friuli Emilia	Marche Lazio	Campania*
36-Manufacture of other means of transport	Piemonte* Lombardia*	Veneto Emilia	Umbria Lazio	Campania
37-Instrument engineering		Trentino* Emilia	Toscana* Marche Umbria	
48-Processing of rubber and plastics	Liguria Lombardia*	Friuli		

* indicates index of patent applications per patenting firm >1

Table 5 **REGIONAL SPECIALISATION (Index of patent applications per industrial employee by sector >1) IN LOCAL PATENTING SECTORS (e.g. sectors wich account for a large share of some regions' patenting activity)**

Sectors	North West	North East	Cent. Italy	South
22-Preparation and processing of metals	Liguria*	Trentino		
24-Manuf. non-metallic mineral products	Liguria*			Abruzzi
41-Food industry			Umbria*	
45-Footwear and clothing industry		Trentino Veneto*	Umbria	
46-Timber and wooden furniture industries		Trentino Friuli	Marche	
50-Building and civil engineering				Campania
83-Acti. auxil. to banking and insurance			Marche	
94- R & D			Lazio	

* indicates index of patent applications per patenting firm >1

Table 6 **REGIONAL SPECIALISATION (Index of patent applications per industrial employee by sector >1) IN SECTORS HIGHLY CONCENTRATED IN FEW REGIONS**

Sectors	North West	North East	Centr. Italy	South
13-Extraction of petroleum and natural gas	Lombardia (66%)		Lazio (33%)	
14-Mineral oil refining	Lombardia (86%)			
16-Prod. distr. electr., gas,steam, water			Lazio (83%)	
21-Extr. prep. metalliferous ores			Lazio (80%)	
26-Man-made fibres industry		Veneto (50%)		Basilicata (50%)
91-Public Administration			Lazio (90%)	
98-Personal services			Lazio (100%)	

-In brackets regional percentage of national patent applications.

Table 7 **INDEX OF INNOVATIVE FERTILISATION CAPABVILITY BY SECTOR AND BY REGION (1970-79)**

Sectors	Piemonte	Lombardia	Veneto	Emilia	Lazio	Italia
25-Chemical	0.85	0.51	0.77	0.79	0.73	0.60
31-Manuf.of metal articles	0.48	0.46	0.40	0.55	0.74	0.43
32-Mechanical engineering	0.45	0.25	0.35	0.21	0.72	0.29
33-Office and data processing mach.	0.57	0.79	---	0.53	---	0.57
34-Electrical engineering	0.47	0.54	0.56	0.57	0.49	0.46
35-Motor vehicles	0.44	0.67	0.74	0.75	0.93	0.58
36-Other means of transport	0.53	0.70	0.71	0.61	0.80	0.56
37-Instrument engineering	0.69	0.62	0.73	0.74	0.61	0.57
48-Proces. of rubber and plastics	0.40	0.47	0.50	0.53	0.88	0.43
83-Business services	0.80	0.45	---	0.85	0.73	0.36

FIG.1 REGIONS ACCORDING TO DIFFERENT INDICATORS OF INNOVATIVE ACTIVITY (l.q. > 1)

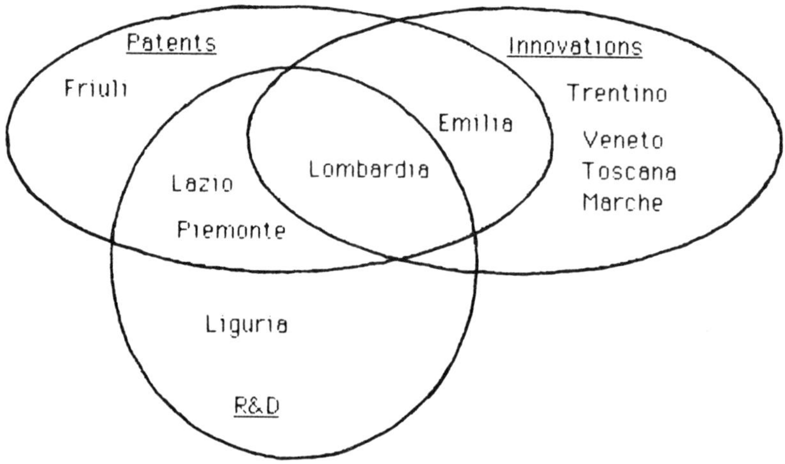

References

Antonelli, C. (1986), L'attività innovativa in un distretto tecnologico, Torino, Fondazione Giovanni Agnelli.

Antonelli, C. (1987), The Determinants of the Distribution of Innovative Activity in a Metropolitan Area, The Case of Turin, Regional Studies, vol. 21.

Archibugi, D. (1987), La struttura brevettuale dell'innovazione industriale in Italia, Un'analisi dei brevetti italiani negli Stati Uniti, Rivista di Politica Economica, no. 2.

Archibugi, D., Cesaratto, S., Sirilli, G. (1987), Innovative Activity, R & D and Patenting, The Evidence of the Italian Survey on Innovation Diffusion, Science Technology and Industry Review, OECD, vol. 2.

Arrow, K.J. (1962), Economic Welfare and the Allocation of Resources for Invention, in: Nelson, R.R. (ed.), The Rate and Direction of Inventive Activity: Economic and Social Factors, Princeton University Press.

Baker, R. (1976), New and Improved Inventors and Inventions that Have Changed the Modern World, British Museum Publication Limited, London.

Belussi, F. (1985), Localizzazione dell'attività di R & S, dei servizi delle imprese, delle reti informatiche e delle reti brevettuali delle imprese, Archivio di studi urbani e regionali.

Boehm, K. (1967), The British Patent System, 1: Administration, Cambridge University Press, Cambridge.

Bound, J., Cummins, C. et al. (1984), Who does R&D and Who Patents?, in: Griliches, Z. (ed.), R&D, Patents and Productivity, University of Chicago Press, Chicago.

CER (1988), L'attività innovativa in Italia: i brevetti nell'industria, Report no. 6.

Cohen, W.M., Levin, R.C., Mowery, D.C. (1987), Firm Size and R&D Intensity: A Re-examination, Journal of Industrial Economics, vol. 35.

Dasgupta, P., Stiglitz, J. (1980), Industrial Structure and the Nature of Innovative Activity, Economic Journal, vol. 90.

Dasgupta, P. (1987), The Theory of Technological Competition, in: Dasgupta, P. and Stoneman, P. (eds.), Economic Policy and Technological Performance, Cambridge University Press, Cambridge.

Dasgupta, P. (1988), Patents, Priority and Imitation or, The Economics of Races and Waiting Games, Economic Journal, vol. 98.

Dosi, G. (1982), Technological Paradigms and Technological Trajectories, Research Policy.

Dossier Brevetti (1986), Università degli Studi di Bologna.

Epifani, E., Giannagostino, C. et al. (1987), Processi di diffusione spaziale delle innovazioni nel sistema urbano italiano (1961-81), in: Tosi,

A. and Cardia, C. (eds.), Il territorio dell'innovazione, Franco Angeli, Milano.

Firestone, O.J. (1974), Economic Implications of Patents, University of Ottawa Press, Ottawa.

Freeman, C. (1982), The Economics of Industrial Innovation, Frances Pinter, London.

Freeman, C., Clark, J.A., and Soete, L. (1982), Unemployment and Technical Innovation, Frances Pinter, London.

Freeman, C. (ed.) (1984), Long Waves in the World Economy, Frances Pinter, London.

Freeman, C. (ed.) (1986), Design Innovation and Long Cycles in Economic Development, Frances Pinter, London.

Freeman, C., and Soete, L. (1986), L'onda informatica, Milano, Edizioni del Sole 24 Ore.

Gilbert, R.J., Newbery, D.M.G. (1982), Preemptive Patenting and the Persistence of Monopoly, American Economic Review, vol. 72.

Griliches, Z. (ed.) (1984), R&D, Patents and Productivity, University of Chicago Press, Chicago.

Griliches, Z., Pakes, A., and Hall, B.H. (1987), The Value of Patents as Indicators of Inventive Activity, in: Dasgupta, P. and Stoneman, P. (eds.), Economic Policy and Technological Performance, Cambridge University Press, Cambridge.

Howells, J.R.L. (1984), The Location of Research and Development: Some Observations and the Evidence from Britain, Regional Studies, vol. 18.

Jawkes, J., Sawers, D., and Stillerman, R. (1969), The Sources of Invention, Macmillan, London.

Kamien, M., and Schwartz, N. (1983), Market Structure and Innovation, Cambridge University Press, Cambridge.

Levin, R., Levorich, K., Nelson, R., and Winter, S. (1987), Appropriating the Returns from Industrial Research and Development, Brooking Papers, 3.

Malaman, R. (1988), L'appropriabilità dell'innovazione tramite brevetti: politiche nazionali e strategie europee, in: Malaman, R. and Ranci, P. (eds.), Le politiche industriali della CEE, Il Mulino, Bologna.

Malecki, E.J. (1983), Technology and Regional Development: A Survey, International Regional Science Review, vol. 8.

Nelson, R.R. (ed.) (1962), The Rate and Direction of Inventive Activity: Economic and Social Factors, Princeton University Press.

Nordhaus, W.D. (1969), Invention, Growth and Welfare: A Theoretical Treatment of Technological Change, MIT Press, Cambridge, Mass.

Onida, F. (ed.) (1985), Innovazione, competitività e vincolo energetico, Il

Mulino, Bologna

Pavitt, K. (1982), Patents and Inventive Activity: A Statistical Exploration, <u>Research Policy</u>, no. 11.

Pavitt, K. (1985), Patent Statistics as Indicator of Innovative Activity: Possibilities and Problems, <u>Scientometrics</u>, no. 7.

Rosenberg, (1976), <u>Perspective on Technology</u>, Cambridge University Press, Cambridge.

Scherer, F.M. (1965), Firm Size, Market Structure Opportunity and the Output of Patented Inventions, <u>Canadian Journal of Economics and Political Science</u>, no. 31.

Scherer, F.M. (1983), The Propensity to Patent, <u>International Journal of Industrial Organization</u>, no. 1.

Schmookler, J. (1966), <u>Invention and Economic Growth</u>, Harvard University Press, Cambridge Mass.

Schumpeter, J.A. (1934), <u>The Theory of Economic Development</u>, Harvard University Press, Cambridge Mass.

Schumpeter, J.A. (1934), <u>Business Cycles</u>, McGraw-Hill, New York.

Soete, L. (1978), <u>Inventive Activity, Industrial Organization and International Trade</u>, Ph.D. Dissertation, University of Sussex.

Soete, L. (1979), Inventive Activity and Firm Size: the Evidence Reconsidered, <u>European Economic Review</u>, no. 12.

Taylor, C., and Silberston, A. (1973), <u>The Economic Impact of the Patent System</u>, Cambridge University Press, Cambridge.

Tosi, A., and Cardia, C. (eds.) (1987), <u>Il territorio dell'innovazione</u>, Franco Angeli, Milano.

PART B
SERVICES

7 The dynamics of producer services

J. N. MARSHALL

1. Introduction

During the last decade and a half there have been profound changes in regional economies in the developed world. Services have increased their relative importance, due to the contraction of manufacturing employment and their continued growth. However, the decline in manufacturing and the growth in services has not taken place in isolation. It is becoming increasingly clear that the growth of service industries can be attributed in part to the changing structure of business organisations and their tendency to contract out services previously carried out in-house (Barcet, Bonamy and Mayere, 1984; Momigliano and Siniscaldo, 1982). Equally important is the fact that services seem to be making an increasingly significant contribution to the competitiveness of industry and this is enhancing business demand for them. In particular, services are playing a critical role in providing the expertise to develop and implement new technologies (Marquand, 1983).

The prominence of service activities and the changes wrought in them by technology have important implications for the nature of uneven development. In many economies service activities are over-represented relative to total employment in a small number of areas close to the capital and other large cities. Examples include the Paris region in France, the South East of the UK and Northern and Central Italy. The continued growth of services may, therefore, enhance spatial disparities (Howells, 1988). Such a view is supported by recent economic change. The strength of services has helped capital cities to weather the recent recession. High growth sectors, such as the financial and business services, are also heavily concentrated in the major conurbations. While services associated with the new technologies, though decentralising from the largest cities, remain over-represented in their hinterlands (Gillespie et al., 1984). In contrast, in less-favoured regions the slowdown in economic growth during the 1970s coincided with a further substantial decine in shipbuilding, iron and steel and heavy engineering. As the scale and breadth of sectoral decline increased new problem regions have emerged. Here the development of business and consumer services has been constrained by the contraction of the manufacturing sector.

There are, however, countervailing tendencies. Dynamic growth of services outside conurbations is widely established (Champion et al., 1987; Coombes and Della Longa, 1987) and new growth regions such as the Southern USA have emerged (Stanback et al., 1981). Reorganisation within industry often using new technology has created more flexible, specialised and decentralised forms of working (Atkinson, 1984; Clutterbuck, 1985) which may create opportunities for a more dispersed pattern of services. For some years it has been argued that changes in telecommunications technology will reduce the friction of distance and increase the attractiveness of peripheral locations (Abler, 1970). There are also persistent demands from professional office workers for improvements in their quality of life and for more rewarding forms of work. Such developments are adding to pressures for the dispersal of services to localities beyond the hinterland of metropolitan regions (Jaeger and Ernste, 1988). These trends lead some to argue that the growth of services will assist provincial areas (Bailly, 1986; Bailly and Maillat, 1986; Planque, 1982; Aydalot, 1984).

While it is relatively easy to describe in such general terms the changes taking place in regional economies, it is more difficult to tease out the precise nature of the processes at work and their geographical significance. Academic analysis is grappling with structural and locational changes that are only partly understood. To analyse service activities, and understand their role in uneven development, it is necessary to proceed beyond the aggregate classification of a service or tertiary sector. This chapter explores the dynamics of producer or intermediate services, that is market services supplying business. The justification for this is that these services are the main contributors to the uneven distribution of service activities.

This chapter considers why producer services are growing unevenly. It suggests that recent changes in producer service location have been brought about by the reorganisation of large firms, market changes and the introduction of new technology. The implications of these developments for the balance of service activities between central and peripheral regions in the economy are discussed.

2. Defining Producer Services

2.1 Definitions of service activity

Economists make a fundamental distinction between goods and services, but the reasons for this are unsatisfactory. While some services are intangible, ephemeral or labour intensive and others involve personal contact, there are frequent exceptions. When services were a small part of the economy it was possible to argue that such ambiguities were relatively unimportant. Activities apparently outside the primary and secondary sectors could be simply grouped into a residual or tertiary sector. However, as services have grown this has become increasingly problematic. A disaggregated analysis of services which makes sense of their diversity is necessary to understand their dynamics.

The seminal works of Greenfield (1966) and Katouzian (1970) have distinguished between services on the basis of the market served, identifying services which (1) serve businesses and those which (2) serve individual consumers. This classification has been developed to distinguish between (3) market and (4) non-market production, recognising the important influence of the state on the service sector. Finally, within private services differences between (5) office and (6) blue collar services have been highlighted (Stanback et al., 1981). Such distinctions represent an important step forward but they tend to treat the service sector as a separate entity. They fail to sufficiently acknowledge that service activities can be carried out by non-specialist producers (most notably in the manufacturing sector) and that changes in the degree of internalisation of such demand have important implications for the specialist supply sector. Any analysis of services must recognise that goods and service production are highly interdependent (Walker, 1985; Bailly et al., 1987).

2.2 The characteristics of producer services

Problems of definition and classification are a major constraint on our understanding of producer service activities (Illeris, 1985; Tauveron, 1974). Their diversity, the limitations of conventional means of classifying them and the variety of approaches to definition limit generalisations about their dynamics.

Producer services are, "services which supply business and government organisations rather than private individuals" (Marshall et al., 1988). They are traded both commercially on the open market and internally within organisations. Some of these services may even be indirectly tradeable through their contribution to other sectors of the economy. Producer services include:

* general management	* legal
* innovation	* technology
* development	* maintenance

* design	* transport
* administration	* comunication
* personnel	* wholesale distribution
* production	* training
* finance	* advertising and marketing services
* computer services	

These producer services are of particular interest because (like manufacturing) "they play a significant role in spatial differentiation.... their demand and supply need not be geographically coincident and they are not solely dependent upon the level of economic activity in an area" (Marshall et al., 1988). These services are most likely to create additional jobs at the regional scale and to avoid displacing other employees in the regional market. Since they also contribute directly to the competitiveness of their business customers it can also be argued that product services (unlike retailing) are part of the productive capacity of an area and influence its long term prospects for economic development.

The essential integration of producer services with other sectors of the economy suggests that an appreciation of the markets they serve is fundamental to an understanding of their development. Research has identified the following market categories of producer service:

1) Services produced by firms for their own use i.e., the internal supply of services within a firm (including non-production activities in manufacturing).

2) Services supplied by specialist producer service firms to meet the externalised demand of other firms for services (i.e., producer service industries in the service sector).

3) Services produced by specialist firms which meet both intermediate and consumer demand. This recognises that producer services in industries such as accountancy and finance are provided alongside services which supply private individuals.

Of course it is still necessary to recognise the different types of expertise offered by producer services. Some offer expertise in goods handling (wholesalers), while others handle information (data processing companies) and some provide support services to personnel in business (employment agencies). It is also important to recognise the different functions performed by producer services; for example, the distinction within information processing services between financial services which are fundamental to the generation and development of business activity and other business services such as advertising, market research and consultancy which help to identify and develop markets or run business efficiently.

It is very difficult to move from such a general description of producer services to a definition and classification appropriate for economic analysis. Research associated with the Producer Services Working Party (Marshall et al., 1988) highlights:

1) The inadequate detail concerning services in conventional government statistical sources.

2) The lack of continuity in the classifications used in government statistics.

3) The failure of conventional sources to take account of the interdependence of services and other activities, and in particular the supply of services both within service industries and in the manufacturing sector.

4) The heterogeneity of service industries and the diversity of the markets they serve which makes classifying them difficult.

5) The resulting confusion in the literature concerning the definition of producer services which clouds any analysis of their dynamics.

There is no easily available or widely acceptable classification of producer services. Indeed, one of the worrying aspects of recent work reacting to the report of the Producer Services Working Party is the continued search for the 'holy grail' of a perfect classification of services (Watts, 1987; Allen, 1988). A more appropriate approach is the development of a small number of different classifications which reflect the variety of service activities, and the different aims of research. To successfully classify and re-classify producer services, however, greater detail on services will be required in government statistical data.

3. Why Are Producer Services Growing?

3.1 Established economic views

The lack of substantial research into producer service activity makes it difficult to answer the tantalisingly simple question posed at the beginning of this section. Economists have neglected the study of intermediate or business demand for services. It was not until the 1930s and 1940s that the economics of service activities was first written. The Fisher-Clark's model of the economy suggested that employment shifted from the primary and secondary sectors to the tertiary or residual sector as a product of increasing affluence and the greater efficiency of the non-service part of the economy. There is no mention in this or related later work of intermediate demand for services (Fuchs, 1968).

Post-industrial analysis in the tradition of Daniel Bell (1974) has accepted these arguments at 'face-value' and highlighted their impact on society. The growth of professional office work associated with a 'service economy' and the relative decline of a blue collar employees, is argued to encourage changes in social values, a decline in industrial conflict and the development of a knowledge-based meritocracy. Such ideas have, however, lost credence during the economic crises of the 1970s, which have indicated that rising affluence for all and a growing demand for services is not predetermined.

A re-interpretation of the original 'post-industrial' concepts has emphasised the role of information technology (see Miles (1985) for a review). It is argued that an 'information society' is developing based on the growing demands of business and consumers for information and advice, and that this will provide the sustained demand for services missing during the 1970s. There is some concern in this literature that the growing use of computer technology will check the development of office occupations, but in general the literature is optimistic. It is argued that computer technology is reducing the cost and increasing the availability of information and thereby facilitating a range of new services which will enhance demand.

These studies do highlight certain facets of the growing demand for service activities, but the analysis is very general. It does not take sufficient account of the different markets services supply. Rather surprisingly producer services and the impact of changes in business demand on services are largely ignored.

3.2 Output, employment and labour productivity

In fact a recent study of the industrial economies (OECD, 1984) highlights the sluggish employment growth in consumer services and the dynamism of producer service industries. There are two possible reasons for this growth in employment in producer services:

1) Employment change reflects a strong underlying growth in output and an increase in business demand.

2) Or, if the growth of demand for services is sluggish, employment growth could largely reflect low rates of labour productivity growth.

We cannot disaggregate the output and employment statistics sufficiently to thoroughly assess these alternative explanations. But evidence from Europe suggests that while output changes broadly in line with employment in public administration, education, health and distribution (retail and wholesale), in services which have a more important business function (e.g., transport and communication and finance and business services) output growth appears to be outstripping employment change (Marshall, 1988). It also seems that productivity growth in market services in the European Community was faster than the rest of the economy during the latter 1970s and early 1980s, and this has been associated with a growth in investment and capital intensity (Green, 1985; Ochel and Wegner, 1987).

Such evidence implies that a growing demand for producer service industry output lies behind their increased importance in the economy. This argument is supported by statistical evidence which suggests that output growth in producer services is under-estimated by the use of employment proxies, and the neglect of quality improvements in the constant output series of government statistics (Barras, 1983).

3.3 The demand for producer services

There have been few empirical investigations of the processes underlying the growing demand for producer service industry output, but recent research indicates that the situation is more complex than simply a growing demand for information and advice. Changes in markets, law and regulation, technology and organisation are encouraging producer service growth (Petit, 1986; Ochel and Wegner, 1987):

1) The globalisation of manufacturing and service markets which exposes firms to competition and opens up new market opportunities encourages the development of a range of financial, civil engineering, consulting, accounting, market research, promotion and selling services.

2) The growing sophistication and differentiation of markets in the developed world, together with attempts by large firms to standardise component production to achieve economies of scale leads to the use of marketing and design services to differentiate products to meet the demands of individual markets.

3) Relatively recently the more competitive economic climate has also encouraged the demand for management consultancy and accountancy services as firms seek to reduce costs and increase efficiency.

4) The increasingly comlex regulatory frameworks which support trade, including import and export regulations, consumer protection and safety regulations, all encourage the development of legal services.

5) Growing government regulation concerning corporate financial practices and disclosure increases the need for accountancy services.

6) Technological changes are also encouraging service demand. The increasing

technological sophistication of new products is enhancing the need for research and development, while ever more sophisticated production process technology requires greater training, maintenance and repair services to support the equipment. Information technology is also offering new possibilities for companies to monitor changes in markets and other businesses, to expand their data processing and retrieval services and thus add to the quality of information available to the firm.

7) Such developments are encouraged by organisational change. The growth of large complex organisations requiring new management techniques and better organisational and communication practices has increased the demand for training, work organisation and telecommunication services. In recent years there have been significant changes in the organisation of these and smaller businesses. Increasingly firms are contracting out their service activities (see Roura, this volume). This shift in producer service provision across the sectors has significant implications for the geography of producer services.

4. Corporate Reorganisation and the Geography of In-house Services

4.1 Established views

A substantial literature argues that corporate structures have a major influence on the location of service activities. Studies suggest a corporate complex of head offices and their associated satellite suppliers in the hinterland of the capital and other large cities underpins the concentration there of service activities (Stanback et al., 1981; Stanback and Noyelle, 1982; Bade, 1983; Semple, 1985; Wheeler and Brown, 1985; Hutton and Ley, 1987; Daniels, 1985; Parsons, 1972; Buswell and Lewis, 1970).

Geographical research links corporate and spatial structure (Crum and Gudgin, 1977; Goddard, 1979). It is argued that the concentration of high-level administrative, technical and research functions at head office attracts them to centres of government, international airports and other major businesses (Westaway, 1974). The important role of the head office in both in-house service provision and the purchase of external services means that its location encourages the development of a local specialist service supply sector. Head offices close to the capital or other large cities place major contracts with locally-based national and international service companies with the capacity to service their operating units (Marshall, 1983).

It is suggested that divisional or intermediate levels in the corporate hierarchy display similar centralising tendencies to head offices, but their location is also influenced by the spatial distribution of the operating units under their control. In contrast, branches specialising in production activities lack the non-production activities of both their parent companies and independent organisations. Nevertheless, they are more evenly distributed than higher levels in the corporate hierarchy, and functions directly serving production tend to be decentralised to the production site. But most of the service requirements of the branch are imported from head and divisional offices (Polese, 1982; Marshall, 1982).

Trends in the location of in-house service production are less certain. One argument, based on work on the growth of large companies by acquisition, suggests a growing spatial concentration of service activities (Goddard and Smith, 1978; Leigh and North, 1978). After acquisition the head office functions of the acquired company are frequently switched to the headquarters of the acquiree, and links with specialist suppliers of services are broken. Given the tendency of acquiring companies to be larger national and international companies based close to the capital or in other large cities, this encourages a growing centralisation of service activity within the space economy. This concentration is intensified by the decline of indigenous firms in established industries in peripheral regions, and their gradual replacement by branch plants from the corporate sector with fewer service requirements. Areas adjacent to the capital may benefit, however, from the localised decentralisation of head offices from urban centres as cost pressures encourage the relocation of administrative and clerical staff.

4.2 Recent research

Recent evidence indicates that this description is too simplistic. It over-emphasises the factors encouraging centralisation by focusing on office activities and placing insufficient weight on the more spatially decentralised blue collar services (Marshall et al., 1988). It ignores the decentralisation of corporate headquarters down the urban hierarchy (Wheeler, 1987). Furthermore, it neglects the variety of corporate structures and in particular significant differences between manufacturing and service organisations. By generalising from the organisation of manufacturing companies it also implicitly assumes that manufacturing industry demand for services is the sole influence on in-house service supply.

Evidence from case studies of service industries suggests that partnership forms of organisation and the absence of large production units reduce the significance of the regional or divisional level in service organisations. The structure of service companies also influences manufacturing industry demand for services. For example, service organisations serve outlying areas from more central conurbations. This intensifies the under-representation of service activities in more peripheral areas and in turn encourages manufacturers to import in-house services into these regions. A realistic view of service demand and supply, then, is that manufacturing and service organisations respond to each other, rather than manufacturing corporate structure determining service supply (Marshall, 1987).

4.3 New forms of organisation

More fundamentally, the established model of corporate-spatial structure provides few pointers to the way in which corporate structures change and develop over time, and the implications of this for the economy. It is largely divorced from the impact of broader economic and technological changes. Yet, in the last decade changes in the economic climate have brought about corporate reorganisation (Brusco and Sabel, 1983; Piore and Sabel, 1984; Atkinson, 1984; Harvey, 1987). In the harsh climate of recession companies have become more conscious of the cost of their administrative overheads, and have sought to cut back, using computer technology to improve the efficiency of their service workers. Companies have also tended to concentrate more on their main lines of business, rather than continuing to build up loosely integrated conglomerate organisations. This has been associated with attempts to make management structures simpler and more responsive to change. To this end accountability has also been increased at lower levels in the corporate hierarchy.

Associated with these developments there has been a growth in the contracting out of services (Rajan, 1987). This trend goes back to the early 1970s in the case of physical distribution, where its reorganisation to form a separate managerial function promoted tighter cost control and encouraged contracting out. More recently, there is evidence that contracting has intensified and spread to white collar functions. Rajan and Pearson (1986), for example, argue on the basis of an extensive employer survey in the UK that out of 700,000 jobs created in distribution, (including on their definition retailing), finance and business services between 1979 and 1985, as many as 300,000 were due to contracting out by non-specialist producers. They expect the externalisation process to increase and between 1985 and 1990 possibly 60% of the job creation in distribution, finance and business services could come from a transfer of service work across the sectors.

However, it is too simplistic to view organisational changes as just an increase in contracting out. Complex and dynamic shifts in the source of service production are taking place, and the boundaries between economic activities are becoming increasingly fluid. Organisations change their use of in-house sources of supply in response to changes in business demand which affect their use of service products, and developments in the expertise, skills and costs of production in their in-house operations and the specialist supply sector. The recent recession has promoted contracting out. It was not always possible for firms to continue their in-house service operations as the company retrenched. External contracts also had the advantage of a clear measure of cost. So firms could monitor more carefully the cost of service production. Where economies of scale were important firms could also reduce costs by employing a supplier, though this benefit could be lost if suppliers possessed a monopoly position.

Changes in management practice have also encouraged the use of external sources of supply. The use of external contractors increases the flexibility of the firm because they provide a buffer of spare capacity to deal with changes in business demand. The constraints imposed by labour legislation on their own employees are also avoided.

There are, however, limits to the contracting out process. Firms retain some services (such as research and development) in-house because they wish to maintain confidentiality. In addition, dramatic market changes and the need to be sensitive to the demands of their customers have encouraged firms to internalise critical market-related services. For some firms the quality of service they provide is also becoming increasingly important to their success, here again internalisation of services is encouraged.

Computer technology has also played a critical part in increasing the fluidity of organisational boundaries. Computer technology, introduced to improve the productivity of service workers, has enhanced the range of tasks they can carry out. This may mean that additional services are conducted in-house, but new demands are also created for external support services.

4.4 Geographical perspectives

Clearly, the conventional description of the role of corporate structure in the space economy with its emphasis on a simple hierarchy of functions and changes in ownership and control is not able to explain the recent dynamics of service location very well. It does not do justice to the diversity of change in service location, nor does it take account of the way in which corporate structures have changed. A new approach is needed.

The geographical implications of corporate reorganisation remain largely uncharted territory. It is likely, though, that the relationship between corporate and urban structure described earlier will have changed (Christopherson and Storper, 1986). The rationalisation of head and divisional offices and the decentralisation of autonomy within companies will alter the spatial balance of in-house service activities within large companies. Simplistic generalisations about the development of services on the basis of the location of head offices or branch plants can, therefore, easily be inappropriate. Some locations may not benefit in terms of employment growth from the location there of head offices, even though the company as a whole is growing. The contracting out of services and in-house productivity exercises may produce a decline of head office employment. In contrast the tendency of companies to decentralise autonomy to their operating units and expand their capacity there could help areas dependent on branches (Marshall, 1988).

Changes in the internalisation and contracting out of services in the corporate sector could also produce important changes in the location of services. The contracting out of technical, consultancy and professional business services by firms as they adopt new technology could produce a leakage of service demand up the urban hierarchy because the suppliers of such services are concentrated near capital cities (Marshall, 1988). However, in blue collar services such as transport and distribution, where the location of production is important, contracting out has resulted in a decentralisation of services to areas with good motorway access to manufacturing and retail outlets. Similarly, while the internalisation of services within non-specialist organisations is likely to reflect the distribution of dynamic companies, potentially at least it does offer some opportunities for the development of in-house services by successful companies in provincial regions (Marshall, 1989).

5. The Changing Geography of Producer Service Supply

5.1 Established views

Consumer Services

Rather like the economics literature, early work on the geography of service supply assumed services supplied consumer markets (see Daniels, 1975; 1985). Following the work of Chirstaller and Losch service location was explained in terms of the volume of demand at a given site, which was determined by the distribution of population and transport costs. This basic approach has been modified to take account of variations in operating costs. The geography of services is argued to reflect the way in which service firms trade-off the revenue generated at a given location against the cost of rents or taxes. Functions requiring a large market area tend to concentrate in central places and to serve lower order centres. The latter contain ubiquidous services which are purchased more frequently, and those services which are particularly sensitive to cost pressures in large cities.

Such work takes insufficient account of the role of producer services in satisfying business demand, which brings with it a non-hierarchical pattern of inter-regional exports, and greater locational flexibility than in consumer services (Beyers and Alvine, 1985; Kirn, 1987). Furthermore, central place theory and its derivatives offer only limited insights into the dynamics of change in producer service location (Parr and Denike, 1970).

Office activities

The geographically mobile office or information processing services have been the subject of most recent research. The literature draws on both behavioural and Weberian theory. Though it considers the impact of factors such as rents, taxes and labour costs, communication and information linkages are regarded as the critical factors in office location.

Early work in the USA described the clustering of offices in large city centres and infered agglomeration economies such as the availability of information in a hurry and lower communications costs to explain this (Foley, 1956). In the 1960s and early 1970s Swedish research (Thorngren, 1970) focused more closely on the communications factor in office location and this approach was applied elsewhere (Goddard, 1975; Daniels, 1975; Alexander, 1979). The work showed that office complexes had a dense network of local linkages and these frequently involved face-to-face contacts. Few of these contacts were very sophisticated, and many were suited to the telephone. Office decentralisation from city centres seemed possible; and in fact during the 1970s evidence of decentralisation began to build-up.

This was not just communication-related decentralisation. The expansion of offices in cities was constrained by lack of space, traffic congestion and higher taxes. Decentralised locations offered lower costs and a captive labour market. However, communications factors provided a constraint. Only more routine functions needing little face-to-face communication were decentralised (Goddard and Morris, 1976). Relocation was also limited because beyond a certain distance savings in rents or labour costs were outweighed by the increased cost of travel for those workers who still needed to attend meetings in the city centre (Goddard and Pye, 1977).

5.2 Information technology and the office

Telecommunications and office location

Telecommunications technology, by altering office communications, can change the economics of producer service location. Early work concentrated on the possibility that multi-party telecommunications and the transmission of pictures would substitute for face-to-face contact and undermine large agglomerations of office activities. As early as 1970 Abler observed, "Advances in information transmission may soon permit us to disperse information-gathering and decision-making activities away from metropolitan centres, and electronic communication media will make all kinds of information equally abundant everywhere in the nation, if not the world. When that occurs, the downtown areas of our metropolitan centres are sure to lose some of their locational advantages for management and government activities" (Abler, 1970).

It was also argued that, by reducing the need for proximity, technology would permit remote working from home and allow workers to more satisfactorily combine domestic and work roles (Webber, 1968). More decentralised operations within organisations would also be possible once it was no longer necessary for supervision to take place face-to-face. A development of this possibility was that peripheral regions might use telecommunications technology to keep in touch with the main information centres in the economy and to more successfully develop administrative work (James, Marshall and Waters, 1979).

While a number of studies have provided evidence that new developments in telecommunications technology can assist firms to decentralise from city centres in the face of cost and congestion pressures (Kutay, 1986; Goddard and Pye, 1977; Goddard, 1975), with hindsight early views seem too deterministic, and to over-estimate the impact of technology. There is a considerable degree of inertia in existing spatial forms (Gottman, 1984). Telecommunications technology can also enhance the role of face-to-face contacts as well as substitute for them. Finally, the technology is a permissive factor (Gottman, 1977; Goddard, 1979). It can be used for a variety of purposes depending on business and government structures (Marshall, 1984).

The merging of computer and telecommunications technology

However, the merging of telecommunications with computer technology, the expansion of data processing and the development of computer networks will have a more profound impact on producer services than telecommunications technology per se (Kellerman, 1984). Information technology has the potential to create entirely new services, to expand the market for existing services, to provide the opportunity for companies to develop new expertise and to change the nature of producer service jobs (Howells, 1988; Hepworth, 1987; Foord and Gillespie, 1985). Notwithstanding the pervasive impact of technology, research has concentrated largely on the impact of the telecommunications infrastructure on the take-up of information and communications technology.

The upgrading of the telecommunications infrastructure from mechanical switching and analogue transmission to electronic switching and digital transmission requires a large investment. As a consequence some nations are developing overlay networks to meet their new transmission requirements. Even for those countries aiming to develop a national Integrated Services Digital Network the infrastructure must be provided incrementally (NEDO, 1987). This means that there are lags in the availability of the new technology, especially in countries where there is an unsatisfied demand for telecommunications facilities or where resources are not sufficient for telecommunications provision.

Studies of the modernisation of the telecommunications infrastructure and the development of new services using it suggest, contrary to the thrust of earlier work, that it will favour the regional concentration of producer services (Gillespie et al., 1984; Gillespie and Hepworth, 1987). The distance shrinking potential of computer technology and the development of distance independent charging for data communications, which could benefit less central locations, are overridden by the uneven introduction of the new technology. Advanced telecommunications infrastructure and services are developed first in large conurbations and in more prosperous regions.

This reflects differences in the demand for telecommunications services. For example, studies of the penetration of the telephone and more importantly telex and network termination points (which give a clearer idea of business demand) in Europe show that penetration levels standardised by population are lower in rural, peripheral and less-favoured regions, and the gap between advanced and lagging regions tends to be greatest in the less-developed countries (Gillespie and Hepworth, 1987). Building on earlier work which suggests that technology facilitates rather than determines change, it is argued that these disparities reflect the economic structure of problem and peripheral regions, and their dependence on traditional and declining forms of economic activity.

Such a conclusion emphasises the importance of the economic and social milieu within which technological change occurs. One feature of this changing milieu is the growing internationalisation of service transactions. It is also frequently argued that the upgrading of the telecommunications infrastructure in the more prosperous metropolitan areas will enhance their position in a world-wide telecommunications marketplace (Moss, 1987).

5.3 Market changes

A growing internationalisation of markets

Trade in services has traditionally been restricted because of the sensitive nature of many financial and business services, and because they are deeply integrated in the system of legal and financial controls in individual states. This is changing, a growth of international trade in services (which is probably understated in conventional trade statistics) has been paralleled by the development of multinational service organisations in a wide range of producer services (Clairmonte and Cavanagh, 1984). The US and some European countries favour greater liberalisation of this trade as a means of creating further exports and jobs. Attempts to develop general principles around which the liberalisation of service trade can be organised, greater scrutiny of barriers to trade in service activities, and pressure for change have been initiated within OECD and GATT. Progress has been slow. However, within Europe the gradual legitimation of professional qualifications throughout the European Community, and greater harmonisation in financial markets, has facilitated the establishment of European service companies throughout the EEC. The move towards a unified internal market by 1992 with minimum barriers to trade will further assist producer service exports and the rights of establishment of service companies.

Liberalisation, de-regulation and growing competition

The pressure to reduce restrictions on service trade are part of a more general trend towards a liberalisation of service markets and a growth of competition (Marshall, 1985; Langdale, 1985; Howells, 1988). The trend is most obvious in financial markets where the more liberal regulation of the New York Stock Exchange has encouraged de-regulation in London. At the government's insistence the London Stock Exchange has permitted new entrants and dual capacity within companies operating in the market, and removed fixed comissions. This is part of a wider restructuring of the financial sector which has reduced restrictions on financial institutions and increased competition. These changes represent an attempt to enhance London's role as the financial capital of Europe, rivalling New York and Tokyo in world markets. A chain reaction of reorganisation has occurred in other stock markets, which includes allowing foreign companies greater freedom to operate in the domestic market, and reducing state restrictions on the operation of financial markets (Keating, 1986).

The merging of the computer and telecommunications industries has also produced regulatory changes in the telecommunications sector, where until recently carriers had a virtual monopoly. The telecommunications sector is central to the strategy of most national governments to develop information technology. A less-regulated environment is believed to be more condusive to rapid network development and innovative service provision. De-regulation and a reduction of state involvement in the US has encouraged more limited liberalisation elsewhere, most notably in Britain and Japan (Cambridge Econometrics, 1985).

Changes in other sectors have been far less dramatic. However, moves to encourage competition and reduce air fares in Europe are gradually taking shape. Privatisation of state organisations, and encouraging them to operate in a more competitive environment are also recurring themes in many countries.

Structural change and an increasing interdependence of markets

Contraction in the manufacturing sector, the growth of new markets for services, developments in information technology and contracting out have increased the interdependence of producer services and other activities.

The recessions in the 1970s brought changes in the demand of manufacturing for services. This was felt most strongly in physical distribution where a reduction in demand was associated with a growth in contracting out. A reorganisation of the distribution sector resulted and a growing interdependence of manufacturing and distributors.

The demand for services associated with information technology (for example, value added telecommunications, computer consultancy and training, and data base services) has grown apace. The demand has been strongest in service industries because of their dependence on information and data processing. This has encouraged a growing interdependence of service activities.

Firms have also been encouraged to diversify into new growing producer service markets, to expand the range of services they offer, to service their existing customers rather better, or reduce their dependence on slowly growing markets. Thus, for example, retailers have expanded into financial services, banks into real estate, accountants into management consultancy, management consultants into computer consultancy and computer service companies into related training and work organisation services (Howells, 1988; Marshall et al., 1988).

Geographical perspectives

The geographical implications of these trends can only be hinted at. One argument is that the trend towards internationalisation, and the concentration of the main multinational operators in agglomerations of producer services will reinforce the position of a small number of world cities (Friedmann, 1986). However, the riple effects of growth may reach widely into the hinterland of the large city via consumer income multipliers (Thrift, Daniels and Leyshon, 1988). In addition, while multinational firms in finance and accountancy establish themselves initially in the capital city, as they become more established, they expand (along with indigenous firms) into other large provincial centres to serve new customers (Thrift, 1988). There are, though, substantial differences between countries in the extent of decentralisation (Daniels, 1986).

A second concern is the growing importance of human capital in the economic development process (Beyers, 1988). Workforce knowledge, information and skills are critical in the development, take-up and use of technology. Firms are also making increasing demands for labour flexibility in producer services, including the use of temporary contracts, part-time labour and contracting out (Manpower, 1986). This means that the quality of a region's producer service workforce, its ability to respond to changes in market and technological demands will play a critical role in its economic development prospects (Kirn, 1987). This would again appear to favour those regions with strong concentrations of producer service activities.

Nevertheless, case study research also shows that information technology creates new opportunities for non-specialist producers to expand their expertise in services, and this offers dynamic provincial firms the opportunity to expand into producer service markets (Marshall, 1989).

6. Conclusions

The changes in producer services described above, including:

* the growing internationalisation of markets
* increasing competition

* the growing interdependence of producer services and other forms of economic activity
* corporate reorganisation
* a greater fluidity of organisation boundaries
* rapid technological change

have produced changes in service location. Research has concentrated on explaining the spatial centralisation of producer services in national economies in terms of the concentration of head offices and internationally tradeable services in the largest metropolitan regions, which is in turn reinforced by technological changes. The dynamics of the growth of services in provincial locations and smaller towns, and the opportunities created by corporate reorganisation and new technology for a more spatially dispersed pattern of producer service growth are much less well understood.

To more fully understand the dynamics of producer services, and the way in which locational changes occur, research is needed which appreciates the interdependence of service and other activities, which identifies shifts in provision between service and non-service organisations, which acknowledges a diversity of corporate structures and provides a more sophisticated account of the response of organisations to external events. An analysis of the integration and externalisation of service activity by organisations and the way locational changes in services evolve from this process appears to present the most pressing need in the study of producer service location.

The changes in the location of producer services brought about by these processes are not predetermined, they will vary between countries in response to a wide variety of factors. The size of a country and the distribution of its population will obviously be significant. The structure of the existing urban hierarchy (the extent to which it is dominated by a small number of cities) and the accessibility of centres within it will also be important. The following factors are likely to be of particular significance in determining the pattern of producer service activities which evolves as a consequence of organisational, technological and market changes:

The role of the state

Government influences the producer service sector through the demands of government departments for intermediate services, and through the pulling power of centres of decision-making. It is clearly no coincidence that countries with federal and decentralised structures of government such as the USA, Switzerland and West Germany tend to have a more evenly distributed producer service sector than countries such as Britain and Ireland which have a unitary system of government, with less autonomy at the local level.

A wide range of government policies including defence, innovation, technology, regional policy and higher education are also critically important for the location of producer services. There has been a shift towards sectoral innovation and technology policies and a reduction in the role of regional economic policy. Unless sectoral policies are well co-ordinated with spatial policy, or regional policy develops a strong service and technology arm, there is a danger that this will enhance spatial disparities. It is well established, for example, that the concentration of 'high tech' manufacturing and services along the prosperous M4 corridor in southern Britain owes much to the cumulative impact of government defence, research and innovation policies. On the other hand, the development of a strong electronics sector with some associated services in Central Scotland shows what can be achieved by a well conceived and executed regional strategy to develop a sector.

A more contentious issue is the impact of action by government to liberalise markets and encourage competition on the location of producer services. We cannot do justice here to the debate concerning whether in the long run competition reduces or enhances spatial disparities. However, there is evidence that moves to increase competition in telecommunications while they may enhance innovation and customer choice also encourage carriers to retreat from the concept of universal service. This need not always necessarily be to the disadvantage of peripheral regions. In Britain, for example, since limited competition has been introduced, British Telecom has reduced tarrifs on long distance business routes. Multi-region companies and firms serving national markets from provincial regions have as a consequence gained. However, competition also encourages new developments in networks and services to concentrate in more prosperous regions where demand is highest. In turn these facilities shape the ability of businesses to adopt new technology, develop new markets and generally adapt to change.

The role of manufacturing

There is a tendency for the debate concerning the role of manufacturing and services in the economy to polarise into two alternative camps each arguing the pre-eminence of one of the sectors in economic growth. Supporters of manufacturing as the engine of growth emphasise the fact that its outputs are traded and the greater scope in that sector for productivity improvements. Those who emphasise the importance of services argue the independence of services from the manufacturing sector and the importance of services as an employer.

An important theme of this paper has been the interdependence of producer services and other activities. A growing demand by manufacturing for services is clearly an important part of the explanation for the expansion of producer services. Changes in electronics production are also encouraging innovation in producer service products (e.g., more customised insurance policies), streamlining production processes (e.g., automated stock control) and the computer service sector is feeding on this growth. The relationship between manufacturing and producer services can also, however, operate in the opposite direction. Producer service inputs such as research and development, marketing and computer services are important to manufacturing competitiveness, and are probably becoming more critical. Manufacturing and service products are also interchangeable as the trend towards 'self-servicing' using goods (such as the TV or motor car) to provide services (such as entertainment or transport) shows.

If follows from this analysis that a healthy manufacturing industry and a prosperous producer service sector are likely to complement each other. It is no surprise, therefore, that those regions with a sluggish producer service sector tend also to be characterised by an over-dependence on declining manufacturing industries and an under-representation of growing service intensive electronics industries.

The importance of infrastructure

Finally, for regions to participate fully in the development of the producer service sector they are likely to require a good infrastructure (broadly defined) for service provision. They are likely to need an up-to-date telecommunications network, an adequately trained workforce in computer technology and related skills, and an attractive environment to encourage locationally mobile professional workers.

References

Abler, R. (1970), What makes cities important? Bell Telephone Magazine, 49:10-15.

Allen, J. (1988), Service industries: uneven development and uneven knowledge, Area 20:15-22.

Alexander, I. (1979), Office Location and Public Policy, Longman, London.

Atkinson, J. (1984), Flexibility, uncertainty and manpower management, Report 89, Institute for Manpower Studies, University of Sussex, Brighton.

Aydalot, P. (ed.), Crise Economique et Espace, Economica, Paris.

Bade, F.J. (1983), Large corporations and regional development, Regional Studies, 17:315-327.

Bailly, A. (1986), Le secteur des services: une chance pour le development local, in: Illeris, S. (ed.), The Present and Future Role of Services in Regional Development, Occasional Paper 74, FAST, Commission of the European Communities, Brussels.

Bailly, A. and Maillat, D. (1986), Le Secteur Tertiaire en Question, Anthropos, Paris.

Bailly, A. et al. (1987), Services and productivity: for a reassessment of economic sectors, The Annuals of Regional Science, 21:45-59.

Barcet, A., Bonamy, J. and Mayere, A. (1984), Les services aux entreprises: problemes theoriques et methodologiques, Commissariat General du Plan, Services des Etudes et de la Recherche, L'economie des services, nouvelle serie no. 9, ler trimestre.

Barras, R. (1983), Growth and Technical Change in the UK Service Sector, Technical Change Centre, London.

Bell, D. (1974), The Coming of Post Industrial Society, Heinemann, London.

Beyers, W.B. (1988), The view from the US, in: Marshall, N.J. et al., Services and Uneven Development, Oxford University Press, Oxford.

Beyers, W.B. and Alvine, M.J. (1985), Export services in post industrial society, Papers of the Regional Science Association 57:33-45.

Brusco, S. and Sabel, C. (1983), Artisan production and economic growth, in: Wilkinson, F. (ed.) The Dynamics of Labour Market Segmentation, Academic Press, London.

Buswell, R.J. and Lewis, E.W. (1970), The geographical distribution of industrial research activity in the UK, Regional Studies 4:297-306.

Cambridge Econometrics (1985), Forecasts for Information Technology in Britain-A Long Term View, Cambridge Economics Ltd., Cambridge.

Champion, A.G. et al. (1987), Changing Places, Edward Arnold, London.

Christopherson, S. and Storper, M. (1986), The city as studio; the world as backlot: the impact of vertical disintegration on the location of the motion picture industry, Environment and Planning D 4:305-320.

Clairmonte, F. and Cavanagh, J. (1984), Transnational corporations and services: the final frontier, Trade and Development 5:215-273.

Clutterbuck, D. (ed.) (1985), New Patterns of Work, Gower, Aldershot, Hants.

Coombes, M.G. and Della Longa, R. (1987), Counterurbanisation in Britain and Italy: a comparative critique of the concept, causation and evidence, Discussion Paper 84, CURDS, University of Newcastle upon Tyne.

Crum, R.E. and Gudgin, G. (1977), Non-Production Activities in UK Manufacturing Industry, Regional Policy Studies Series 3, Commission of the European Communities, Brussels.

Daniels, P.W. (1986), Foreign banks and metropolitan development: A comparison of London and New York, Tijd Econ Soc Geog, 77:269-287.

Daniels, P.W. (1985), Service Industries: A Geographical Appraisal, Methuen, London.

Daniels, P.W. (1975), Office Location: An Urban and Regional Study, Bell, London.

Friedman, J. (1986), The world city hypothesis, Development and Change 17:69-83.

Foley, D.L. (1956), Factors in the location of administrative offices, Papers and Proceedings, Regional Science Association 2:318-326.

Foord, J. and Gillespie, A.E. (1985), Reorganisation, new technology and office jobs, Discussion Paper 75, CURDS, University of Newcastle upon Tyne.

Fuchs, V.R. (1968), The Service Economy, National Bureau for Economic Research, Columbia University Press, New York.

Gillespie, A.E. et al. (1984), The effects of new information technology on the less-favoured regions of the Community, Regional Policy Studies Collection 23, Commission of the European Communities, Brussels.

Gillespie, A.E. and Hepworth, M.E. (1987), Telecommunications and regional development in the information society, Newcastle Studies of the Information Economy, Working Paper 1, CURDS, University of Newcastle upon Tyne.

Goddard, J.B. (1979), Office location in urban and regional development, in: Daniels, P.W. (ed.) Spatial Patterns of Office Growth and Location, Wiley, London.

Goddard, J.B. (1975), Office Location in Urban and Regional Development, Oxford University Press, Oxford.

Goddard, J.B. and Smith, I.J. (1978), Changes in corporate control in the British urban system, Environment and Planning A 10:1073-1084.

Goddard, J.B. and Pye (1977), Telecommunications and office location, Regional Studies 11:19-30.

Goddard, J.B. and Morris, D. (1976), The communications factor in office decentralisation, Progress in Planning, 6, 1, Pergman, Oxford.

Gottman, J. (1984), The Coming of the Transnational City, Institute of Urban Affairs, University of Maryland.

Gottman, J. (1977), Megalopolis and antipolis: The telephone and the structure of the city, in: Pool I de S. (ed.) The Social Impact of the Telephone, MIT Press, Boston.

Green, M. (1985), The development of market services in the European Community, the United States and Japan, European Economy 25:69-96.

Greenfield, H. (1966), Manpower and the Growth of Producer Services, Columbia University Press, New York.

Harvey, D. (1987), Flexible accumulation through urbanisation: reflections on 'post-modernism' in the American city, Antipode 19:260-286.

Hepworth, M. (1987), Information technology as spatial systems, Progress in Human Geography 11:157-179.

Howells, J. (1988), Economic, Technological and Locational Trends in European Services, Gower, Aldershot, Hants.

Hutton, T. and Ley, D. (1987), Location linkages and labour: the downtown complex of corporate activities in a medium size city, Vancouver, British Columbia, Economic Geography 63:126-141.

Illeris, S. (1985), How to analyse the role of services in regional development, in: Bannon, M.J. and Ward, S. (ed.) Services and the New Economy: Implications for Regional Development, Regional Studies Association, Dublin.

Jaeger, C. and Ernste, H. (1988), Europe: A Way Beyond Fordism? Paper presented at the Annual Conference of the IBG, ETH, Zurich.

James, V., Marshall, J.N. and Waters, N. (1979), Telecommunications and Office Location, CURDS, University of Newcastle.

Katouzian, M.A. (1970), The development of the service sector: a new approach, Oxford Economic Papers 22:362-382.

Keating, G. (1986), The market for financial services in Europe, in Promethee, CEPII and CEPS, Europe in the New International Division of Labour in the Field of Services, FAST, Commission of the European Communities, Brussels.

Kellerman, A. (1984), Telecommunications and the geography of metropolitan areas, Progress in Human Geography 8:223-346.

Kirn, T.J. (1987), Growth and change in the service sector of the US: A spatial perspective, Annals of the Association of American Geographers 77:353-372.

Kutay, A. (1986), Effects of telecommunications technology on office location, Urban Geography 7:243-257.

Langdale, J. (1985), Electronic funds transfer and the internationalisation of the banking and finance industry, Geoforum 16:1-13.

Leigh, R. and North, D.J. (1978), Regional aspects of acquisition activity in British manufacturing industry, Regional Studies 12:227-245.

Manpower, (1986), Flexible Manning in Business, Manpower, London.

Marquand, J. (1983), The changing distribution of service employment, in: Goddard, J. and Champion, A. (ed.) The Urban and Regional Transformation of Britain, Methuen, London.

Marshall, J.N. (1989), Corporate reorganisation and the geography of services, Regional Studies, 23:139-150.

Marshall, J.N. (1988), Producer Services and the Manufacturing Sector in the West Midlands, West Midlands Enterprise Board, Birmingham.

Marshall, J.N. (1987), Industrial linkage and regional development, in: Lever, W. (ed.) Industrial Change in the UK, Longman, London.

Marshall, J.N. (1985), Business services, the regions and regional policy, Regional Studies 19:333-363.

Marshall, J.N. (1984), Information technology changes corporate office activity, GeoJournal 9,2:171-178.

Marshall, J.N. (1983), Business service activities in British provincial conurbations, Environment and Planning A, 25:1343-1360.

Marshall, J.N. (1982), Linkages between manufacturing industry and business services, Environment and Planning A, 14:1523-1540.

Marshall, J.N. et al. (1988), Services and Uneven Development, Oxford University Press, Oxford.

Miles, I. (1985), The Service Economy and Socio-Economic Development, paper presented to UNCTAD, SPRU, University of Sussex.

Momigliano, F. and Siniscaldo, D. (1982), The growth of service employment: a reappraisal, Banca Nazionale del Lavoro, Quarterly Review 142:269-306.

Moss, M. (1987), Telecommunications, world cities and urban policy, Urban Studies 24:534-546.

NEDO (1987), IT Futures. IT Can Work, Information Technology Economic Development Committee, National Economic Development Organisation, London.

Ochel, W. and Wegner, M. (1987) The role and determinants of services in Europe, Occasional Paper 132, FAST, Commission of the European Communities, Brussels.

OECD (1984), The contribution of services to employment, Employment Outlook, September, 39-54, Paris.

Parr, J.B. and Denike, K.G. (1970) Theoretical problems in central place theory, Economic Geography 46:568-586.

Parsons, G. (1972), The giant manufacturing corporations and balanced regional growth in Britain, Area 4:99-103.

Petit, P. (1986), Slow Growth and the Service Economy, Frances Pinter, London.

Piore, M. and Sabel, C. (1984), The Second Industrial Divide, Basic Books, New York.

Planque, B. (1982), Le Development Decentralise, Litec, Paris.

Polese, M. (1982), Regional demand for business services and inter-regional service flows in a small Canadian region, Papers of the Regional Science Association 50:151-163.

Rajan, A. (1987), Services: A New Industrial Revolution? Butterworths, London.

Rajan, A. and Pearson, R. (1986), UK Occupation and Employment Trends to 1990, Butterworth, London.

Semple, K. (1985), Toward a quarternary place theory, Urban Geography 6:285-296.

Stanback, T.M. et al. (1981), Services: The New Economy, Allenheld, Osmun, Totowa, NJ.

Stanback, T.M. and Noyelle, T.J. (1982), Cities in Transition, Rowman and Allenheld, Osmun, Totowa, NJ.

Tauveron, A. (1974), Le tertiaire superieur moteur du developpment-regional, L'Espace Geographique 3:169-178.

Thorngren, B. (1970), How do contact systems affect regional development? Environment and Planning A, 2:409-427.

Thrift, N. (1988), The City of London: Southern Fortune or Northern Misfortune, paper presented at the Annual Conference of IBG, University of Bristol.

Thrift, N., Daniels, P.W. and Leyshon, A. (1988), Location Behaviour of Large Professional Producer Service Firms in Britain, Report to ESRC, University of Bristol.

Walker, R. (1985), Is there a service economy? The changing capitalist division of labour, Science and Society 49:42-83.

Watts, H.D. (1987), Producer services, industrial location and uneven development, Area 19:353-355.

Webber, M.M. (1968), The post city age, Daedalus 97:1091-1110.

Westaway, E.J. (1974), The spatial hierarchy of business organisations and its implications for the British urban system, Regional Studies 8:145-155.

Wheeler, J.O. (1987), Fortune firms and the fortunes of their headquarters metropolises, Geografiska Annaler 69B:65-72.

Wheeler, J.O. and Brown, C.L. (1985), The metropolitan corporate hierarchy in the US South, Economic Geography 61:66-78.

8 Services and new information technologies – producer services

J. R. CUADRADO ROURA

1. Introduction

The increasing importance of services in the economy is clearly evident, even taking into account the reservations which arise from the point of view of statistics. The initial interpretations of this growth in tertiary activities were undoubtedly over-simplistic. Nevertheless, over the last two decades, and especially since the mid 1970s, the analysis of the growth of services and their internal dynamics has been considerably enriched, making way for interpretations which link this growth to changes in the production system (such as de-industrialization and growth of small businesses), to certain social changes (population shifts, changes in consumer behaviour, and part-time employment) and of course to technological change. As a result of this we are now facing questions which are less generalized than those which refer to the growth of such a heterogeneous sector.

The dynamics of employment in services, variation in functions and occupations, factors which explain changes in demand and supply of different services, the development of innovations in services, and the role of these activities at national, regional and local levels have all been the subject of numerous research projects and publications.

Within the wide range of questions to be taken into consideration, we shall concentrate on the specific area of the relationship between services and technological change. Moreover, since our intention is to refer to the most recent past and to what present trends indicate for the future, the technological changes referred to will be almost exclusively those in the development of new information and communications technology (NIT), we shall restrict the scope of the discussion to one specific field of services, namely producer services.

This term has given rise to doubts and differences among writers and statisticians (producer services/business services, **without** or **plus** transport and distribution activities). Here we are dealing with business services including producer services, such as R&D, maintenance, engineering, etc., services related to purchasing, producers' sales and distribution, management services, the various types of advisory services, financial services, information and communications services, logistics, hygiene and safety in production activities. In some cases we will refer to some of these specific services, but attention will be paid above all to the more general features, problems and questions.

In the follwoing pages I shall try to present certain topics which have been studied in recent literature and research on the impact of technological change (essentially NIT) on services, especially those services which companies demand and use, whether to produce goods or to offer other services. Sections 2 and 3 will be given over to highlighting some coincidences surrounding the effects of new information and communications technology on the production of goods and services. Section 4 will deal more specifically with the case of services to companies, and Section 5 will analyze the important question of service location. Finally, the article will conclude (Section 6) with observations on services to companies and regional development.

This article has been prepared from a review of the extensive literature mentioned at the end. The text itself, however, only includes a limited number of references, for ease of reading. It is necessary to point out that many of the ideas set out in this text come from a much larger, theoretical and empirical work on the case of Spain and the EEC, produced in collaboration with Professors M. González, to whom I am particularly grateful, and C. del Rio.

2. New Information Technology and its Impact on Production of Goods and Services

The latest technological advances have been concentrated in a certain number of areas which will almost certainly influence the whole production system: microelectronics and computers, telecommunications, automation, lasers, biotechnology, energy and new materials. Of all these, those which fall into the broad generic category of information technology (electronic communications, computers and their derivations) would seem to have the widest and most immediate repercussions, since apart from their horizontality they can be combined with other technological advances.

There is agreement that the emergence of NIT and the internationalization of the economy, along with other factors (growth in incomes, the appearances of new occupational and labour models, etc.) have had an enormous effect on the production of goods and services. Great changes have already taken place, or are developing (Table 1) in the following fields:

- The combination and composition of products and services (**what** is produced).
- Markets (**who** the product or service is produced for).
- Location of economic activities (**where** the production takes place or where the service is performed).
- Production or service processes (**how** the production is carried out or how the service is performed).

This profound transformation in the economy has been accompanied by a considerable increase in service activities. We shall briefly analyze the most important changes.

a. Transformation of products and services

The factors mentioned above have changed the composition of products in our economies. Goods and services are increasingly no longer required individually, but as integral parts of a product package. So very often it is only in combination that the goods and services contribute to a greater satisfaction of the demand. Services facilitate the use of goods, the maintenance of goods and a certain style of using goods. Consequently, the relative importance of services in these product packages seems to be increasing constantly (Bessant et al., 1985).

The demand for services is also influenced by the fact that new types of services are being offered through the use of information technology such as automatic cashpoints, credit cards, airline and car rental reservation systems. These services can stimulate demand, and at the same time promote self-service activities as a substitute for market services.

Furthermore, due to the spread of NIT, mass-production of highly standardized goods is being replaced by the production of custom-made goods in response to changing market conditions (differentiation of products by size, efficiency, colour, accessories, etc.).

Also, the type of product produced by European economies is affected by the changing international division of work, since the more advanced developing countries have made progress in mass-production of standardized goods (textiles, simple consumer goods, basic chemicals, steel, etc.), which also affects the needs and the demand for technical, marketing and other similar services.

b. Market transformation

Market transformation is another feature of the 'new economy' resulting

Table 1: EFFECTS OF NIT ON THE ECONOMY

INFORMATION TECHNOLOGY

TRANSFORMATION OF PRODUCT OR SERVICE

- Growing complementary relationship between goods and services (integrated packages).
- Increase in service functions
- Development of new services:
  - Stimulation of demand for market services
  - Substitution of demand (growth in self-service activities).
- Differentiation of services instead of mass services (standardized services).
- Greater importance of technical, R&D and marketing services.

TRANSFORMATION OF MARKETS

- Internationalization of markets:
  - Growth of planning and organizational capacity
  - Increased demand for information services, market research, etc.
- Increased commercialization of services.

TRANSFORMATION OF SERVICE LOCATION

- Internationalization of goods and service industries:
  - Increase in planning and organizational capacity
  - Increased demand for information and advisory services.

TRANSFORMATION IN SERVICE PROVISION

- Growing flexibility in provision.
- Provision of large and small scale services.
- Greater demand for skilled labour.
- Increase in service activities within industrial companies.
- Reintegration of services.
- Automation of provision of old and new services.
- Growing importance of non-material investment.

OTHER FACTORS:

- Increased returns
- Occupational and labour models
- Time "budgets"
- Institutional factors

primarily from the application of NIT. Information on international markets can be easily transmitted, so that a world market is gradually developing. Due to the interdependence between goods and services, the internationalization of goods markets often requires the internationalization of service markets, and vice versa. Thus exportation of goods is normally followed by the sale of services such as training and maintenance contracts.

This internationalization of markets is facilitated by the growing commercial viability of services (Rada, 1985). Advances in communications technology and data processing and storage have made it possible to produce services at any time for consumption at a later date. Therefore, any service product which can be broken down into electronically coded pieces of information can, theoretically, be provided reliably and immediately anywhere in the world at relatively low costs. It must, however, be pointed out that there are several difficulties involved in putting such an information services into practice (Feketekuty and Hauser, 1985).

The gradual formation of a world market exposes domestic industries, both manufacturing and services, to international competition. Therefore companies must be more aware of the market situation and of innovations. So in order to reduce uncertainty about demand and about the competition, different types of services are being promoted: market research, promotion, legal and financial advice, financial services, etc. (Petit, 1987).

c. Internationalization of the economy

The internationalization of service companies has grown considerably in recent years, especially in the U.S., and to a lesser extent in Europe. There are various explanations for this process. Direct foreign investment in service companies may complement the internationalization of manufacturing industries. Multinationals often find it more efficient to buy services (insurance, banking, accountancy...) from companies which can provide these services worldwide and with uniform quality, rather than to deal with several suppliers in different places.

Contrary to the situation in the manufacturing industry, the formation of networks in service companies is not restricted to the use of unskilled labour. Along the lines of the model created by IBM, large corporations are now trying to use specialized agencies (engineering, R&D and other professional fields) in several countries simultaneously. For example, American companies are moving their R&D laboratories, design departments and software development to countries such as Israel and India. These departments work closely with services of other companies located in Europe and the U.S. (Ginzberg et al., 1986). In short, this internationalization process is stimulating demand for different types of service, such as planning, organization and infirmation, within multinationals.

d. Changes in production processes

The application of NIT has one effect which has far-reaching consequences in the production process: manual work on the product and simple mechanical operations are becoming less important. Production and processing of information is fast becoming the backbone of the modern production system. As factories automate, they need fewer workers on the production line and more to programme computers, design robots and develop software to run CAD/CAM technologies. In other words there is a growing use of knowledge and information as inputs in the production of goods and services.

According to Piore and Sabel (1985), a flexible type of production based on small-series manufacturing will predominate in the future. In this way production will be flexibly adapted to meet changes in demand. In this new production process 'non-material investments' (software development, advanced training, etc.,) becomes more important in comparison with 'material investments'.

3. Consequences of the Introduction of NIT in Services: some Basic Aspects

A study of growing literature on this subject shows certain coincidences in the fundamental aspects. The main consequences attributed to the introduction of NIT in services, whose clearest repercussion is in transportability of information, are as follows:

- **Increased productivity** in a considerable number of service fields.
- Greater **market transparency.**
- **Breaking down of boundaries** between sectors.
- Changes in **entry restrictions.**
- **Greater internationalization** of services.

These consequences are not mutally exclusive. In the majority of cases they are even common, although it is possible to separate them, as many authors do, in order to gain a better understanding of the changes which are taking place. We must also add two of the most important effects of NIT on services, the effect on **employment and production methods,** and the effect on the **location** of services companies. The latter will be dealt with in greater detail in Section 5. Employment is already covered in other contributions.

To give a detailed account of each of the five consequences of NIT on services in general would require a lengthy discussion which would also limit the attention we wish to give to business services. We shall therefore restrict ourselves to a few brief remarks.

* **Increase of productivity in service production**

The first effect of transportability involves the production of the services themselves. In services, thanks to the introduction of new technologies, there is a reduction of transaction costs, particularly in fields such as banking, insurance and other information-intensive activities. Many services can now be carried out without staff 24 hours a day from different places, including home. The application of office automation equipment and particularly correctly operated on-line systems increase productivity in a whole group of areas which until now had not been formalized and were difficult to rationalize. This is the case of professional and management work, for example.

Although it is clear that technological advances substantially increase productivity in numerous services, it is well known that the gauging of productivity in this sector is a difficult question which is open to debate. De Bandt (1985 and 1988), Denison (1973), Freeman and Soete (1985), Kendrick (1982 and 1986), OECD (1987), Petit (1984 and 1987) and Ray (1986) present various arguments. There are few empirical studies, and methodology differs so much as to prevent comparisons. It is also evident that services have a quality aspect which improves substantially with technological advances, and which is difficult to measure. Only ad hoc methodologies for different activities can shed a little light on the subject.

* **Market transparency**

A transparent market is that in which all the necessary information on the goods to be transacted in the market is available. This is essential for competitors, industrial or individual consumers, since a wide range of choices are on offer. Transparency occurs for goods and services more than for producers of information per se (for example, a consultant's report or a market research study). Information does not behave like 'normal merchandise'; this has led to considerable discussions on the nature of information and the economic theory required to threat it as 'merchandise'.

Data bases increase the amount of information concerning alternative sources, prices and quality, and thus choice is widened and market transparency increased. This in itself has important consequences, since it reduces the response time to changing market conditions and reduces entry restrictions to markets, where it is necessary to spread information about a given product or service quickly and widely. One example of this can be seen in the information terminals in the stock market; great operators practically do work all over the world. This type of system is successful when the time factor is essential and most of the information provided is perishable. In the future there will be systems which offer both perishable and more permanent information, particularly with the increased use of local cable networks and a combination of powerful processing and storage technology.

The transparency of the different markets will increase with the appearance of more transaction systems and on-line data bases. The effect of more transparent markets on the international economic system is still unclear. There are substantial positive effects in at least two areas. One is the greater choice for the domestic consumer (company or individual), and the other is that there is an easy, cheap way for producers to expose their goods on the market, thus reducing entry restrictions.

* **Breaking-down of boundaries between sectors**

The classic case is the fusion of bankign, financial services and retailing. In Europe and the U.S. many department stores and supermarkets use interactive sales points systems and have their own credit cards, which reduces the number of operations carried out by the banks.

Whereas the above case refers to the breakdown of boundaries between different service providers, the same thing happens between industry and services. One classic case here is the computer industry and the telecommunications industry, which were clearly different in the past, but which are now becoming one. Computer companies are offering telecommunications services or international or local networks, while those companies offering telecommunications services are entering the computer field.

The breakdown of boundaries between sectors confirm the way that goods and services are becoming intermixed in many fields. This has important implications for economic policy, particularly in the area of service regulations. It also increases the complexity of data accumulation procedures and the distinction between goods and services in international trade. One of the most important effects of this convergence of sectors is an increase in scale economies through vertical integration. The trend in most areas is to offer total packages that are a combination of goods and services. This is valid for the area of services and more and more so for manufacturing. Consequently, this process will concentrate activities and vary entry restrictions in many fields.

* **Changes in entry restrictions**

There are two main changes in this field, one where the entry restriction for service providers is reduced, and the other where entry restrictions are increased for mass-consumer services or where industry intermixes its offer of goods with services.

Entry restrictions are reduced when the telecommunications infrastructure is used to sell and/or distribute services. Connection to a network may make costly investment in branch offices or localized services unnecessary. The cost of the network may in turn be shared by all the users, suppliers and consumers. Through telecommunications clients can be attended to without the physical presence of the service provider. This is particularly true for information-intensive services such as software or data base companies which can enter markets through telephone lines.

Although the growth and diversification of services could take place due to fewer entry restrictions, it is also true that certain services enjoy larger scale economies both nationally and internationally; for example, insurance companies and financial service companies. When goods are intermixed with services the entry restriction is considerably greater, since 'captive markets' can be created when purchase and/or consumption of goods is linked to service consumption.

* **Internationalization of services**

The consequence of the process of service trnasportability mentioned previously is inevitably related to international aspects. The internationalization of services increases as more services become marketable through improvements in communications, and as more services are incorporated in goods, and finally, as services are held on physical supports (cassettes, video and the like) or substituted for goods. Despite the varied effects of technology on the internationalization of services, research has concentrated mainly on the role of data transmission in old and new services. This is not surprising, since data transmission between countries is the clearest example of change in world trade. Acknowledgement of this change has led some countries to review their policy towards data transmission in order to maximize domestic profits and to avoid possible adverse effects.

The majority of developed countries have followed a policy of promoting the offer of information-intensive services through special government programmes, financial incentives and R&D.

The second type of reaction has been the explicit regulation of the mechanism of data flow as part of a general information policy aimed at promoting the production of information goods and services. Such a policy is equivalent to a programme of substitution of service importation, similar to the policies of substitution of manufactured goods imporatation used in the past. The only clear example in this field is Brazil, which has been extensively studied by J.F. Rada (1987).

To sump up, the development of telecommunications will greatly condition the trends towards the internationalization of services. Growth is expected to accelerate due to increased connections via satellite, cable TV and cheaper telecommuniations rates. These factors and the concentration of information resources will generate as yet unknown long term economic effects.

With regard to the general repercussions of the introduction of NIT in services, we shall subsequently highlight a few very interesting points:

***First,** services are becoming increasingly important. In evaluating the relative comparative advantages, it seems that the organic link between the production of goods and services is a key element. In this context, the loss of comparative advantage in the production of goods could affect demand and the type of services required by an economy. The reverse can also occur due to the multiplying effect of services on the production of goods. There is a need for additional study in order to clearly establish the precise mechanisms of these links and how they are affected by technological change.

***Second,** the greatest impact of NIT on services lies in the better transportability of many services. This creates conditions for the birth of new services, and above all opens up the possibility of international exchange of services through telecommunications systems. This phenomenon is especially important in world trade.

***Third,** the international situation of services is significantly favourable to the developed countries. This appears to be true for world trade and also for the growing differences in equipment and infrastructure necessary to increase productivity and which facilitate transportability of services. The internationalization of services, either through telecommunications or through more modern, more economic production, creates problems for developing countries, problems which are much more complex than those experienced during the first stages of industrialization.

It appears undeniable that the above can be transferred to the case of different regions within the same country, as well as that of those regions integrated in a large supranational organization, i.e., the EEC. The literature agrees on this point, or rather in the diagnosis. However, there is a lack of research and experience in the possible actions to be taken, which clearly must be carried out on several levels: local/regional; national; and supra-national (i.e., EEC level). The 'Strategique 4 dossier' of FAST II (Brussels 1987) proposes a taxonomy of possible action and fields of responsibility by levels.

4. Business Services and Technological Change: Externalization Versus Internalization

One of the features which characterizes the expansion of services in recent years are the changes which are taking place in their internal composition. Most of the more recent studies (Illeris, 1986; Giarini, 1987; Cuadrado and González, 1987; Cappellin, 1986; Petit, 1987; Bailly and Maillat, 1988; and Barcet and Bonamy, 1988, among others) highlight the basic feature of the last decade as being the development of modern services, where essentially business services are usually included, and a relative fall-off in more traditional services. Not only in terms of employment (where several observations would have to be made), but in terms of contribution to GNP and in terms of growth rates.

There is a certain amount of agreement in pointing to two factors which explain, to a different degree, the advance of business services: the appearance of **new tasks** in the organization of production and the possibilities of using productiong processes **outside** the company to carry out these new tasks. This duality does not mean that business services always necessarily fulfill new functions. In this regard we can observe four types of change which lead to the creation and development of new tasks:

- The changes that have taken place in the product markets (internationalization; increased competition; diversification versus mass-production, etc.), which create a need for more services related to market analysis, financing formulas, production logistics, design, and the introduction of innovations.
- The changes that have taken, or are taking place in the fields of commercial, labour, fiscal and consumer-protection legislation, which make the framework within which companies have to work much more complex, thus creating an increasing need for advice.
- The development of new technologies, which affect the increase in handling of very different types of information (technical, market,

accountancy, commercial), broadening the need for service functions within the company and the demand for services outside.

- The spread of new types of management. This is to a great extent linked to the introduction of new information technology which allows old systems to be altered or replaced, both in production and marketing as well as in management and control tasks.

The result of these four changes which we have distinguished, although they tend to coincide to some extent in some cases, is to be seen in the exampansion at present being experienced by services **within** companies (**service funtions**, as described by Gershuny and Miles, 1983) and also **outside**. This is more clearly reflected by the growth figures for some service activities.

These growing **needs** for services in manufacturing companies, modern agricultural concerns, and many companies which are part of the services sector (hotels, health, commerce) enable the new functions which satisfay those needs to be carried out inside the company or meet demand from outside. As different studies show (e.g., Chandler, 1977), the pattern of internalizing functions is usually justified in the exploitation of the synergy inherent in a hierarchical organization, at least in large companies. The externalization of functions is, on the other hand, an attempt to take maximum advantage of the scale economies generated by the expansion of external markets (Stigler).

4.1 Externalization of functions

The principle of market expansion proposed by Stigler (1951) to explain the externalization of certain functions explains **a priori** the recent development of the intermediate demand for services. The activities of a company fulfill various functions. If the scale of production is increased, these different functions offer different scale economies. According to Stigler, at a certain level of production it is preferable to resort to the outside market and externalize the functions which offer the largest scale economies. Stanback and others (1981) correctly point out that it is not so much a question of scale economies as a matter of task specialization. On the one hand it is difficult to discover scale economies at the level of functions, and on the other, the specialization of tasks in a given function reveals the possibilities of externalizing that function. The development of business services is based precisely on the capacity of businessmen to ascertain which specialization within the company can be externalized.

Nevertheless, Petit (1987) sees these specializations as being more closely linked to the organization of production than to the scale. Therefore, a certain way of organizing personnel policy of the accounts department in a company will allow part of the process of salary payment or auditing to be carried out externally. Having made the distinction between specialization (which is essentially a question of organization) and scale economies (which are related to the volume of production), the factor of the size of the compnay no longer works in the same way. One interpretation of Stigler's principle could lead us to draw the conclusion that the increase in the size of a company, expressed in volume of production, increases the use of external services or even the setting up of an external production process (open to outside users). However, another interpretation suggests that, on the contrary, a company of any size can use external services as soon as it reaches a certain level of organization. In this context, the scale of production in a large company may permit the possibility of internalizing a given function completely, but not the reverse.

These different ways of interpreting Stigler's principle affect the conclusions that can be drawn regarding the development of demand for business services. The increase in services provided to companies is, then, the result of an accumulative process in which the spread of new organizational practices facilitates the expanstion of business services, and in which the development of this provision of services accelerates the spread of habitual management practices. Under these conditions, the distribution by sectors of service users is due to a great extent to the level of the development of service provision and the spread of modern management practices which, as we have seen, are closely linked to the spread of new technology within the service industry.

Consequently the frequent use of business services in the tertiairy sector is linked to the importance of small and medium-sized businesses, and also to the development of the offer of services and to the spread of new technology, which are especially susceptible to externalization in these sectors. Thus the arguments put forward by Chandler and Stigler, seen from this angle, are to a great extent complementary.

4.2 Internalization of new functions

Chandler (1977) emphasizes the development of the different types of hierarchical organization which still explain the might of large companies. His work helps to clarify the development of the demand of companies for external services and the gradual structuring of the offer of business services. In the first type of organization that can be determined, and which appeared in the United States at the beginning of this century, the large company was divided into departments according to its main functions (personnel, research and methods, production and general administration). As a result of this first type of centralized organization -the U-shaped structure of Williamson's formulation- the company satisfied most of its needs internally (vertical integration). The rigidity of this very centralized organization tended to increase with the size of the company (Williamson, 1981). In the period between World Ware I and World War II another type of organization appeared which integrated the diversification of production processes, creating product divisions (M-shaped structure). These divisions, which reproduced in their own sphere the unitary functional division of the U-shaped structure, enjoyed greater independence regarding product management; consequently it was easier to use external production processes. This positive correlation between a structured internal division of tasks and the use of external services seems to have spread to small and medium-sized businesses. Barcet and Bonamy (1983) observed that the use of business services by French companies increased with the functional division of tasks (in the case of those companies with U-shaped management structures). From this point of view the demand for business services is more a complement to the reorganization of the internal activities of a company than a substitute.

This marked increase in the use of these business services took place between 1960 to 1975, especially in large American and European companies (Caves Report, 1980; Williamson, 1981; Chandler, 1981). Nevertheless, the fact that the pace of reorganization in small and medium-sized businesses is slower perhaps contributes to the maintenance of a constant increase in the demand for business services. The very development of service provision is a factor which contributes to this evolution. Thus, the large companies which are already present in the distribution , hotel, restaurant and car rental sectors also demand service activities such as consultants, data management and research. The standardizing of service production, which is without doubt a decisive factor in the development of these chains, allows small and medium-sized businesses greater facility of access to intermediate services. Therefore, this restructuring of service supply explains to a great extent the relatively high level of demand for business services in the services and consumer goods sectors which have a low degree of concentration.

In this way, the principles of industrial organization put forward by Chandler and Stigler go a long way to explaining the appearance and development of business services. The increase in demand for these services is the result of three successive changes in the production process:

- Changes in the organization of large companies.
- The structure of the offer of services to producers.
- The functional reorganization of the administration of small and medium-sized businesses.

As Petit (1987) points out, in reference to various studies, the above analyses based on scale economies or on complementary aspects, are to a greater or lesser extent related to purely technical aspects of production. No explicit mention is made of the interdependence of the social and technical aspects of production organization. However, the nature of industrial relations plays an important part in decisions whether to internalize the production processes, or whether to use external services. This criticism made by Williamson (1981) of Chandler's studies on American companies is even more justified in the case of European economies, where labour laws are usually more restrictive. According ot the sector, the size of the company and the degree of trade union activity among the workforce, working conditions and salaries can respond more or less rapidly to variations in the general level of economic activity or to changes in the nature of the product. Therefore, companies in which the organization of the internal work market is rather rigid can benefit from conditions in the external work market.

From this point of view the externalization of certain tasks is aimed mainly at increasing flexibility in the use of labour, which leads us to consider the problem of adjustment costs (the costs of adjusting production conditions to market fluctuations). All this brings us to the conclusion that decisions to externalize or internalize production activities do not depend solely on purely technical considerations.

There can be no doubting the **strategic importance of business services** at present, an importance which is determined by the following facts:

- In some industrial sectors the demand for business sectors represents a considerable proportion of added value.
- The strategic value specialized services have for companies. The services of highly-qualified specialists help the decision-making centres of modern companies in many areas. Stanback (1980) pointed out that in fields like commercial strategy, engineering, financial administration and international law specialized companies work better than the internal services of the different divisions of the modern company. This facilitates the diversification of products and markets.

At the same time, external specialists offer possibilities of adaptation to small and medium-sized companies subject to strong foreign competition in their domestic markets.

- The importance of business services also lies in the capacity of service companies to solve temporary strategy problems or to help industrial companies in the processes of structural adaptation.
- Finally, as McKenzie and Smith (1986) showed for the case of the United States, contracting external services allowed industrial companies to maintain or increase their international competitiveness between 1969 and 1985. Had these companies not contracted all sorts of services externally, they would have had to incur higher direct costs and would have imposed higher costs on the services they provided to other companies. In consequence, the U.S. industrial sector would not have increased production to the extent it did. Therefore, total employment in the combined industrial and services sectors was perhaps higher in 1985 than it would have been if part of industrial employment had not transferred to the services sector.

To sum up, business services have experienced a continuous period of growth. The externalization of tasks not directly related to the production process represents in fact a phase in the organization of production systems for two reasons:

- It allows medium-sized companies access to a functional division of labour which has been used by large companies for decades.
- It allows large companies to diversify their production in accordance with new strategies for the development of multiple products on a worldwide scale.

The importance and amplitude of this double change in the division of labour has been possible thanks to the introduction of new information technology in the business services sector, which has facilitated the externalization of tertiary production. Consequently, new information technology facilitates this trend towards the externalization of tertiary functions, and also allows small companies to benefit from this same trend.

The empirical studies available are still insufficient, and do not cover the evolution of a large number of countries, but those studies which have been carried out support many of the claims we have tried to summarize.

5. Some Notes on the Location of Business Services: Greater Centralization?

One thing which is perfectly clear is that one cannot generalize when referring to the factors which explain the location of services. There are obvious and substantial differences according to the type of services (public/private, and in any case there are differences according to activities).

Having made this general statement, which is in some ways applicable to the case of business services, the considerations which follow only refer to these services, and we shall omit any reservations and specific observations which would be necessary in a more extensive examination of the subject.

The importance of business services is determined basically by a qualitative element: their strategic nature. Within the process of the tertiarization of western economies, the dynamics of advanced services linked to production is the fundamental guide which allows us to distinguish between the 'assistance' element (the growth of a services sector with the functions of an 'employment hospital') and the 'advanced' element (tertiarization as a product of the introduction of new technologies and changes in production and in the international division of labour).

In studying the process of the physical location of business services we are answering three basic questions:

* The first refers to the degree of integration of the structure of production in the area studied, its dependence on other sevice centres and, consequently, its position in the metropolitan and regional production hierarchy.
* The second deals with the level of competitiveness of companies, the introduction of new management techniques and the processes of segregation of services.
* The third, to a certain extent a summary of the first two, deals with the existence of economic and business dynamism, and therefore with growth prospects in terms of production and employment.

The objective is to ascertain whether we are seeing the establishment of new foundations for economic growth, with an expansion of the tertiary sector which benefits industry and allows technologically advanced development thereof, or whether we are faced with an involved process where services take on a new social function - absorbing part of the excess labour force of industry, giving rise to a possible poverty spiral. In this case the industrial decine of a city or region would be generating a compensatory growth of the tertiary which, by putting pressure on the economic use of certain resources would expel other production activities and accentuate the economic decline.

5.1 A first reflection

The process of division of labour that determines the birth of new services may be the result of a process of horizontal disintegration between production phases, similar to that which occurs in the industrial sector, or of a process of differentiation in the demand for services.

In the first of these instances, differentiation may mean that more modern services need a greater optimum dimension, as a result of which this new activity would tend to be located in areas of greater concentration/ accessibility in order to increase its market. Inthis case the proximity of urban areas means a larger market and lower transport costs, while in more remote areas the cost of transport cannot compensate the segregation of the service. The result is that new services tend to **concentrate in central areas**, and that even services of origin (internalized in the company or external) may suffer **redimensioning** due to the fact that some of the functions fulfilled by traditional services are absorbed by the new structure.

The second factor in the process of division of labour may lie in the differentiation of services brougth about by an increase in **innovation, higher-quality** production of widely-spread services. In this case an increase in demand could bring about the appearance of a superior service which could not be produced in certain areas due to its production cost at a certain level of demand. This seems to be the case of the variations which occur in different sizes of city, especially in production services: the higher the population, the more new services which have managed to find a threshold.

Consequently, in the case of strongly integrated areas, the appearance of new needs deriving from increased demand tends to be concentrated, especially in production services, in the areas with a higher density of services.

These patterns of **concentration** may, however, suffer variations in the case of certain types of service. Certain 'position returns' in peripheral quality areas can have similar central effects when the level of communi- cations is sufficient, and on the contrary, agglomeration expenses (cost of land, saturation of communications networks, etc.) can force services which occupy a great deal of space (public centres, large stores) to peripheral locations or make them heavily dependent on the efficiency of transport networks.

This process of concentration is also subject to other complex factors affected by the peculiarities of the spatial structure of the services, which complicate the concentration process.

In the first place, unlike industrial activities, services linkages are carried out through communications networks or personal contacts. In most cases the merchandise is information (transmitted by networks or personal contact) or management. This makes the services' sphere of influence highly complex, and the spheres of attraction, and consequently the minimum operating thresholds, vary substantially.

Also, spatial structures which allow one or other linkage do not share the same infrastructures. The need for frequent personal contact in a wide sphere necessitates proximity to modern communications centres: large cities with airports, good transport connections, etc.; while other services which require more sporadic contact can manage without the degree of flexibility mentioned, and their location will depend on the efficiency (or the mere existence) of modern communication networks. We must also take into account the fact that the threshold of need for contacts is not defined for ever for a given service. Modern techniques of long-distance communication have recently allowed a great deal of personal contacts, which were considered essential, to be eliminated (closed circuit television, telephone transmission of pictures, etc.).

In principle these techniques work in favour of the spatial spread of services, **but there is no guarantee that this happens in practice.** Indeed, the improvement in accessibility (physical or of information) seems to favour the spread of services, but in practice the indications are that this spread is extremely selective and does not occur in all cases. The problem lies in the fact that certain improvements in accessibility through new information technology tend to increase the sphere of attraction of central services, whose market is growing as the costs of transport (of people, merchandise or information) decrease.

Another factor which favours **concentration** is the possibility of access to skilled labour, which is concentrated inthe centres of large cities, either because of the existence of specialized training structures (universities, specialized centres) or because of the quality of the urban environment (greater density and quality of available services, residential areas).

The different concentration factors tend to produce a feedback effect, and together they act cumulatively one on another, generating a social specialization of space, based on prestige, which is reflected in the high value of land in these areas of advanced services.

These considerations of the 'spontaneous' tendency towards the concentration of services, particularly new or advanced services, independent of the technological possibilities fo the spread of services, are of great importance in planning regional policy. In fact, in the absence of a definite policy of decentralization or spread of services, the latter will tend to concentrate in a small number of areas, reproducing and worsening the current shortage of services in peripheral regions (and/or cities).

5.2 A subject for debate

A number of studies (in Italy, France, Canada, United States, Great Britain, for example) emphasize the trend towards a geographical concentration of more advanced production services. Others (Switzerland, France - B. Planque (1982, Aydalot (1979 and 1984) - Holland) point to a decentralization within the existing centralization. Molle and Van Holst (1984) referred to the coexistence of tendencies towards centralization and decentralization of service activities in the EEC.

The studies available perhaps do not allow completely clear statements to be made. This was done of the conclusions reached by the studies carried out on services as part of the FAST II programme (EEC), with studies on France, Italy, Greece, Spain, Denmark, Great Britain, Germany, Holland and Belgium, although the methodologies, and above all the areas studied were not comparable from the point of view of gleaning ideas which can be considered generalized.

The first thing to be said, then, is that this subject is one which demands more research: work is already being carried out (in Spain, and apparently in France, Italy and Germany, at least for some regions).

The spatial factor in production services is extremely complex. There are at least three relevant spaces: the space **where the services is provided**; the one in which **demand is generated**; and the one **where the service supplier is located.** The granting of greater importance to one or another of these depends on the factor to be studied. The first is fundamental when it comes to analyzing the effects induced by the spatial structure of public services (for example, the fact that a court or a railway station is in a particular place is extremely important when deciding on a location for a law firm or a postal service company. The second allows us to gauge the degree of development of a production fabric, and the third defines the relative situation of each zone within the metropolitan and regional hierarchy.

Understandably, detecting the location criteria used by companies is never simple. Even less so in the case of services, which have widely differing qualitative and quantitative spatial requirements. In the case of location factors in business services, specific studies must be examined. For example, in the Comunidad Autonomia de Madrid (1986) a study was carried out on the spatial structure of production services in the city. There is considerable unanimity in pointing to the cost of premises as a decisive factor. This unanimity is not at all surprising, since in most cases the companies dealt with were small ones which did not require large investments; inthis context the study shows that the expense involved in purchasing or renting premises is an extremely important factor, and minimization of this expense is fundamental.

It would, however, be superficial to conclude that the main criterion for location in the production services sector is the cost of premises. The fundamental question would remain unanswered: of all the premises in this price range, why did you choose that one? From this new standpoint, according to the study, the cost of premises appears more as a restriction, and what should be examined are the discrimination criteria used.

The first of these is the closeness to the company owner's home. Apart from the obvious reduction in transport expenses, this facilitates quick access to information. Another criterion is the existence of good communication systems in the area. The complexity of the space element in production services makes the existence of efficient communications and/or information networks paramount to development. In some activities there is a third factor: closeness to the place where the service is to be provided. For example, the study shows that proximity to a railway station was a deciding factor in the location of postal publicity companies in the area.

It is striking that among the criteria detected there is nothing directly related to closeness to the client or the existence of a large market nearby. It is usually said that one of the characteristics of service companies is their closer proximity to demand, although in advanced production services this is less evident. The complex nature of the space factor in these services requires 'proximity' not to one, but to several areas. If we add their metropolitan, and in many cases suprametropolitan nature, the result is a space hierarchy which assigns advanced services to high-quality central areas, and traditional services to more dilapidated areas with worse communication facilities, with a great diversity of conditions to be found between these two extremes.

In pure advanced services, the study concludes, their domestic scope means that the relevant space criterion is closely linked to the function of representation. Their preferred location, therefore, is in high-quality areas where there is an already established, strongly interrelated tertiary fabric.

There is no doubt, as Cunha and Racine (1984) pointed out, about the consequences of the location processes of business services, since the proven **centralization** of these services activities must necessarily lead to an inequality in the spatial distribution of functions; this could possibly bring about an increase in regional disparities, since tertiary production activities are typified by a preference for the best equipped areas, which works to the detriment of the peripheral regions.

6. Business Services and Regional Policy

As Wadley (1988) emphasized, it can be said that until very recently those in charge of regional policy have had a dubious, or rather prudent, attitude towards the services sector. Marquand (1980) placed the main types of reigonal policy into four categories according to the importance attached to the services sector:

1. Regional policies exlcusievely devoted to tertiary activiites (in whole or in part).
2. Those which include the tertiary sector more or less under the same conditions as the secondary sector (but often with special restrictions or reservations).
3. Those which explicitly exclude tertiary activities, which are thus seen as inferior.
4. Those which in fact have national scope, but which have considerable influence on the spatial distribution of services among the various regions.

In his study on the EEC countries, the author showed that regional policies applied to the tertiary sector in Belgium, Denmark, Germany, Italy and Holland only differed in small ways from those applied to the manufacturing industry (type 2). In France, Luxembourg, Ireland and the United Kingdom there are programmes specially aimed at service activities (type 1), but the U.K. also applied type 3 measures. The main policies of national scope which have repercussions on the services sector are those which refer to the development of tourism and the decentralization of the administration.

It has been said, and not without reason, that the services sector should receive more limited aid than the industrial sector. It has been shown, for example, that services contribute only slightly to the improvement in welfare, that their productivity is lower, that they do not always allow exports to develop, but rather their objective is to satisfy the needs of the local population, and that therefore, in facilitating the creation of companies in this sector, there is always the risk of harming existing companies; in short, that really mobile activities deserve much more help than service activities..

In this context, it must be pointed out that the Report to the OECD (Wadley, 1988) refutes all these argumetns. There is no reason to deny aid to service activities in those regions where manufacturing industries receive aid, except for certain services provided to consumers who already have a sufficient offer. Nevertheless, it is recommended that those responsible for regional policy establish a certain distinction between different types of service. Services provided to companies (and office activities) have a tendency to concentrate in certain regions and areas, especially in large cities which have well developed infrastructures and communications networks, while consumer services are usually more clearly aimed at the local population.

The problem of the development of business services is undoubtedly an important aspect of regional policy which must be considered at the present time. In this sense, the demand for greater efficiency in the search for markets, in obtaining financing, in perfecting production functions, and in the search for flexible forms of management give business service activities a prime role in regional plans (Cunha and Racine, 1984). A coherent regional policy for promoting intermediate and peripheral reigons must consequently attempt to decentralize (or ensure the development of) these tertiary activities which are decisive factors in the implantation or renovation of industries, in the strengthening of the possibilities of innovation, and equally decisive in the enriching and diversification of the endogenous potential of regional economic systems (Sweeny, 1988; Stöhr, 1988; Landaburu, 1988; Cappellin, 1988; Cuadrado, 1988).

The application of a policy of this type appears to be closely linked to the development and spread of NIT. It seems that the lines of action suggested by various experiences in EEC countries should cover four basic fronts where simultaneous action is recommended:

- The **development of technical infrastructures** required by services; among these are: telecommunications infrastructures, traditional communications, those required for personal services and real estate infrastructures ('Advanced Business Service Centres').
- The establishment of a **set of measures** which facilitate the functioning

of the activities; e.g., organizational, legal transfer of technology, provision of service centres and data banks, among others.

- Action aimed at **stimulating supply and demand** of services in relation to the needs of the primary and industrial sectors; mainly in strategic services (R&D; innovation), aid for small and medium-sized businesses, grouping of service companies and clients according to needs.
- Action aimed at improving and increasing available **human resources**: specialized regional institutes; research/industry co-operation; training in new fields; co-operation with universities, etc.

It is impossible here to go into an assessment of the action that has been taken in individual countries, almost always without an overall approach. Some actions have not yet reached the stage of maturity necessary for an assessment to really be made. Others never come to fruition. In the documents related to this subject drawn up within the FAST II programme (summary in: Dos. 4, 'Services, Technologies avancées et regions', Brussels, 1987) it is concluded that the policies have not yet had the expected impact on the location of service activities and on their dynamic role. This is mainly due to shortage of resources, but in many cases due to the **lack of co-ordination** between national and regional or local authorities in decision-making on such essential questions as the provision of infrastructures (telecommunications), information networks.

In this section we must also conclude that more research is needed into the achievements and weaknesses of the experiments carried out to date. The most desirable approach would undoubtedly be a comparative international approach.

References

Amin, A. and Goddard, J. (1986), Technological Change, Industrial Restructuring and Regional Development, London, Allen & Unwin.

Antonelli, C. (1984), Multinational firms, international trade and international telecommunications, Informations Economics and Policy 1.

Aronson, J.D. and Cowhey, P.F. (1984), Computer, Data Processing and Communication Services, University of Michigan, October.

Aydalot, P. (1979), L'entreprise dans l'Espace Urbain, Economica, Paris.

Aydalot, P. (1984), Crise Economique et Espace, Economica, Paris.

Bailly, A.S. and Maillat, D. (1988), Le Secteur Tertiare en Question, Economica, Paris.

Bailly, A. et al. (1984), Tertiare moteur et developpement regional: le cas des petites et moyennes villes, Revue d'Economie Regionale et Urbaine, no. 5.

Bannon, M. and Blair, S. (1985), Services Activities, the Information Economy and the Role of Regional Centers, University College, Dublin.

Barbier, J.M. (1981), Le Quotidien et son Economie, CNRS.

Barcet, A. (1988), The Development of Tertiary Services in the Economy, Labour Market and Employment, The Service Industries Journal, vol. 8, no. 1.

Barcet, A. and Bonamy, J. (1983), La demande de services industriels, CGP.

Barcet, A. and Bonamy, J. (1988), Services et transformation des modes de production, Revue D'Economie Industrielle, I Trimestre.

Barcet, A. et al. (1985), Changement Technologique et Developpement des Services aux PMI, Lyon.

Barras, R. (1984), Growth and Technical Change in the UK Service Sector, TCCR-83-05, London.

Baumol, W.J. (1985), Information Technology and the Service Sector: A Feedback Process?, Fishman-Davidson Center for the Study of the Service Sector, Discussion Paper, no. 7, June.

Baumol, W.J. (1986), Productivity Policy and the Service Sector, in: Inman, R. (ed.), Managing the Service Economy: Prospects and Problems, Cambridge, C.U.P.

Bell, D. (1974, 1976), The Coming of Post-Industrial Society, London, Heinemann.

Bertrand, O. and Noyelle, T. (1985), L'evolution des Emplois Tertiaires, CEREQ, Paris.

Bessant, J. and Cole, S. (1985), Stacking the Chips, London, Rowman & Ablanheld.

Bessant, J. et al. (1985), IT Futures, What Current Forecasting Literature Says about the Social Impact of Information Technology, London.

Black, F. (1986), The Future for Financial Services, in: Inman, R. (ed.), Managing the Service Economy: Prospects and Problems, Cambridge, Cambridge Univ. Press.

Boisson, P. (1984), Informatique et emploi, Rapport au Conseil Economiqe et Social, February, Paris.

Bonamy, J. (1988), Services: New Opening for Employment, The Service Industries Journal, vol. 8, no. 1.

Bozeman, W.E. (1985), Ordenadores y Cálculo en Educación: una Instrucción, Fundesco.

Braibant, M. (1982), Le tertiaire insaisissable?, Economie et Statistique, June.

Branscomb, L.M. (1986), La tecnología de la información como igualador económico, Perspectivas Económicas, no. 4.

Braun, E. et al. (1979), Micro-electronics: Secondary and Tertiary Changes in Technology and Society, Comisión de las Comunidades Europeas.

Brender, A. et al. (1980), Etats Unis: croissance, crise et changement technique dans une economie tertiaire, Economie Prospective Internationale, Paris, CEPII, Documentation Française, no. 5.

Bressand, A. and Distler, C. (1985), Le Prochain Monde, Paris.

Buttner, O. and Mouriaux, M.F. (1983), Le developpement des services marchands aux enterprises: mouvement d'extériorisation ou croissance autonome?, Centre d'Etudes de l'Emploi, Paris.

Camagni, R. and Rabellotti, R. (1986), Innovation and Territory: The Lilan Heigh-Tech and Innovation Field, University of Paris.

Cappellin, R. (1986), The development of service activities in the Italian urban system, in: The Present and Future role of Services in Regional Development, S. Illeris.

Cappellin, R. (1988), Opciones de política regional en la CEE, Papeles de Economía Española, no. 35.

Cassia, C. (1987), Gestión de las implicaciones laborales de las nuevas tecnologías, in OCDE.

Castells, M. et al. (1986), Nuevas Tecnologías, Economía y Sociedad en España, Alianza Editorial, Madrid, 2 volumes.

Castells, M. et al. (1986), El Desafío Tecnológico. España y las Nuevas Tecnologías, Alianza Editorial, Madrid.

Commissariat General du Plan (1983), Services consommés par le secteur productif, La Documentation Française, November, Paris.

Comunidad Autonoma de Madrid (1986), La Gran Fábrica del Sur Madrileño,

Madrid.

Cossette, A. (1982), La Tertiarisation de l'Economie Quebecoise, C. Morin, Quebec.

Chamoux, J.P. (1980), L'Information sans Frontière, La Documentation Française, Paris.

Chandler, (1977), The Visible Hand: The Managerial Revolution in American Business, Harvard University Press.

Cuadrado, J.R. (1988), Políticas regionales: hacia un nuevo enfoque, Papeles de Economía Española, no. 35.

Cuadrado, J.R. and González, M. (1987), El Sector Servicios en España, Orbis.

Cuadrado, J.R. and C. Del Rio, (1989), Structural Change and the Service Sector in the OECD, The Services Industries Journal, vol. 9, no. 3.

Cunha, A. and J.B. Racine, (1984), Le rôle des services aux entreprises dans une société post-industrielle: technologies nouvelles et décentralisation, Revue d'Economie Regionale et Urbaine, no. 5.

Curnoy, M. (1986), Educación y Ordenadores, Fundesco.

Daniels, P. (1978), Service sector office employment and regional imbalance in Britain, ESG, no. 69.

De Bandt, J. (1985), La productivité dans les activités de services: sens et non sens; Programme mobilisateur, Technologie, Emploi, Travail, Cahier 1, June, Paris.

De Bandt, J. (comp.) (1985), Les Services dans les Sociétés Industrielles Avancées, Economica, Paris.

De Bandt, J. (1988), Le débat sur la productivité dans les services: des problèmes mal posés, Revue D'Economie Industrielle, I Trimestre.

Del Rio, C. (1988), Dinámica y distribución espacial de los servicios en España entre 1960 y 1985, Papeles de Economía Española, no. 34.

Denison, E. (1973), The Shift to Services and the Rate of Productivity Change, Survey of Current Business, 53, October.

Donges, J.B. (1986), Inversión Productiva, Nuevas Tecnologías y Creación de Empleo, Cámara de Comercio e Industria de Madrid.

Dordick, H.S. (1987), Las incipientes sociedades de la información, Perspectivas Económicas, no. 2.

Ecalle, F. (1986), La Revolution Tertiaire aux Etats-Unis, La Documentation Française.

Faulhaber, G. (1986), Deregulation and Innovation in Telecommunications, Fishman-Davidson Center for the Study of the Service Sector, Discussion Paper, no. 14.

Faulhaber, G.R. et al. (1986), The Impact of Information Technology on the Service Sector, Ballinger.

Feketekuty, G. and Hauser, K. (1985), Tecnología de la información y comercio de servicios, Perspectivas Económicas, no. 4.

Fischer, M. and Nijkamp, P. (1988), Tendencias actuales de los mercados de trabajo regionales: una comparación internacional, Papeles de Economía Española, no. 35.

Fontaine, C. (1988), Le developpement des activités de service aux entreprises et dans les entreprises, Revue D'Economie Industrielle, I Trimestre.

Freeman, C., J. Clark and L. Soete, (1985), Desempleo e innovación tecnológica, Ministerio de Trabajo y Seguridad, Madrid.

Freeman, C. and L. Soete, (1983), Cambio tecnológico y políticas de ajuste, Papeles de Economías Española, no. 15.

Freeman, C. and L. Soete (eds.) (1987), Technical Change and Full Employment, Oxford, Basil Blackwell.

Fuchs, V. (1965), The Growing Importance of the Service Industries, NBER, New York.

Fuchs, V. (1968), The Service Economy, NBER, New York.

Gershuny, J. (1978), After Industrial Society. The Emerging Self Service Economy, MacMillan.

Gershuny, J. (1982), Social innovation, change in the mode of provision of services, Futures, December.

Gershuny, J. (1983), Social Innovation and the Division of Labour, Oxford, Oxford University Press.

Gershuny, J. and Miles, I. (1983), The New Service Economy, London, Frances Pinter.

Gershuny, J. (1987), The Future of Service Employment, in: Giarini, O. (ed.), The Emerging Service Economy, Oxford, Pergamon Press.

Giarini, O. (ed.) (1987), The Emerging Service Economy, Oxford, Pergamon Press.

Giarini, O. (1988), Les nouvelles conditions du progrés économique: de la rigidité de l'offre à l'economie de service, Revue D'Economie Industrielle, I Trimestre.

Gierch, H. (1985): The present economic situation in Europe and future trends, in: Gibb, J. (ed.), Science Parks and Innovation Centres: Their Economic and Social Impact, CEE.

Ginsberg, A. (1979), Good Jobs, Bad Jobs, No Jobs, Cambridge, Mass., Harvard University Press.

Ginzberg, E. et al. (1986), Technology and Employment, Concepts and

Clarification, Boulder, London.

Glowinski, A. et al. (1980), Telecommunications en l'an 2000, Dunod, Paris.

Grey, R. (1983), Trade Computer Services, Royal Bank of Canada, Montreal.

Guy, K. (1987), The UK Tertiairy Service Sector, in: Freeman, C. and Soete, L. (eds.), Change and Full Employment, Oxford, Basil Blackwell.

Hill, T.P. (1977), On goods and services, Review of Income and Wealth, December.

Hirschhorn, L. (1988), The Post-Industrial Economy: Labour Skills and the New Mode of Production, The Service Industries Journal, vol. 8, no. 1.

Holmerstrom, B. (1986), The Provision of Services in a Market Economy, in: Inman, R. (ed.), Managing the Service Economy: Prospects and Problems, Cambrdige, Cambridge University Press.

Illeris, S. (1986), The present and future role of services in regional development, FAST, no. 74.

Inman, R. (ed.) (1986), Managing the Service Economy: Prospects and Problems, Cambridge, Cambridge University Press.

Instituto de Estudios Economicos (1981), Nuevas Tecnologías y Empleo. Revista del Instituto de Estudios Económicos, no. 1.

Kane, E.J. (1984), Technological and Regulatory Forces in the Developing Fusion of Financial Services Competition, Journal of Finance, vol. 39, no. 3.

Keeble, D. et al. (1985), The urban-rural manufacturing shift in the European Community, Urban Studies, no. 4.

Kendrick, J.W. (1982), Measurement of Output and Productivity in the Service Sector, ARA/Wharton Conference 'The Future of the Service Economy', Philadelphia.

Kendrick, J.W. (1986), Measurement of Output and Productivity in the Service sector, in: Inman, R. (ed.), Managing the Service Economy: Prospects and Problems, Cambridge, Cambridge U.P.

Kranzberg, M. (1986), La era de la información: ¿evolución o revolución? Perspectivas Económicas, no. 3.

Krommenacker, R. (1986), The Impact of Information Technology on Trade Interdependence, in: Inman, R. (ed.), Managing the Service Economy: Prospects and Problems, Cambridge, Cambridge U.P.

Krommenacker, R. (1987), Services and Space Technology: The Emergence of Space Generated, Highly Integrated Goods and Services (IGS), in: Giarini, O. (ed.), The Emerging Service Economy, Oxford & Pergamon Press.

Landaburu, E. (1988), La política regional de la Comunidad, Papeles de Economía Española, no. 35.

Langeard, E. and P. Eiglier, (1988), Le couple produit-service dans l'offre

globale de services aux entreprises, Revue D'Economie Industrielle, I Trimestre.
Lefebvre, N. (1981), L'exteriorisation des fonctions tertiaires, Cahiers de Laboratoire d'Economie Sociales, October.

Lesournes, J. and M. Godet, (eds.) (1985), La Fin des Habitudes, Seghers, Paris.

Malsot, J. (1984), Services et croissance: perspective à l'horizon 2000, CGP, La Documentation Française, Paris.

Marquand, J. (1979), The Service Sector and Regional Policy in the United Kingdom, CES, London.

Marquand, J. (1980), Le Role des Activités Tertiaires dans la Politique Regionale: Analyse Comparative, CEE, DI no. 7.

Missika, J.L. et al. (1981), Informatisation et Emploi, menace ou mutation?, Informatisation et Société, no. 11, La Documentation Française, Paris.

Molle, W. and B. Van Holst, (1984), Services et regions; une analyse des developpements a long terme des services dans les regions de la CEE, Revue d'Economie Regionale et Urbaine, no. 5.

Momigliano, F. and O. Siniscalco, (1983), The Growth of Service Employment: A Reappraisal, Banca Nazionale del Lavoro, Quarterly Review, no. 142, September.

Nicholas, T.I. (1984), La Banca y la Tecnología, Papeles de Economía Española, no. 19, Madrid.

Niosi, J. (1988), Du nouveau dans des services internationaux, les multinationales de l'ingenierie, Revue D'Economie Industrielle, I Trimestre.

Nixdorf, (1986), Presente y futuro de los sevicios bancarios y de las entidades financieras, Informe.

Nora, S. and A. Minc, (1978), L'information de la société, La Documentation Française, Paris.

OCDE (1977), La politique de l'innovation dans le secteur des services. Definitions et profil des facteurs fonctionnels, Roneo, Paris.

OCDE (1985), Perspectivas del empleo. Ministerio de Trabajo y Seguridad Social, Madrid.

OCDE, (1986), Restructuration Regionale, Paris.

OCDE, (1987), Crecimiento del empleo y cambio estructural, Ministerio de Trabajo y Seguridad Social, Madrid.

Ochel, W. and M. Wegner, (1987), Service Economies in Europe, London, Pinter Publishers.

Petit, P. (1984), Automatisation des Services: le Cas des Services Bancaires, CEPREMAP, Paris.

Petit, P. (1987), El crecimiento lento y la economía de los servicios,

Ministerio de Trabajo y Seguridad Social, Madrid.

Petit, P. (1988), Tertiarisation, croissance et emploi: quelles nouvelles logiques?, Revue D'Economie Industrielle, I Trimestre.

Piore, M. and Sabel, C. (1985), The Second Industrial Divide, New York.

Planque, B. (1982), Le Development Decentralisé, Litec, Paris.

Planque, B. (1983), Innovation et Development Regional, Economica, Paris.

Rada, J.F. (1984), Advanced Technologies and Development: Are Conventional Ideas about Comparative Advantage Obsolete? Trade and Development, no. 5, Geneva.

Rada, J.F. (1985), The International Division of Labour and Technology, Geneva.

Rada, J.F. (1987), Information Technology and Services, in: Giarini, O. (ed.), One Emerging Service Economy, Oxford, Pergamon Press.

Ray, G.F. (1986), Productivity in Services, National Institute Economic Review, February.

Revell, J. (1986), Effects of New Technology on the Operations of Financial Institutions, The World of Banking.

Rothwell, R. and W. Zegveld, (1981), Industrial Innovation and Public Policy: Preparing for the 1980s and the 1990s, Frances Pinter, London.

Ruyssen, O. (1988), Nouveaux services et renouveau des services, Revue D'Economie Industrielle, I Trimestre.

Sabolo, Y. (1975), The Service Industries, OIT, Geneva.

Sema-Metra Conseil (1986), Services to the Manufacturing Sector: A Long-Term Investigation, Paris.

Shelp, R.K. (1981), Beyond Industrialization: Ascendancy of the Global Service Economy, Praeger, New York.

Shelp, R.K. (1985a), Entrepreneurship in the Information Society, Heritage Foundation.

Shelp, R.K. (1985b), Tecnología de servicios y desarrollo económico, Perspectivas Económicas, no. 4.

Shelp, R.K. et al. (1984), Service Industries and Economic Development: Case Studies in Technology, Praeger, New York.

Singleman, J. (1979), From Agriculture to Services, Sage Publications, Beverly Hills.

Skinner, W. et al. (1982), The Impacts of New Technology: People and Organizations in Service Industries, New York and Oxford.

Smith, A.D. (1972), The Measurement and Interpretation of Service Output Changes, National Economic Development Office, London.

Soete, L. and C. Freeman, (1987), Nuevas tecnologías, inversión y crecimiento del empleo, in OCDE.

Spero, J.E. (1985), International Trade and the Information Revolution, Harvard University Press.

Stanback, T.M. (1980), Understanding the Service Economy, John Hopkins University Press, Baltimore.

Stanback, T.M. et al. (1981), Services: The New Economy, Allanheld & Osmun, New York.

Stigler, G.J. (1951), The Division of Labor is Limited by the Extent of the Market, Journal of Political Economy, no. 59.

Stigler, G.J. (1965), Essays in the History of Economics, Chicago University Press.

Stöhr, W. (1988), La dimensión espacial de la política tecnológica, Papeles de Economía Española, no. 35.

Sweeney, G.P. (1988), Innovación, tecnología y reorientación del desarrollo regional, Papeles de Economía Española, no. 35.

Thwaites, A. and R. Oakey, (1985), The Regional Economic Impact of Technological Change, Frances Pinter, London.

Wadley, D. (1988), Estrategias de desarrollo regional, Papeles de Economía Española, no. 35.

Williamson, O.E. (1981), Emergence of the Visible Hand, Harvard University Press.

9 Knowledge technologies and the evolution of the service sector

T. R. LAKSHMANAN

1. Introduction

Services have replaced production of goods as the dominant mode of economic activity in the industrialized affluent economies. The 'typical' picture of a modern economy is not one of factories engaged in the fabrication of goods but that of a society engaged in the production and consumption of services - a society of dining, of travelling, of learning, of medical visits and healing, of business and legal consultancies, of transmission and reception of various messages and of leisure activities.

While the nature and dimensions of the changeover to an economy of service activities (that currently account for the greater portion of the national product in advanced societies) is beginning to be noticed, the vitality and dynamism of the service sector has received little, if any attention, in the literature (barring a few exceptions such as Fuchs, 1968; Gershuny, 1978; Stanbeck et al., 1981). This neglect of the service sector reflects a long-term bias against services among economic analysts of various stripes from Adam Smith through Marx. Veblen, to Baumol (1967), and Walker (1985). Adam Smith viewed services as unproductive. Marx adopts a similar position when he relegates into the 'superstructure' all activities which are remote from material production. Also, he suggests:

> Services are consumed the moment they are produced.... (services) do not exist as a utility different from the process, a use-thing which does not function as an article of commerce, does not circulate as a commodity after it has been produced. (cited by Burger, 1970, p. 74).

Further, there have been arguments that in the service sector scale economies are limited, substitution of captial for labor is not possible in many cases and that there is a 'cost disease' (Baumol, 1967).

Such points of view have discouraged serious analyses of service production as well as the development of statistical concepts and measurement so that the available picture of the service sector is somewhat sketchy and fragmented. The result has been a general neglect of the vitality and dynamism of the service sector.

A fundamental characteristic of most services is the participation and modification of the consumer in the service production. Consequently, the efficiency and quality of the service delivery depends to a great degree on the quality of all agents -producer and consumer- participating in the production. To facilitate such joint production, the content of work is transmitted among the production agents as information on the purposes and modalities of the service rendered. Such information in general (acquired or accumulated) together with the information flows among various economic actors and institutions performs for the services the same role that raw materials perform for goods production. The economy is an organization for linking on the one hand, individuals by interaction through technical processes of production and distribution and on the other, by communication networks of information flow necessary for coordination and control among the individuals. Gradually over time, there has been a qualitative change, in which the cerebral labor (of design, coordination and control) has become more significant than manual or physical labor as a key factor of production.

The objective of this paper is to conceptualize the role that information and knowledge technologies may play in the evolution of the service sector. It is part of a large multi-dimensional study underway on the dynamic transformation of the service sector in the industrialized affluent societies. The paper opens in Section II with a survey of the major characteristics of the service sector. It focuses on the central attribute of the service activity, namely, that consumers of the service are intrinsic participants in the production process. To produce a service, the consumer and the producer must interact for the delivery of the service to be complete. Thus the context within which the knowledge technologies of service production are brought to bear on the problems of the consumer is essentially a process of transaction between the producer and consumer. This transactional process involves not only the exchange of information between the producer and consumer, but also the utilization of the consumer in the production of service output.

Service activities differ from one another in the nature of their producer-consumer interface, specifically in the levels and types of information exchanged and the degree and type of participation of the consumer in service production. Indeed, we present a classification of service sector activities into four groups, based on the nature of this producer-consumer interface, which affects the context and scope of the technology of service production. Since each category of service activity differs from the other in the quality and quantity of information flow, the type of knowledge technologies available and the environment of service production, the classification can provide clues to the likely future technological behavior of these sectors. In particular, how can the development in information technologies (IT) be exploited to improve the efficiency and quality of information generation and use in service activities, thereby, improving the productivity of service activities? We report briefly on the dimensions of change in these four service sectors in the U.S. in the decade of industrial structuring following the energy price shocks of 1973.

We proceed in Section III to conceptualize the nature of relationships between knowledge, technology and transformation of the service sector from the perspective of technical change as an evolutionary process. We view the emerging knowledge technologies as a social-technical system. Such a system comprises of two parts: a set of **intrinsic** elements that indicate its new possibilities and qualities of experience and a set of **contingent** elements of the system that define the individual, institutional and macro responses to the new possibilities. We suggest that the process of interaction between these two elements of the system, as shaped by the rise of the new communication infrastructure leads to the patterns of diffusion of technical change in the service sector (Lakshmanan, 1987).

We suggest further that a variety of structures that facilitate diffusion of technical changes develop out of this interactive process. Such structures take the form of (a) a variety of (private and public) producer and consumer services (to reduce transactional and adjustment costs) and (b) of some institutional innovations. We conclude by linking such diffusion structures to different phases of progressive evolution of the service sector production technologies. For example, the first class of changes may pertain to more efficient ways of carrying out older production tasks as in the use of peripheral activities such as automated personnel, billing and financial systems. A second level of change pertains to the change in the core technologies of service production through the use of IT.

2. Technologies in the Service Sector

2.1 Service sector technologies: some highlights

The enormous diversity of activities ecompassed in the service sector has often led to loose definitions and ascription of attributes to the sector that are both misleading and analytically unhelpful. Typical of such characteristics attributed to the sector as a whole are its low rates of productivity growth, its labor intensity, low capital intensity and its major role in the slowdown in productivity in the overall economy in the last 15 years.[1] Since these characterizations are not accurate, we engage in a definitional digression.

The first attribute of service activities is that they involve a change in (or for) the consumers or the goods they possess. Three types of such changes in the asset structure of the consumer during service production can be noted (Garn et al., 1976). In one type, there is a change in the tangible asset structure of the consumer and provider of services (e.g., retail services, repair and maintenance services for durable goods, economic development activities). In a second type of service activities, there are changes in the human capital of the consumers (e.g., education and training, counseling and health care services). The third type of activities include primary brokerage activities (e.g., referral programs and real estate sales) where the output appears as change in the personal assets of clients who are providers and clients who are cosumers of the service. However, a few services involve change neither in the consumer nor in their goods. For example, rental services allow persons to make use of goods without acquiring them - a change in accessibility created for the consumer without ownership of the asset providing the accessibility. Thus the change process involved is **in** or **for** the consumer and confers on him or her **time, place, and form utility** (Riddle, 1985).

The second general attribute of service activity is that consumers are intrinsic participants in the production process. This distinguishes the production of goods from that of services. In goods production, resources -labour, capital, energy and materials- are combined. (Figure 1).

---

[1] An analysis of 145 industry groups in terms of capital stock per work hour shows that the service industry groups formed nearly one-half of the 30 groups in the top 2 deciles and the bottom 3 deciles of industry ranking by capital intensity did not contain any service industries (Kutscher and Mark, 1983). Further the range of productivity growth noted in the service sector is not significantly different from the range among goods producing sectors (Kutscher and Mark, 1983). The growth of services cannot possibly be the cause of the recent slowdown in productivity in the larger economy (Fuchs, 1985).

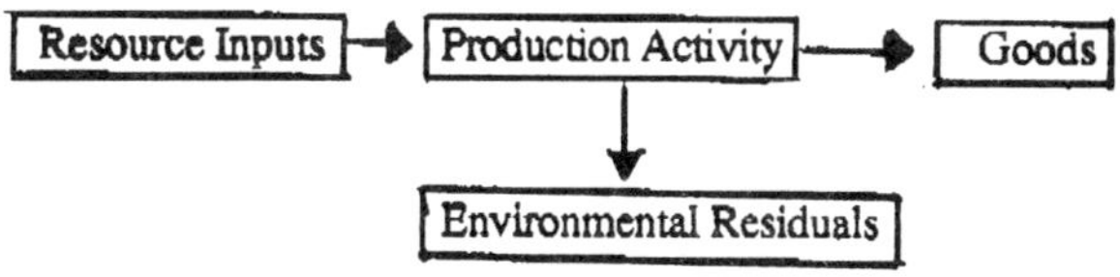

Figure 1 Production of Goods

A production unit is subject to a variety of opportunities and constraints on its actions. However, the transaction between the producer and consumer will not directly affect the level of output. The asset positions of the seller and buyer are altered during sales but the output is not affected by the attitude or behavior of the seller or buyer. This enables the analyst to view goods producing units as self-contained units for analysis of efficiency.

This is not the case with production of services (Figure 2), where

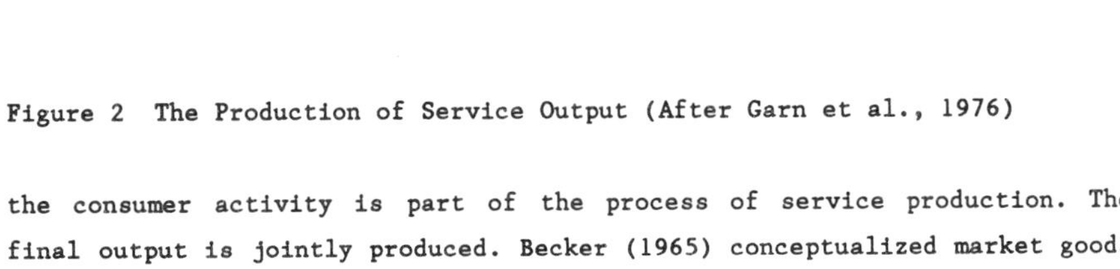

Figure 2 The Production of Service Output (After Garn et al., 1976)

the consumer activity is part of the process of service production. The final output is jointly produced. Becker (1965) conceptualized market goods as intermediate goods that individuals combine with their time and skills to produce final products or commodities (e.g., meals, entertainment, recreation, etc.) they consume. Becker's notions have been extended by social indicator researchers (Garn et al., 1976) so as to view consumers as producers in the realm of service production. For example, the provision of quality health services requires the participation of the patient in the form of producing accurate medical histories, cooperating in the treatment etc. In other words, the efficiency of the service production cannot be solely attributed to the activities of the producer. The consumer's actions bear upon the efficiency and effectiveness of the services.

We briefly characterize the nature of service provider - consumer interface along two dimensions. First, in terms of the four **functions** the service client can perform (Normann, 1984):

- **Specification** of the service: e.g., by participation in the diagnosis of the problems by providing health data, specifying items in a menu of services that are desired as in a restaurant, etc.
- **Self Service:** where the consumer is used as primary input without interaction with the producer (e.g., retailing, automatic banking, etc.). The idea is to reduce the labor costs of service.
- **Coproduction:** where the client carries out part of the production (e.g., client's staff doing data collection and some analysis for market analysis; residents participating in housing services delivery; the student contribution to education, etc.).
- **Development** of the services in a professional organization (e.g., in advertising, auditing, consulting).

Third, service technologies are typically 'knowledge technologies' that are defined as 'a set of ideas that comprise of the goals for work, its functional importance and the rationale for the methods employed' (Thompson, 1967; Dubin, 1968). By contrast, the typical features of manufacturing technologies associated with materials handling, tool guidance, and energy supply are functions of equipping and sequencing of workflow activities. The typical organization of good production is a combination of mental labor, machinery, and infrastructure in a linear conversion of raw materials passing through into output - a process often described as a Fordist labor process.

Fourth, an open-system logic that is consistent with the transaction process is vital so that there is considerable workflow and task uncertainty (Mills and Moburg, 1980). Different services handle these uncertainties in various ways: by altering the type, quality, generality and instrumentality of information exchanged. In some cases, the conversion process can be routinized, in others, the consumer can be restricted to a discrete set of options. In still others, it could be complex, ill-defined strategic technologies. All of these variations provide a continuum of service sector technologies: at one end, restricted service model - narrow technical repertoire, high degree of routinization, low levels of professionalism and the technical core of service productions fully insulated from the consumer (as in goods production) (e.g., fast food restaurants, credit bureaus, department stores). At the other end are activities where service workshops have high level of expertise, high levels of non-routine activities - both as inputs to service production, and in the process of service delivery (e.g., doctors, university educators, etc.).

Fifth, the context within which technologies are brought to bear on consumers' problems is also different in a service production. The intangibility of service outputs and the transactional nature of the service production leads to situations where the quality of the service delivery is a key component of the service output. Part of the service production technology is knowledge technology that assists the worker in these transactions with the consumer (in some cases, the service provider may attempt to help the consumer in socially reconstructing perceptions of the value of service outputs!).

Given such differences between the technologies in goods production and services production, there is need for caution in the use of common measures of inputs and outputs and related measures of productivity among the two sectors. Second and more important, is the need to develop measures of service technology that relfect the different quantities, levels of generality and instrumentability employed at the producer-consumer interface. Since the service activities are often consumer-intensive, better measures of this type can be helpful in organizing the information exchange at the producer/consumer interface as a way to enhance productivity. We explore the nature of this information exchange in a broad framework in the next part of the paper. Prior to that, we highlight below some selected findings on the service sector from ongoing analysis of a large database on the U.S. households between 1973 and 1984 (Lakshmanan, 1988).

2.2. Changing patterns of service technologies in the U.S.

Given these patterns of producer-consumer interface and our interest in service technologies, we proceed to a classification of service activities that highlights the differences among the activities in the **type, generality**, and **instrumentality of information exchanged** at the producer-consumer interface during service production. Such a classification will throw light on the relative potential of information technologies and their diffusion in various sectors. We present now such a classification that may be appropriate for analyzing technical change. It builds on the ideas of

Mills and Moburg (1980) and Baumol (1985) and is based on the nature and intensity of producer-consumer contact and the types of skills and knowledge brought to the service by the clients.

We recognize four classes of services:

(i) Quasi-industrial Services (QI)

(ii) Routine Interactive Services (RI)

(iii) Task Interactive Services (TI)

(iv) Personal Interactive Services (PI)

(i) Quasi-industrial Services (QI)

The term Quasi-industrial Services refers to a class of services whose technologies are closest to those of goods production. They are complementary to goods production and include the greater part of the Browning and Singlemann (1975) category of distributive services. Examples would be transportation, communications and wholesale activities. Such organizations operate in stable environments and the information resources exchanged at the producer-consumer interface are direct and simple. Since the consumer's demands are known and the producer can develop technologies that are amenable to economies of scale and considerable capital/labor substitution. New physical technologies for materials assembly, and digital communication and intelligence information technologies for automated warehousing operations are broadly applicable and hence the appellation Quasi-industrial for these services.

(ii) Routine Interactive Services (RI)

In these activities, the direct interaction between the service employee and the consumer is limited and the information transmitted is relatively limited. However, that information input is vital to the production function since the customer knows what he or she wants. An example of this is the interaction between a bank teller and a customer. Examples of such activities are retail outlets, fast food restaurants, banks, etc.

These activities occur in environments with low uncertainty. The customer needs and production technologies are well known. The service provider is less involved withproduction and more in dispensing them. Specifying in advance responses to consumer demands, there is a tendency to standardize activities and set up large-scale operations, thereby increasing productivity. The consumer may be incorporated in the self service mode and the information exchanged is simple in nature. This group includes services using self service e.g., retail and personal services amenable to 'industrialization' (Levitt, 1976) e.g., fast food restaurants. Customer involvement is minimal and is often completely removed in some predictable environments by employing appropriate equipment (ATM's in banks).

(iii) Task Interactive Services (TI)

The term task interactive pertains to the technologies in task accomplishment that are paramount in these activities. Examples are legal firms, engineering and architectural firms and marketing and advertising organizations. The focus is on the encounter between the provider of the service and the consumer and the tasks or skills needed to provide the services (Mills, 1986). The firms in this sector possess a body of knowledge that is relevant to the problem to be solved.

Moderate to high degree of uncertainty characterize the environment of these activities. The technologies to solve these task-oriented problems is known, but large amounts of information and efforts are expended before the customer's needs are known. The consumer is aware of his problem but looks to the expertise of the provider. Further, the uncertainty exists because of the uniqeness of the client's needs and the contingencies in specific situations, which may prevent the imposition of precise standards on service performance. Since TI service environments are inherently uncertain, the producer-consumer interface tends to be concentrated and periodic. If a client requires litigation, the lawyer would work out the feasibility of the solution, though the outcome may be beyond his control. Thus there may be much uncertainty in such services.

(iv) Personal Interactive Services (PI)

The environments on these services are both dynamic and complex. The consumers are unaware or imprecise about their problems or solutions. Thus the service producer and consumer jointly strive to diagnose the problem. The body of knowledge that exists is such that the cause-effect relationship between what the service producer does and the outcome of the action is not sound. It is in this category that 'professionals' and professional organizations originally develop. Examples of such services are health care, phychological consulting, welfare agencies, etc.

These activities operate in dynamic and complex environments. The complexity derives from the environmental uncertainty since consumers are often imprecise or unaware of their problems (e.g., health provision, welfare servicese, etc.). The health professional will utilize considerable knowledge technology in diagnosing the problem. The consumer participates in this diagnosis by providing information. Each service episode may refer to an entirely different set of problems so that the diagnostic process is continuous. Further, the knowledge technology used to address a problem may have limited predictive power. For example, a welfare worker's efforts at assisting a client to be self-sufficient may not work. Thus there is much scope for creativity.

What follows now is a selective brief summary of a separate ongoing statistical analysis of the dimensions of change in these four service sectors in the U.S. by census region and SMSA (Lakshmanan, 1988c).

The growth of the service employment and the four component sectors between 1973 and 1984 is evident in Table 1. Only the quasi industrial service, comprising of transportation, communication, and wholesaling grows slower (29%) than the sector as a whole (44%). All sectors that involve some level of producer - consumer interaction grew faster. The sectors that grew fastest are those requiring the minimal routine interactions (routine interactive - Q.I.) and those requiring the most intense and complex interactions (personal interactive - P.I.) services. These are also the two sectors where female workers have outnumbered male workers and the pace of feminization is quickening. The task interactive services maintain more or less the same level of male-female ratio. The quasi industrial services, like the manufacturing sectors are still largely male preservers, though increasingly less so.

By contrast, the average wages (in 1984) show a slight decline between 1973 and 1984 both in services and manufacturing sectors. This slight decline in average wage in services reflects the 10% drop in wages in the relatively low skill routine interactive services and very slight increases in the other three sectors. Female workers experience wage increases in all sectors, though from a smaller initial wage. Male wages drop both in R.I. and T.I. services and slightly increase elsewhere. The overall drop in service sector wage thus reflects the increasing participation of lower paid (but relatively improving wage wise) female workers and the drop in male wages.

Summing up, in the decade of rapid industrial restructuring following the energy price shocks of 1973, the service sector in the U.S. has grown, particularly so in 2 component subsectors: (a) that are most amenable to standardization and minimal and routine customer information exchange (routine interactive) and (b) the dynamic and complex personal interactive services (P.I.) where the potential for application of a variety of knowledge technologies is maximal. Second, these are also the traditionally dominant female worker sectors that are becoming more so. Third, the average real wage has dropped in this period, reflecting the 10% drop in the wages

Table 1

Employment and Wages in Services and Manufacturing Sectors in U. S., 1973 and 1984 (all workers)*

Sector	Employment (in 1,000)		% Change	Wages (in 1984 dollars)		% Change
	1973	1984	1973-1984	1973	1984	1973-1984
Quasi Industrial	5,124	6,596	29	$13,567	$13,897	2
Routine Interactive	16,949	25,408	50	9,166	8,244	-10
Task Interactive	9,300	13,477	45	8,067	8,277	2
Personal Interactive	16,541	24,359	47	10,672	11,053	4
All Services	47,756	68,768	44	9,940	9,721	-2
All Manufacturing	19,729	21,315	8	11,981	11,694	-2

* Computed from census data on households

in routine interactive services. Another way to view the situation is that this reflects the increasing participation of lower paid female workers and the dropping male wages.

3. Knowledge, Technology and Transformation of Services: a Conceptualization

Technical change is a very complex social process, which covers a continuum of inventive efforts stretching from initial conceptualization, to establishment of technical feasibility (invention), to commercial feasibility (innovation), and finally to the subsequent diffusion into the larger economy.

The theoretical approach adopted in this paper, is to view technical change as a process of trial and error, as the accumulation of small, often, individually minor improvements of the production process. To paraphrase Braudel (1985, 1986) between the **inventio**, which represents the technically possible, and **usurpatio** (application) there is often a significant interval when for various reasons - economic, geographic, social of psychological- society has not yet attained the necessary degree of receptivity. Thus technology change and diffusion is a historical process, a non-linear process in which there is no single onward movement, but many actions and reactions, many changes in gear. New possibilities and qualities of experience thrown into relief by a major new technology are engaged interactively by the often-conflicting economic, social and political interests to produce a technical 'choice' or outcome. It is this interactive process between the **intrinsic** attributes and qualities of experience of a new technology and the **contingent** individual, institutional, and macro response of society that we suggest leads to the evolution of what may be termed as a new socio-technical system. We delineate here the elements of such a socio-technical system in the case of the emerging intelligence and knowledge technologies.

The contemporary stage of technical evolution is characterized by introduction into the various activities of the socio-technical system, enlarged capacities to substitute, perfect and amplify human information processing capacities - in the production and consumption of both goods and services. These enhanced technological capacities for distribution of information and knowledge quickens the tempo of creation of knowledge, which in turn is embodied in further technology and so on. We elaborate this interactive process of knowledge generation and technical evoluation particularaly as shaped by the rise and evolution of the new communication technologies which are greatly extending information (text, video, voice, and data) across space and time with great speed, permanency and fidelity.

Next, we attempt to sketch the intrinsic attributes of the emerging knowledge technologies in this framework. We follow this by identifying the contingent aspects of or the nature of socioeconomic responses to knowledge technologies at the micro, meso, and macro levels.

Then, we suggest that a variety of structures for technical diffusion emerge from the interaction between the intrinsic and contingent attributes of the knowledge technologies. These structures take the form of a variety of (private and public) producer and consumer services, that facilitate the transaction processes and adjustment processes in the knowledge technology system. Such services (e.g., management, consulting, engineering, organizational, etc.) facilitate the introduction of newer knowledge technology into the various economic activities. We conclude this section by linking services for diffusion of technology to different phases of evolution of the service sector production technologies.

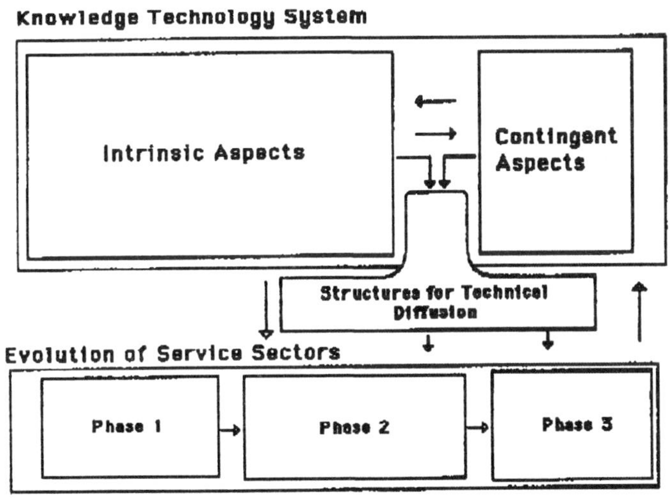

Fig. 3 An abbreviated model of the evolution of the service sectors

Figure 1 provides a broad outline of such a model of the role that knowledge technology system plays in the evolution of the service sector. We elaborate on the content and scope of this model in the rest of this section of the paper.

3.1. Technical change as an evolutionary process

Theoretical inquiry into the process of technical change evidences two approaches (Elster, 1985). The first view is one of technical change as a rational choice of the best innovation among a set of feasible changes in technology. Neoclassical and Marxian theories of technical change adopt this rational actor approach, although they impute different goals to the technological entrepreneur. In neoclassical theory, profit maximization is the object and in Marxian view innovation is power in the class struggle. Schumpeter goes beyond the straightjacket of profit maximization and focuses on creativity, disequilibrium and the dynamic character of capitalism. All these formulations, however, focus on the earlier part of technical change processes, specifically from conceptualization of an idea to its innovation.

The second approach looks at this process more completely, including the spread and diffusion of innovations into the larger economy. This approach is the 'evolutionary' approach that view the technology diffusion process historically and focuses on the factors that affect the pattern and spread and diffusion of innovations in society (Usher, 1954; Rosenberg, 1976; David, 1975; Nelson and Winter, 1982; Braudel, 1985, 1986).

We characterize the emerging information technology as a knowledge technology system comprising of two components: (i) a technology-economic element whose **intrinsic** characteristics typify the newly possible innovations and (ii) a social-institutional action subsystem where the adjustments, responses, decisions, and actions on the new technology are made **contingently** determining thereby the level, the speed and pathways of wider technology diffusion.

While these **intrinsic** properties (e.g., flexible, non-linear growth, increasing intellectualization, non-hierarchical diffusion of knowledge, etc.) of the new knowledge technologies have the potential for opening the door to a new future, intentions do not determine outcomes. As the new technology defines the new ceiling of the possible, social opportunities for choice are multiplied. The imperatives of the new technology have to contend with forces that influence human choice among alternative pathways of transformation to the new 'paradigm' of work and lifestyle. These latter forces derive from the interplay of economic, social and political interests represented by the inherited socioeconomic structures. Such inherited structures have evolved over a previous period time in response to the requirements of earlier techno-socioeconomic paradigm (e.g., based on mass production of low cost energy intensive materials with a linear (Fordist) and hierarchical production process) and are very likely to exhibit properties that are not very receptive to a new technology (Blackburn, Coombs and Green, 1985). Such properties of the decision - action system that act as barriers to change are the **contingent** attributes of the new technology.

The intrinsic dynamism of the new technology comes up against the barriers of the prior action system, so that the exact outcome cannot be determined in advance. To understand how technical change takes place, it is necessary to grasp not only the **intrinsic** and the **contingent** aspects of technology, but also their interrelations, since the final outcome results from the interaction between these two dimensions of change. The action system of the inherited structre operates at the individual (micro), institutional (meso), and social (macro) levels - each level forming a hierarchy inthe formal sense, with each higher level being made up of systems from lower level (Figure 4). The diffusion of the intrinsic attributes of the new technology is contingent on whether complementary, mutually supporting economic, and social adjustments are made at all three levels. If such innovations include new management approaches, new institutions, new sets of incentives to promote mobility in factor and geographic space, newer work organizations, supporting physical infra-structures and the like, a 'virtuous circle' (in contrast to 'vicious circle') results. Wide diffusion of the new technology takes place.[2]

3.2. Knowledge, technology, and communication infrastructure

If the technologies in the service sector represent a systematic exploitation of intelligence and knowledge, insights into their further evolution would require better understanding of the interactions between intelligence, knowledge and technology. We attempt here to sketch briefly some recent research findings on such interactions, in order to structure our inquiry into service sector evolution (Carpenter, 1974, 1978; Bell, 1979; Innis, 1951; Mitcham and Huning, 1986; Levinson, 1986, 1988; Rophol,

---

[2] Such a view of cumulative causation has been elaborated in structural terms by the French Regulation School (e.g., Aglietta, 1979). This school identifies a structural whole of economic, social, institutional and ideological arrangements that can be labelled as an 'accumulation regime' characterized by a specific mode of economic organization and social regulation. A succession of such regimes characterizes the development of capitalism: examples of such changes with associated shifts in the space economy are the heavy industry at the turn of the century, the mass production of consumer durables between the 1920s and 1960s.

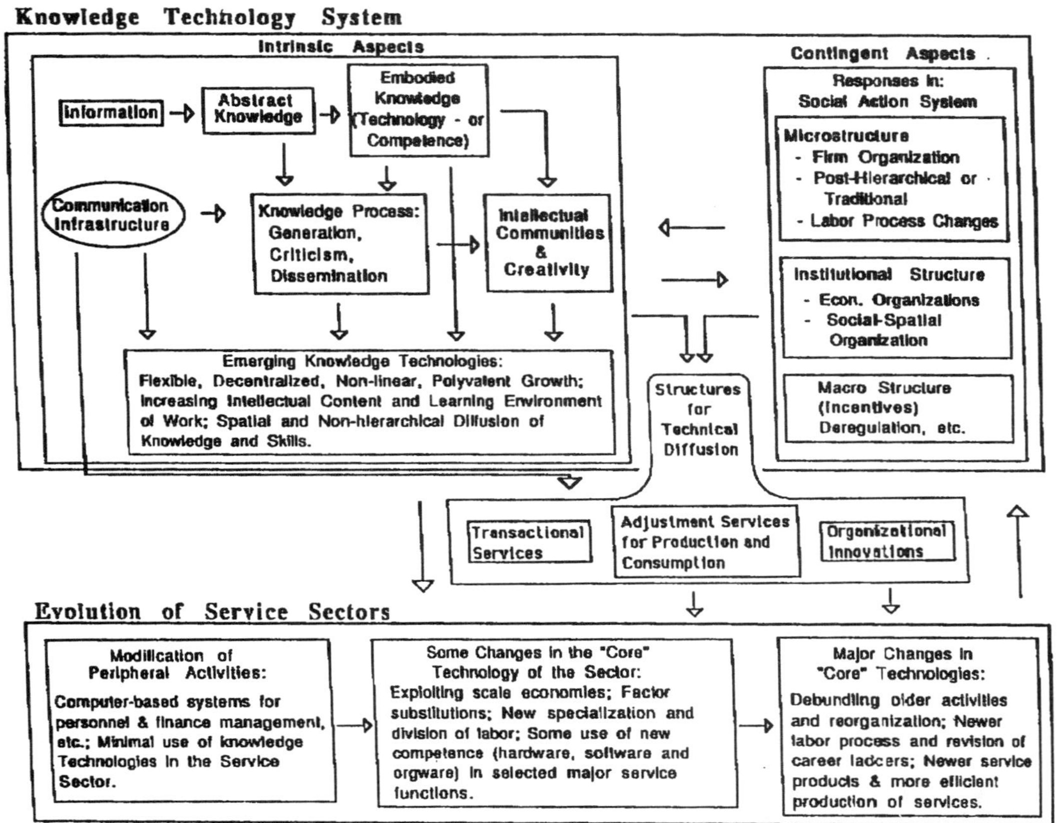

Fig. 4 A MODEL OF THE ROLE OF KNOWLEDGE TECHNOLOGIES IN THE EVOLUTION OF SERVICE SECTORS

1982, 1986).

Knowledge originates through structuring intelligence obtained from the external world. Technology embodies such knowledge in a variety of forms: hardware, software or orgware. It is possible to visualize a continuum of technology that begins with artisan skills at one extreme and progresses through greater and greater systemization and formalization of technique to high technologies. The continuum must be viewed as a progression from primarily prescriptive rules toward increasingly descriptive statements which are themselves ultimately systemized within a comprehensive scientific theory (Carpenter, 1979).

The progressive rationalization of technology in recent times is achieved through an increasing dependence on discursive knowledge. In scientific practice and knowledge generation two features -the manipulation of the physical environment as a new path to knowledge (vita activa), and the ability to abstract from experience- have been crucial in the last 3 centuries. In the last century, such scientific knowledge has been used especially in chemical and electrical industries to guide rational manipulation for practical purposes, (in other words, technological developments). Thus, substantive and methodological insights of scientific knowledge have not only led to useful physical technologies and goods such as autos, air planes, nuclear energy, lasers and the like but also novel commodities such as information, that generate and permit new demands for services such as management and control.

If knowledge supports the development of technology in this manner, the reverse is also true. Technologies not only embody knowledge but oftenhelp generate and disseminate further knowledge by their operations. There are many technologies that are directed not so much to reshaping the world but rather toward expanding the knowledge, which when embodied in other technologies reshape the world and so on. If technology is embodied knowledge, technology applied to the pursuit of knowledge is a second-order or metatechnology, namely technology working toward its own creation (Levinsson, 1986). If first order or general human technology modifies the external, physical world, this second order technology transforms the internal, mental world of our intellect and knowledge.

Technology contributes to the growth of knowledge in three ways (Levinson, 1988):

- As embodied knowledge, technologies contribute to knowledge growth by being durable manifestations and workable implementations of knowledge, in other words as unintended libraries of knowledge that have survived some encounter with the environment.

- Perception enhancing technologies. Telescopes and microscopes are such examples that expand the sensory experience, transform the infinite and the infinitesimal to human scales, thus providing intelligence not accessible to our senses. Such intelligence is structured by our cognitive faculties into useful knowledge.
- Cognition enhancing technologies. Computer greatly speed the sorting and classification of numerical information (in much the same way numbers do for real experience). By vastly expanding the quantity of data that can be handled, bigger chunks and smaller details of the external world are opened to human cognition. 'Computers have signifincantly expanded the domain of tractable complexity... (making possible extensive exploration, validation and refinements that would have been literally impossible' (Baker, 1977).

The latter two categories are the meta cognitive devices since their operation corresponds to the key phase of knowledge generation process. The three stages of this latter process are **generation, criticism,** and **dissemination of knowledge.** The second and third class of meta cognitive technologies listed above facilitate each of these three stages of knowledge generation process. Technology provides the tools used in knowledge generation process. Knowledge develops through the sifting of ideas through criticism and testing, and the dissemination and preservation of ideas that have stood such tests. While the initial generation of an idea may be a solitary activity, criticism and dissemination are social or communicative. Such social activities as criticism and dissemination require technological assistance in terms of capacity to transfer and exchange information between and among individuals. Such capacities are provided by communications media (Figure 2).

The primary goal of communications media is not the production of representations of reality, but the transmission of such representations, regardless of their truthfulness or correspondence to reality to as many people as possible as accurately as possible (Innis, 1950, 1951; McLuhan, 1964; Levinson, 1988). These two objectives -plurality and accuracy- have been sought through the ages by communications media. Unlike the earlier media of speech and writing, which required abstraction, recent developments in communications have permitted plurality, high degree of realism, and permanency in transmission of ideas. Developments in electricity, electronics, photography and computing in recent decades facilitate the transfer and permanent 'record' of information, with high degree of realism, and at the speed of light.

The new communications infrastructure reflecting a blending of computing and telecommunications not only sends information across space, but also retains or preserves sound, motion and color. The traditional goals of extensions across space and time and fidelity to the original are being realized. The speed, permanency, and realism with which initial ideas and observations are transmitted from one to many individuals defines a **new context** for criticism and dissemination of knowledge - thereby influencing the number and variety of minds, both across space at one time and through time, that can facilitate the development of these ideas and the eventual utility of such social cognitive efforts (Levinson, 1988). The distribution of information by the rapidly evolving communications infrastructure is thus as central to the growth of knowledge as in the technological enhancement of perception and mental processing of the two cognitive metatechnologies discussed earlier. The computer as part of the communication infrastructure (as disseminator of printed words and data) is as revolutionary of communications as it is for primary cognition.

There is another remarkable novel development in that th einformation technologies transmitted over the telecommunications infrastructure affect further development of knowledge and the evolution of economic activities based on such technologies. First, the ISDN (Integrated Services Digital Network) not only enhances our cognitive processes as suggested above, but also **externalizes** such cognitive processes. Levinson (1988) claims that the potential such technologies offer in terms of juxtaposing or sequencing (text, video and voice) information across time and space can stimulate knowledge development processes in novel ways.

3.3. Intrinsic and contingent aspects of knowledge technologies

The evolutionary view of knowledge technologies presented so far from the recent literature on philosophy of technology, has its echo in the regional science literature.

Anderson and Johanssen (1984) exploring the bases of knowledge driven economic growth make an important contribution in this regard by recognizing four types of knowledge and production. These types, in increasing order of complexity, are:

1. **Information** (or data): elemental units with a limited structure that can be aggregated, disaggregated and transmitted.
2. **Knowledge**: structurally ordered information.

3. **Competence:** which is knowledge embodied within hardware (machinery, instruments, etc.) software and orgware (institutions, organizations, etc.) and,
4. **Creativity:** which is the concept of the highest order, implying a capacity to order and reorder information with the aid of a knowledge system.

We have already discussed the progression from information or intelligence on the external world to abstract knowledge and to its eventual embodiment as competence.

Knowledge is of two types: the first is the facts of the domain - the widely shared knowledge obtained from books and journals. The second is the heuristic knowledge, which is obtained from experience, good practice and good judgement in the field, getting a feel for the problem, learning when to go by the book and when to skip the rules. In this manner, a repertory of rules of thumb are 'heuristics', that combined with the domain of knowledge, produces experts.

Another perspective on knowledge would distinguish between generality, power, and instrumentality. Generality an be viewed as the 'number' of the aspect of a solution. It indicates the **scope** of a solution or the class of problems that can be handled by that solution (Malvey, 1985). Power is the 'size' quality of the problem, the degree of fit or explanation achieved by a solution. Only for the situation of a near-perfect knowledge can we have both general and powerful solutions - mostly narrow, technical problems. At the other end are broad problems where the 'solutions' have low power and low generality.

Up until very recent decades, the nature of technology was such as to promote such exchanges within small geographic areas, to specific physical locales. The more profound developments of such locale-based creative developments are associated with intellectual communities such as the Vienna Circle, and the Bloomsbury Group. The special circumstances for intellectual incubation (that comprises the initiating thinking, creative community or 'localized collective mind') have operated in specific physical locales.

The development of the apparently complementary activities of jet plane travel and teleconferencing promises to stretch the operative intellectual circle to a larger geography, possibly global proportions. In other words, the number and variety of minds that can be marshalled to nurse an idea through early growth can increase enormously. Kantorovich (1983) claims that our more significant theories are 'collective world pictures', resulting from pooling mental abilities beyond an individual's abilities. Under the new realities of communication infrastructure, with spatial separation or distribution of intelligence, knowledge, and processing, knowledge production also becomes a world production. Very likely, the number of such intellectual or creative communities are likely to proliferate.

The operation of these processes is leading to the creation of information technologies that have the following inherent or intrinsic characteristics:

- The most important aspect is flexibility. Potential regarding both size and capacity to adapt product or service mix to changing market patterns is great. The change toward the 'economics of scope' reflects the trend toward input-adaptable, market-adaptable, flexible technologies.
- The arrival of spatially organized information networks permits decentralized information processing that permits debundling an economic activity into separate components that can be later reintegrated.[3]
- The nature of these technologies require much greater interactive processes within firms and with customers. This creates the need for knowledge at various levels in a non-hierarchical manner and new styles of work and learning experiences.

The response to these intrinsic attributes in the action system occurs at various levels. At the micro level, the firm may respond either in promoting the newer non-hierarchical patterns of intra-firm exchange thereby promoting synergistic growth of the firms' activity, or frustrating such developments by maintaining the older hierarchical structures (Zuboff, 1988). Institutional responses may display similar patterns of response. Since the new flexible technologies relfect a new **relative cost structure**, there is a need to revise the existing overall structure of public and private incentives at the macro level. Developments such as deregulation of some industries, government programs at trade promotion are examples of such positive responses. As noted earlier, the outcome will be contingent on interaction between such intrinsic and action system responses. We now turn to the special structures that grow out of such interactions and facilitate

---

[3] There are instances of American construction companies that allocate their resources efficiently by utilizing American managers, British insurance, Japanese finance, Swedish building materials, Indian engineers and Korean workers in a Saudi Arabian project.

technical diffusion.

3.4. Structures for technical diffusion

The economic activities have been increasing their productivity over the last two decades by significantly extending their markets (e.g, to global markets and new product markets) and by increasing the division of labor. As markets expand and a finer division of labor takes place, two classes of costs are additionally incurred. These are transaction costs and adjustment costs. The emergence of these costs stimulate the generation of new producer services which reduce these costs and thereby augment manufacturing productivity.

Transaction costs represent the costs of brining together the supply and demand sides of an ever increasing market and circuitous production process. Such costs include for example: Transportation and Communication costs, Measurement Costs, Risk Costs, Uncertainty Costs, and Coordination and Control Costs. Corresponding to each type of such transaction costs, a producer service activity develops to lower such costs (Wiliamson, 1908).

Adjustment costs are incurred in increasing the supply of resources that have long gestation periods. Examples of such resources are highly skilled labor and sophisticated capital goods. To reduce these costs, a variety of educational, engineering, consulting, management, and strategic planning services have emerged. Such services play a crucial role in facilitating the transition to higher productivity in the goods sector.

These two classes of services provide a variety of professionals - engineers, consultants, lawyers, accountants, and a variety of newer experts incorporating integrative skills. The latter group are new professionals who develop expertise at the interface of two or three domains, e.g., medicine and information technology, consultant engineering and strategic management, and developers of new multidisciplinary training programs. Such professionals facilitate the interaction between producers of new technology and users of such technologies by being the bridge group producing intermediate services that help the transition to the new.

Another important component is the social development of new innovations in the structure of incentives or institutions. Such innovations change the relative cost picture, more in favor of emerging technologies, and thus promote the new innovation (Lakshmanan, 1986).

3.5. A model of diffusion

We suggest a 3-stage evolution of the service sector under the impact of these service sectors:

1) In the first phase, information technology takes place in the **peripheral** activities of administration and management. For example, in the health sector, the early phase of knowledge technology diffusion was the use of computers to substitute for routine low skilled labor in the activities peripheral to health care (e.g., administrative, payroll, and other financial information activities).

2) In the second phase, a variety of knowledge technologies are implemented or promoted by the transactional and adjustment services. Barriers to general diffusion still remain. In the second phase, two sets of changes take place: a) cost reduction activities that involve cheaper ways of doing older hospital operations, e.g., capturing economies of scale by aggregating individual health providers into group practices, or HMOs, etc.; substituting cheaper factors of production such as nurse labor for doctor's labor, or outpatient care for inpatient care; b) new service technologies, exploiting the emerging knowledge technologies in medicine.

3) In the third phase, major changes in the core technology of service production take place because of prior organizational innovations that permit, over time, debundling older activities and reorganization, and because of newer career ladders and labor processes that promote the development of newer service products and more efficient production of services. This phase has hardly begun in the health care sector. When it happens, the new technologies will be widely disseminated.

We cannot predict the specific patterns of evolution. However, our brief survey of evolving technical diffusion patterns in the U.S. indicate the **plausibility** of this sort of interaction process between the intrinsic and contingent aspects of medical knowledge systems as they are mediated by the evolving structures of diffusion of technical change. We plan to follow through on the ongoing U.S. study by an inquiry into the health care sector in the different institutional context of Sweden.

In the long run, the next level of change involves the exploitation of the network attributes of distributed information systems in the development of new services. This would involve not only debundling and reorganizing existing services (as a form of cost competition) but also the development of entirely new service products (dynamic competition). However, a prerequisite for the latter development is often a set of complementary institutional innovations (Lakshmanan, 1987). Such innovations could be in the social organization of the particular service or in the labor process of the sector. We attempt briefly in this section to link the types of knowledge available through IT and the phases of changes in the technology and organization of service production in the form of a sequence model of service sector evolution.

References

Aglietta, M. (1979), A Theory of Capitalistic Regulation, N.L.B. Publishers, London.

Anderson, A. and Johansson, B. (1984), Knowledge Intensity and Product Cycles in Metropolitan Region, International Institute for Applied Systems Analysis (IIASA), Working Paper, Laxenberg, Austria, 44 pages.

Anderson, J.G. and Jay, S.J. (1987), Use and Impact of Computers in Clinical Medicine, Springer Verlag, London.

Baumol, W. (1985), Productivity Policy and the Service Sector, in: Inman, R.P. (ed.) Managing the Service Economy: Prospects and Problems, pp. 301-317, Cambridge University Press, London.

Blackburn, Ph., Coombs, R. and Green, K. (1985), Technology, Economic Growth and the Labor Process, St. Martin's Press, New York.

Braudel, F. (1985), The Structures of Everyday Life, Perennial Library edition, Harper & Row, New York.

Braudel, F. (1986), The Perspective of the World, Perennial Library edition, Harper & Row, New York.

Browning, H.C. and Singlemann, J. (1975), The Emergence of a Service Society, Springfield, VA, National Technical Information Service.

Carpenter, S.R. (1974), Models of Knowing and Technological Action, Philosophy Today, Summer, pp. 162-68.

Daniels, P.W. (1985), Service Industries: A Geographical Appraisal, London, Methuen.

David, P. (1970), Technical Choice, Innovation and Economic Growth, Cambridge University Press, Cambridge, M.A.

Davis, L.E. and North, D.C. (1971), Institutional Change and American Economics Growth, Cambridge University Press, Cambridge, UK.

Elster, J. (1985), Explaining Technical Change, Cambridge University Press, Cambridge, UK.

Faulhaber, G., Noam E. and Talsey, R. (eds.) (1986), Services in Transition, Ballinger, Cambridge, MA.

Freeman, Ch. (1987), The Case for Technological Determinism, in: Finnegan, R., Salaman, G. and Thompson, K. (eds.) Information Technology: Social Issues, pp. 3-18, Hodder and Stoughton, London.

Fuchs, V. (1968), The Services Economy, National Bureau of National Research and Columbia University Press, New York.

Garn, H.A. (1973), Public Services on the Assemby Line, Evaluation, vol. 1, no. 2.

Garn, H.A., Flex, M.J., Springer, M. and Taylor, J.B. (1976), Models for Indicator Development, Urban Institute, Washington, D.C. (Paper 1206-17).

Gershuny, J.I. (1978), After Industrial Society: The Emerging Self Service Economy, MacMillan, London.

Gillespie, A.E. and Hepworth, M.E. (1986), Telecommunications and Regional Develoment in the Information Society, New Castle Studies of the Information Economy, Working Paper no. 1, University of Newcastle upon Tyne, October.

Holmstrom, B. (1985), The Provision of Services on a Market Economy, in: Inman, R.P. (ed.) Managing the Service Economy: Prospects and Problems, pp. 183-213, Cambridge University Press, London.

Kutscher, R.E. and Mark, J.A. (1983), The Service Producing Sector: Some Common Perception Reviewed, Monthly Labor Review, April 21-24.

Lakshmanan, T.R. and Chatterjee, L. (1986), Technical Change and Metropolitan Evolution, Regional Science and Urban Economics, February, pp. 7-30.

Lakshmanan, T.R. (1987), Technological and Institutional Innovations in the Service Sector, paper presented at the symposium on Research and Development, Industrial Change and Economic Policy (RICE), University of Karlstad, Sweden, June 21-26.

Lakshmanan, T.R. (1988a), Infrastructure and Economic Transformation, in: Andersson, A., Nijkamp, P., Batten, D. and Johansson, B. (eds.) Advances in Spatial Theory and Dynamics, North Holland, New York.

Lakshmanan, T.R. (1988b), The Evolution of the Service Sector: Role of Information Technology, paper prepared for presentation at the 27th Meeting of the Western Regional Science Association, Napa Valley, California, February.

Levenson, I. (1985), Services in the U.S. Economy, in: Inman, R.P. (ed.), Managing the Service Economy: Prospects and Problems, pp. 80-102, Cambridge University Press, London.

Levinson, P. (1986), Information Technologies and Evolution, in: Mitcham, C. and Huning, A. (eds.), Philosophy and Technology II, pp. 29-47, D. Reidel Publishing Co., Boston.

Levitt, Th. (1976), The Industrialization of Service, Harvard Business Review, September.

Mills, P.K. and Marguiles, N. (1980), Toward a Core Typology of Service Organizations, Academy of Management review, 5:2, pp. 255-265.

Mills, P.K. and Moburg, D.J. (1982), Perspective on the Technology of Service Operations, Academy of Management review, 7:3, pp. 467-78.

Mitcham, C. and Huning, A. (1986), Philosophy and Technology II, Boston University Studies in the Philosophy of Science, Reidel Publishing Co., Boston.

Nelson, R.R. (1986), Institutions Supporting Technical Advance in Industry, American Economic Review, 76:2, pp. 186-89.

Nelson, R.R. and Winter, S.G. (1982), An Evolutionary Theory of Economic

Change, Harvard University Press, Cambridge, MA.

North, D.C. and Thomas, R.P. (1970), An Economic Theory of the Growth of the Western World, The Economic History Review, 23:1, pp 1-17.

North, D.C. (1981), Structure and Change in Economic History, W.W. Norton, Co., New York.

Perez, C. (1985), Microeconomics Long Waves and World Structural Change: New Perspectives for Developing Countries, World Development, 13, 3, pp. 441-463.

Petit, P. (1986), Slow Growth and the Service Economy, St. Martin's Press, New York.

Porat, M.V. (1982), Information, Communication and Division of Labour, in: Jusswale, M. and Lamberton, D.M. (eds.), Communications, Economics and Development, pp. 75-81, Pergamon, London.

Stanbeck, T.M. Jr., Bearse, P.J., Noyelle T.J. and Karased, R.A. (1981), Services: The New Economy, Allanheld, Osmun, & Co., Totowa, NJ.

Stanbeck, T.M. and Noyelle, T.J. (1982), The Cities in Transition, Allanheld, Osmun, & Co., Totowa, NJ.

Usher, A.P. (1954), A History of Mechanical Inventions, Harvard University Press, Cambridge, MA.

Williamson, O. (1980), Transaction Cost Economics: The Governance of Contractual Relations, Journal of Law and Economics, March.

Zuboff, S. (1988), In the Age of the Smart Machine, New York, Basic Books.

PART C
GENERAL OVERVIEWS

10 Technogenesis: incubation and diffusion

D. J. F. KAMANN AND P. NIJKAMP

1. Introduction

Spatial dynamics has become a focal point of research in regional science. In this context, it is generally taken for granted that certain areas (e.g., large cities) are relatively more favourable generators of technical change than others, although these other areas may sometimes be more flexible adopters of new technologies. Thus both the incubation of new technologies (including new technological regimes or paradigms) and their diffusion (via a physical or non-physical communication network) deserves closer attention.

In the framework of economic restructuring metropolitan areas are often considered as large efficient production and information processing systems. Such 'urban fields' (cf. Pred, 1977) usually encompass a major share of the economic activities in a country and may even be regarded as the focal points of the interregional network across nations. The role played by metropolitan regions in such a network may be quite diverse and may vary from areas accommodating large scale industrial complexes to centres for business and governmental decision making, R&D or information processing. Their evolution does not only reflect stages of fast growth and stagnating maturity, but also obsolescence and decline, while over time they usually exhibit dynamic patterns of specialized roles in national (and international) systems. When these systems change, the specialization of a region may become less competitive or even out-dated, while inertia may prevent a revival of its activity pattern.

Metropolitan regions usually evolve with different growth rates and in different directions. This process has strong spatial and demographic repercussions. Migration and intra-urban relocation of households and firms, however, are not only reactions to the economic development, but have also their own dynamics, caused inter alia by population cycles for households (see Rima and Van Wissen, 1988) and by spatial evolution (see De Jong and Lambooy, 1985) or product life cycles (see Ciciotti, 1984, and Kamann, 1985a) for firms. Some evolutionary patterns are common to all industrial nations and thereby common to metropolitan regions, even if they differ in many other respects. In general, spatial dynamics take the form of activity expansion (or contraction) and of construction of new physical elements such as housing and industrial sites, infrastructure and transportation facilities. The evolutionary process itself has certain general consequences which may to a large extent be similar for all regions; more and more space becomes occupied by buildings, facilities and other extremely durable structures; the activity density increases especially in central locations etc. As time goes by, the inertia incorporated in this fixed structure affects the competitive power of an area, so that a process of decline may start. In case of regional specialisation, the social fabric and culture of the area may show the same inertia hampering new ideas and creativeness (see Storper, 1986).

Consequently metropolitan regions seem to develop from a young to an old structure which may have a long life but which may also extend its vitality through renewal processes. In this context, it has often been observed that a metropolitan area may act as a birthplace for new technologies in the form of new products and production (see for instance Hoover and Vernon, 1959). Studies of such urban incubator phenomena indicate that technologys and knowledgesintensive industries locate in new places with a preference for regions with a rich variety of education, research and cultural opportunities (cf. Davelaar and Nijkamp, 1987).

Spatial dynamics is often exhibiting a link between the stages of metropolitan development (defined in terms of population or jobs) and economic development. The early urban concentration phase can usually be observed together with an early industrialization phase, where people and jobs concentrate in the already existing urban areas, characterized by the best infrastructure available at the time. As industrialization proceeds and per capita incomes rise, the demand for new housing as well as for private gardens, etc. leads to suburbanization. As the network of public and private infrastructure increasingly covers the entire country, urban areas appear to lose their comparative economic advantages and jobs begin to decentralize. In the post-industrial society, desurbanization tends to become a widespread phenomenon. In the context of such an urban life cycle hypothesis, the desurbanization phase should be followed by reurbanization, a trend which cannot as yet be firmly established statistically, although there are some signals which indicate that urban renewal in old city centres already leads to increases in the economic activity (gentrification). A further analysis of urban life cycles in the framework of 'long waves' theory can be found among others in Nijkamp (1985) and Nijkamp and Schubert (1985), where it has been conjectured that the evolution of metropolitan systems does not necessarily run parallel to that of national economic systems, although (metropolitan) systems may have many dynamic key factors in common. Thus, to a large extent, metropolitan areas may be regarded as a family of closely related "species".

The previous remarks on spatial dynamics imply also that the role of urban agglomerations in the process of regional and urban transformation is not always the same. For instance, it has been asserted by Malecki (1985) and Malecki and Nijkamp (1988) that - although there is apparently a strong tendency toward agglomeration-oriented and agglomeration-induced new activities the innovative capability of large cities tends to decrease in favour of medium-sized towns. Although small-scale enterprises (notably business services) may still find a favourable seedbed in large cities, many modern high-tech based firms tend to look for a location outside the large cities (the urban sprawl of enterprises). Especially the improved transport and communication infrastructure favours this disperson to the rim of agglomeration areas (see Kamann, 1986a). Apparently, in many geographical concentrations a situation of saturation (caused by diseconomies of scale) has come into being. Thus cities seem to pass through a life cycle in a way analogous to that of normal products. Various older cities may face an economic and social structure dominated by convential (old-line) industries, leading to a structural spiral downward movement. Economic restructuring favouring new entrants in the existing structures and urban revitalization based on strategic R&D, are here key elements for ensuring a new rejuvenation potential of large cities. Those cities which are most creative (and hence offer the highest relative development potential for new activities) will become the winners in such a competitive evolutionary process. See for a representation of fluctuating spatial patterns also

FIGURE 1 *The fluctuation of R&D intensity and market share of an area*

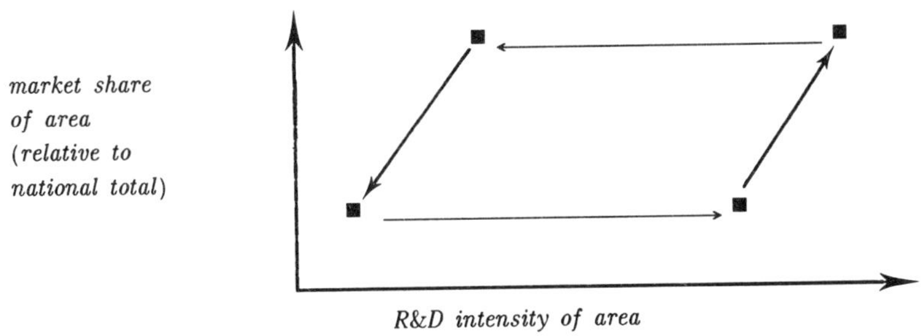

It is taken for granted in the present paper that the key mechanism in explaining the behaviour of these "species" of urban areas in market-oriented countries over time is spatial competition. The growth of urban areas can then be explained by dynamic forms of spatial competition, both between urban areas at the national level and/or at the international level. An example of international competition of urban areas is the competition between major seaports such as Rotterdam and Antwerp or the competition between large agglomerations such as Singapore and Hong Kong.

It should be added, however, that the competitive position of an area does not only depend on its locational qualities, but also on the market performance of actors concerned and their degree of connectedness in a broader communication network, through which diffusion of innovations, acquiring of knowledge, adoption of new products or processes, and rapid transfer of commodities can be ensured. Thus the spatiotemporal development of a country is also strongly determined by the efficiency of a physical network and the formal and informal networks of its actors, while the competitive position of a city is co-determined by its degree of nodality and connectivity in a network. An observation which is in perfect agreement with the traditional growth pole concept à la Perroux.

In the present paper, both the incubation and diffusion aspects of innovations of new technologies will be dealt with. First, in section 2 a brief review of elements of incubation (or nursery) theory will be given, with particular emphasis on the urban seedbed potential regarding innovations. Next, it will be argued that various models which have been developed in the recent past are often - despite their theoretical merits- not very satisfactory from the viewpoint of explaining entrepreneurial innovative behaviour. Then in section 4 various elements from diffusion and adoption theory regarding technological innovations will be presented. Here we will focus on the behaviour of actors in a network, on the attributes of such networks and on some methods for visualizing the relationships shaping a network. It will again be concluded that their behavioural content is not always very impressive. Finally, some policy implications and research recommendations will be discussed.

2. Incubation

Spatial patterns display a strong tendency toward concentration and agglomeration. Apparently, scale economies seem to emerge at places marked by a high information intensity, a high level of competence, a rich academic and cultural milieu, a high potential for external communication (including international connections), and a dynamic entrepreneurial climate. Schumpeterian entrepreneurship and spatial dynamics are essentially two sides of the same coin (cf. Suarez-Villa, 1988, and Thomas, 1987). Bruno and Tyebjee (1982) cite even 12 factors as essential for the environment for entrepreneurship:

venture capital availability
presence of experienced entrepreneurs
technically skilled labour force
accessibility of suppliers
accessibility of customers or new markets
favourable governmental policies
proximity of universities
availability of land or facilities
accessibility to transportion
receptive population
availability of supporting services
attractive living conditions

Clearly, several of the above mentioned factors are simply conditions common to most, if not all, large urban regions. For example, availability of land and facilities, accessibility to transportation, suppliers and customers, and attractive living conditions tend to be features found in virtually any major metropolitan region. Thus urban areas seem to provide important seedbeds for innovative entrepreneurial behaviour, an idea which is not only supported by the growth centre theory, but also by the incubation theory. To a large extent both conceptual frameworks - though developed in different cultures - are equal, although the incubation concept - in contrast with the growth centre concept - places more emphasis on the nursery conditions of new - often small-scale - enterprises.

Since Hoover and Vernon (1959) claimed that large cities provided external scale economies for small new firms (e.g., cheap working space, joint cost sharing of overhead facilities, face-to-face contacts etc), an avalanche of literature has been published on the incubation concept. It was claimed that the external economies provided by other firms in the same city did not only lead to a geographical clustering of activities, but gave also birth to many new firms (cf. Leone and Struyk, 1976). In this context, Cameron and Johnson (1969) stated: "For firms offering a new product the key objective is to maximize the contacts with the new market for a given working capital and fixed capital investment. The central area may be optimal for this since it contains a concentration of highly accessible potential customers, allows distribution costs to be minimal and offers a readily available pool of labour and other inputs, and possibly cheap property as well" (p.259).

In the past decade the question has also been studied whether large cities provide a high incubation potential for innovative activities (cf. Malecki, 1979, and Norton, 1979). Whether or not the city may act as a generator of innovations, depends of course on the specific seedbed conditions for innovative behaviour. In general, the following factors are regarded as highly important for the creation of innovations:

* creative entrepreneurial environment
* presence of research institutes that may act as breeding places
* presence of a highly-skilled labour force
* public support for R&D activities of starting innovative firms
* availability of venture capital
* presence of a stimulating and innovative entrepreneurial climate
* availability of inexpensive areas for new innovative entrepreneurs
* access to information
* international accessibility

It should be added that the accumulation of a critical mass of production in an industry serves also as incubator for new firm generation (cf. Rees, 1969). In particular, the specialization trend implies that incubation forces are relatively strong in particular subsectors.

Economic-technological spin-offs appear to be most common in large urban areas. It is in such places that a sufficient number of potential entrepreneurs are present, as well as other 'environmental' factors that encourage entrepreneurship. Shapero (1971) studied technical company formation in relatively small American countries over a 28-year period. The variables most highly correlated with firm formation tended to be those related to city size and agglomeration, such as manufacturing employment, educational expenditures, and income.

Furthermore, it must be stressed that not all R&D generates new spin-off firms. The state of the local industry's technology must be sufficiently unstandardized, preferably with multiple market niches, and the barriers to entry by new firms must be low (see also Bollinger et al., 1983). Even so, it would seem that the European experience regarding branch plants with their lack of local linkages (Kamann, 1978) and with public sector R&D in peripheral regions has led to very low levels of entrepreneurship (Cooke, 1985).

It is noteworthy that the location of innovative firms and the geographical spread pattern of innovations has to be seen against the background of a product life cycle phenomenon. New technology firms are-according to Davelaar and Nijkamp (1987) increasingly dividing their activities into routine (or standardized) and non-routine (or innovative) operations. The non-routine (mainly R&D-oriented) activities tend to be concentrated in only a few locations marked by significant scale economies such as a good geographical accessibility (see e.g. Ciciotti, 1984; and Kamann, 1986a). For instance, Hekman (1980) indicates that American computer firms tend to maintain their innovative activities in only a few regions (like California, Texas or Massachusetts). Standardized production and assembly operations are either moving into small towns or peripheral areas or into low wage Third World countries (cf. also Bluestone and Harrison, 1982).

Next, non-routine activities in the new technology sector rely heavily on skilled and professional labour input, so that also the quality of the residential climate (including socio-cultural amenities) becomes a major locational motive for high technology firms (see also Brotchie et al., 1985). Similar results were found by Oakey (1981) in a study on the British instruments sector, who came to the conclusion that skilled workers largely determined the location of production. This result was supported by Malecki (1984) and Oakey (1983) who observed that locational preferences of technical personnel exert a large influence on the location decisions of R&D, as this personnel appeared to attach a high priority to cultural, educational and employment opportunities in urban areas.

With regard to routine activities, especially of multi-plant corporate corporations, it is evident that low-skill labour is still the main input. In as far as low-skill employment is abundantly present in various regions, it is mainly the presence of tax heavens, premiums, cheap land and a low wage level which are determining the locational pattern of these standardized activities (see Hansen 1980). It should be added, however, that these activities may be fairly capital intensive, so that aging and life cycle processes of capital stock may also exert a significant long-term impact on industrial location patterns of new technology firms.

In addition, it is noteworthy that a large concentration of new technology activities may lead to congestion phenomena, especially if innovative firms create spin-off effects which lead to a rise in routine activities. Premus (1982), for instance, observed that in recent years there is a tendency of American new technology firms to move from the Sunbelt states to the Mid-West due to bottleneck factors (such as high wage rates, high land rents, insufficient areas for industrial expansion, high local taxes and traffic congestion). Clearly, this 'crowding out' phenomenon may also be related to the firm's position in a product cycle. Similar results were found for Scotland by Cross (1981), for the Netherlands by Hoogteijling et al. (1985) and Wever (1984), and for Germany by Wettmann (1983).

Another (important) locational determinant of the new technology sector is its orientation towards an accessible communication and information network (see also Thwaites, 1982), so that this sector is either located in nodal points of a physical communication infrastructure or in areas near research and educational institutes (Levy, 1983). This may lead to job hopping, for instance, in the Silicon Valley.

Finally, another relevant component of an innovation infrastructure of the new technology sector is the availability of venture capital (Rothwell, 1982), especially in those countries which are marked by regional variations in the provision of venture capital. In small countries, however, it is plausible to assume that regional differences in venture capital are less pronounced, so that this is not a location specific factor (although it may be a generic determining factor for new technological innovations in the country as a whole). In the foregoing sections we have described, more or less extensively, the incubating hypothesis with regard to the role of urban agglomerations as a seedbed for both new firms and innovative activities.

Testing the above mentioned hypothesis (with regard to new firms) raises several problems (see also Fagg, 1980). A first problem is related to the fact that we have to define the geographical area which is supposed to induce an 'incubation function'. Should the area to be considered consist of the core region (Hoover and Vernon, 1959, and Vernon, 1960), the area just adjacent to the core region (Buit, 1970, and de Ruijter, 1978) or the whole agglomeration (Lambooy, 1984)? In case we intend to identify certain 'breeding places of new activities' in a national context, it seems reasonable to consider the whole agglomeration as a 'breeding place'. In an intra-urban analysis this viewpoint is not sensible of course. When we compare the incubation scores of agglomerations, large urban areas and remaining regions, we have to take into account the sectoral specialization (see Kamann, 1988a). Some areas have a higher birth potential for specific activities (e.g. services) than others.

A second problem is the development of a measurable criterion by means of which we can decide whether or not a certain (part of an) agglomeration offers an incubator function. Then the question becomes how many new firms (absolutely or relatively) does a (part of an) agglomeration have to generate (attract) in order to speak of such a 'seed-bed function'?

Finally, it should be noted that rigorous empirical tests on the incubation hypothesis are still fairly rare (see Davelaar and Nijkamp, 1987). In Davelaar and Nijkamp (1988) a more comprehensive model based on a Partial Least Squares (PLS) approach has been developed in order to test the existence of this hypothesis in the Dutch context. Here a distinction was made between the innovation potential (the capacity of industrial firms to generate and adopt innovations), the innovativeness (the realized performance of industrial firms in terms of technology generation, diffusion or adoption) and the selection environment (the set of indicators reflecting the regional production environment). By means of a multivarite PLS model for micro innovation data of Dutch entrepreneurs it was concluded that a spatial product cycle could be identified in which the more routine-oriented activities were shifted away from the main agglomeration in the Netherlands toward the periphery.

3. Urban Incubation Models

In recent years various types of dynamic urban incubation models have been developed. Some of them are relatively simple and only based on the assumption of economies of scale during a first stage of urban evolution, followed by diseconomics of scale in a later phase. Examples can be found in Dendrinos and Mullally (1983) and Nijkamp (1987). Both types of models will be discussed briefly in subsections 3.1. and 3.2., respectively. A more comprehensive model on urban dynamics and innovative behaviour can be found in Blommestein and Nijkamp (1986). This model will be treated in subsection 3.3.

3.1 A dynamic urban Volterra-Lotka model

The urban Volterra-Lotka Model has been described and advocated by Dendrinos and Mullally (1984). Cities may exhibit fluctuating patterns depending on their (dis)economies of scale and their economic performance or incubation potential. The authors argue that many urban areas in the U.S.A. exhibit stable dynamic behaviour, which can be depicted by spiralling sink type paths recorded on two key variables of aggregate metropolitan dynamics: relative population and per capita income. Two differential equations, defined over the ratio of urban to national population size and over the ratio of urban to national average income, are shown to describe an urban area's dynamic behaviour in accordance with the Volterra-Lotka predator-prey model from population dynamics.

The urban Volterra-Lotka model has the following form:

$$\dot{x}_i = x_i(-\overset{0}{\alpha}_i - \overset{1}{\alpha}_i x_i + \overset{2}{\alpha}_i y_i)$$

and (3.1)

$$\dot{y}_i = y_i(\overset{0}{\beta}_i - \overset{1}{\beta}_i x_i)$$

Subscript i stands for urban area i. Variable x represents relative urban population size, and variable y represents relative per capita income in the city concerned. Assuming some plausible values of the parameters a and b, the dynamic trajectory of a city may be simulated.

Various interesting empirical results for American cities were found by Dendrinos and Mullally (1983). These findings suggest a general validity of the urban Volterra-Lotka model. Its major advantage is of course its flexibility, so that the model is able to generate a wide spectrum of dynamic behaviour of a city (including stable or unstable trajectories). A major disadvantage, however, is the lack of a rigorous econometric and statistical test procedure, so that the validity of results cannot be judged in a proper way. The lack of micro data once more is not very appealing, as then entrepreneurial behaviour cannot be satisfactorily dealt with.

3.2 A dynamic urban quasi-production function

In the second type of model for urban evolution the assumption is made that the growth (or incubation) potential of a city depends to a large extent on the available R&D resources. In this framework a simple quasi-production function has been developed in Nijkamp (1987).

A quasi-production function is a generalized production function, which in addition to traditional production factors such as productive capital, labour and landuse also incorporates infrastructure capital (or public overhead capital) and R&D (or innovative) capital. Infrastructure capital- as a complement to private productive capital - serves to enhance the efficiency of entrepreneurial activities, while R&D capital (both private an public) aims at favouring the innovative potential. Then the following quasi-production function for a certain area (region or city) may be assumed:

$$Y = f(C,S,R) \tag{3.2}$$

where

Y = regional share in the national production

C = regional share in the national directly productive capital

S = regional share in the national social overhead capital

R = regional share in the national R&D capital.

By assuming a Cobb-Douglas specification, one may rewrite (3.2) as follows:

$$Y = \alpha C^{\beta} S^{\gamma} R^{\delta} \tag{3.3}$$

where the parameters β, τ and δ reflect production elasticities. These elasticities are assumed to be positive on the range (Y^{min}, Y^{max}). Y^{min} reflects a minimum threshold level of the regional production volume which has to be reached before a self-sustained growth will take place, while Y^{max} reflects a bottleneck level (or maximum capacity level), beyond which congestion factors lead to a negative marginal product. Consequently, the following conditions hold:

$$\begin{aligned} &\text{if } Y \leq Y^{min}, \text{ then } \beta,\gamma,\delta = 0 \\ &\text{if } Y \geq Y^{max}, \text{ then } \beta,\gamma,\delta \leq 0 \end{aligned} \tag{3.4}$$

By assuming now a time-dependent quasi-production function, the shifts in the regional share of the national production volume can be written as:

$$\Delta Y_t = (\beta \tilde{C}_t + \gamma \tilde{S}_t + \delta \tilde{R}_t)\ Y_{t-1} \tag{3.5}$$

with

$$\Delta Y_t = Y_t - Y_{t-1} \tag{3.6}$$

and:

$$\tilde{C}_t = (C_t - C_{t-1})/(C_{t-1}) \tag{3.7}$$

$$\tilde{S}_t = (S_t - S_{t-1})/(S_{t-1}) \tag{3.8}$$

$$\tilde{R}_t = (R_t - R_{t-1})/(R_{t-1}) \tag{3.9}$$

The economy reflected by (3.5) will exhibit a stable growth path without structural changes within the range (Y^{min}, Y^{max}). The lower limit Y^{min} is in the present context of innovation and capacity limits less interesting, so that we will focus our attention mainly on the effect of the bottleneck value Y^{max}.

This bottleneck value reflects congestion phenomena due to too high a concentration of productive capital in a certain area leading to diseconomies of scale, environmental decay, and inefficient land use. Beyond Y^{max}, each additional increase in C will have a negative impact on the regional production share. Such a situation of a negative marginal product of capital can be represented as:

$$\beta_t = \hat{\beta}\ [(Y^{max} - \kappa Y_{t-1})/Y^{max}],\ \kappa \geq 1 \tag{3.10}$$

where $\hat{\beta}$ represents the (fixed) production elasticity of C on the range (Y^{min}, Y^{max}). Consequently, the adjusted production elasticity has become a time-dependent variable. Similar relationships may be assumed for τ_t and δ_t, so that substitution of β_t, τ_t and δ_t into (3.5) yields the following result:

$$\begin{aligned} \Delta Y_t &= (\hat{\beta}\ \tilde{C}_t + \hat{\gamma}\ \tilde{S}_t + \hat{\delta}\ \tilde{R})\ (Y^{max} - \kappa Y_{t-1})\ Y_{t-1}/Y^{max} \\ &= \tilde{Y}_t(Y^{max} - \kappa Y_{t-1})\ Y_{t-1}/Y^{max} \end{aligned} \tag{3.11}$$

with:

$$\tilde{Y}_t = \hat{\beta}\tilde{C}_t + \hat{\gamma}\tilde{S}_t + \hat{\delta}\tilde{R}_t \tag{3.12}$$

$\tilde{Y}_t$ may be regarded as the rate of change in the original quasi-production function. Relationship (3.11) is essentially a Volterra-Lotka-type model, which has often been used in population biology, for instance, to describe predator-prey relationships, notably in the May-type of models.

Relationship (3.11) has some interesting features: models of this type - despite their mathematical simplicity - may exhibit a remarkable spectrum of dynamical behaviour, such as stable equilibrium, stable cycles, stable cyclic oscillations, and chaotic trajectories with a-periodic (but bounded) fluctuations. The behaviour of such a model is determined by the initial conditions of the system and by its growth rate (depending on $\tilde{Y}_t$), but in principle this model is able to generate a wide variety of dynamic growth patterns. Consequently, in a spatial context long-term fluctuations depend on the initial values of a spatial system and its growth rate (which is codetermined by the production elasticities of production capital, overhead capital and R&D capital).

The growth rate, however, is a time-dependent variable, which can also be controlled by (private and public) policy measures. If the model is used in the framework op optimal control theory, generalized geometric (signomial) programming algorithms can be used to identify optimal controls.

A next step may be to introduce an additional relationship for R&D investments, given the assumption that R&D may serve as a tool to remove bottlenecks (the so-called depression-trigger hypothesis). Then we may hypothesize the following relationship, as soon as an area has reached its critical bottleneck level Y^{max}:

$$\tilde{R}_t = \hat{R}_t(Y_{t-1} - \pi Y^{max})/Y^{max} \qquad (3.13)$$

where $\tilde{R}_t$ is the rate of change in R&D capital beyond the value Y^{max}. Substitution of (3.13) into (3.11) yields:

$$\Delta Y_t = [\tilde{Y}_t + \hat{\delta}\hat{R}_t(Y_{t-1} - \pi Y^{max})]/Y^{max} + (Y^{max} - \kappa Y_{t-1})Y_{t-1}/Y^{max} \qquad (3.14)$$

with:

$$\tilde{Y}_t = \hat{B}\tilde{C}_t + \hat{\gamma}\tilde{S}_t \qquad (3.15)$$

The latter relationship is a nested dynamic model. This model may exhibit even more complicated dynamic growth patterns, depending on the superimposition over two dynamic phenomena. The perturbations caused by the bottleneck factors may be neutralized or reinforced by R&D investments, depending on the fine tuning of new technology investments and spatial fluctuations. Thus in conclusion, incubation models of the May type do not ensure a stable spatial equilibrium.

3.4 A comprehensive dynamic urban Verhulststype model

In the present subsection an illustrative model based on a Verhulst specification will be used as a framework for treating urban dynamics in a spatial system (see for details Blommestein and Nijkamp, 1986). The fundamental growth equation for city i is supposed to be:

$$\dot{x}_i = \alpha x_i \, (N + \sum_k \varepsilon^k v_i^k - x_i) - \beta x_i \qquad (3.16)$$

where α is the birth rate of new urban entrepreneurial activities, β the death rate of existing activities, N the initial physical-economic carrying capacity for economic activities of the city, v_i^k the volume of new activities in sector k generated in city i (measured in appropriate units), and ϵ^k the impact of new activities in sector k on the growth of city i. Thus the expression $\Sigma\epsilon^k v_i^k$ indicates the capacity growth in the original volume N of city i due to the introduction and implementation of new activities k.

Next, the growth of these new activities in sector k in city i may be represented as follows (see also Allen et al., 1981):

$$\dot{v}_i^k = \eta\, v_i^k\, (e_i^k - \delta^k v_i^k) \tag{3.17}$$

where η is the growth rate of these new activities, e^i the volume of employment (or, in general terms, production factors) that might potentially be generated in sector k in city i (i.e., a ceiling for new urban activities), and δ^k a market threshold coefficient in sector k.

In addition, one may assume:

$$e_i^k = \mu^k d_i^k \tag{3.18}$$

where d_i^k is the demand for the products generated by sector k in city i, and μ^k a (constant) parameter linking the effective demand for k to their potential employment opportunities (usually, $\mu>1$).

Besides, the total demand in city i generated by residents of other cities j is in case of absence of spatial competition equal to:

$$d_{ij}^k = \lambda^k x_j/(p_{ij}^k)^v \tag{3.19}$$

where p_{ij}^k is the c.i.f. price of a unit of product from sector k, produced in i and shipped to residents in j; λ^k and v are just normal reaction parameters. Now the price p_{ij} is supposed to depend on communication costs between cities i and j as follows:

$$p_{ij}^k = p_i^k + \phi^k \delta_{ij} \tag{3.20}$$

where p_i^k is the f.o.b. price, δ_{ij} the distance between i and j, and ϕ^k the unit communication cost.

Next, one may introduce spatial competition between cities on the basis of an attractiveness indicator p_{ij}^k for sector k which incorporates urban facilities and price levels of sector k:

$$a_{ij}^{k} = \rho \; n_i \; / \; (p_{ij}^{k})^{\theta} \tag{3.21}$$

where aij is the relative attractiveness of city i for residents of city j, n_i the share of facilities is city i, while θ and ρ are standardisation parameters.

Consequently, the demand in city i generated by households outside i is co-determined by the relative attractiveness of city i, so that the urban demand equation may be adjusted as follows:

$$d_{ij}^{k} = \lambda^{k} \; x_j \; a_{ij} \; / \; (p_{ij}^{k})^{v} = \lambda^{k} \; x_j \; \rho n_i \; / \; (p_{ij}^{k})^{v+\theta} \tag{3.22}$$

It is easily seen that the total sectoral demand can be directly calculated from (3.21), while also (dis)economies of scale may be incorporated. By substituting now (3.22) into (3.18), followed by a substitution of (3.18) into (3.17), equations (3.16) and (3.17) describe a highly non-linear dynamic evolution (of the May type) for a spatial system composed of competing regions, which might lead eventually to competitive exclusion (see also Johansson and Nijkamp, 1986). Thus various types of dynamic behaviour ranging from stable growth to chaotic behaviour may emerge, depending on the initial conditions and the various parameters of the system. Various simulation experiments for similar types of urban incubation models can be found in Nijkamp et al (1988).

3.4 Concluding remarks

The previous selection of dynamic urban incubation models is of course rather selective, but nevertheless leads to some important conclusions. Finally, a rigorous test of such dynamic models has not been achieved yet, mainly because dynamic micro data (or longitudinal data sets) do hardly exist in this field. Thus to some extent most models are too aggregate to test the inderlying micro processes of urban dynamics. Simulation experiments may be helpful of course but do not lead to statistically solid inferences.

In addition to the incubation potential of urban areas, also the second attribute of technogenesis, viz. the role of its actors vis-à-vis diffusion and adoption, has to be given adequate attention. This will be done in section 4.

4. Networks

4.1 The 'network' approach

The previous sections focussed mainly on locational and institutional features of the selection environment of actors and dealt with the question whether a spatial differentiation in these elements would lead to spatial differences in new firm formation and innovative behaviour. The present section focusses attention on the individual actor in a network of relations with the various actors in his environment. We will in particular concentrate on diffusion and adoption phenomena and hence emphasize network links (Kamann, 1988a, 1988b). This set of network relations contains actors that are specific for the product/market combinations of an actor and the industrial organisation the actor is part of.

Networks are increasingly becoming popular in the literature. There is a growing awareness in economics and geography that firms should not be seen as individual organisms that live their own lives independently from other actors in their economic, social and cultural environment. Waves of merger activities and take-overs have resulted in noticable effects and disruption, not only in financial circles, but also in local employment situations and welfare (Flynn and Taylor, 1986). Technological change, e.g. in telecommunications information technology -, has resulted in locational changes of parts of multinational organizations and the contents of the activities involved (Kamann, 1985a; Hepworth, 1987; Nijkamp, 1987) Government efforts to stimulate innovative industries, and the growing interest for autonomous growth potential, territorial industrial complexes (Premus, 1985; Levitt, 1985; Stohr, 1985) and local initiatives (Boekema and Verhoef, 1987) have induced a great deal of interest in the networks of relations between firms, including the diffusion of innovations (Pavitt, 1984).

When reviewing the available theories in regional science, we find that Perroux (1955) already dealt with networks of firms in his concept of the Growth Pole. Unfortunately, his theory about clusters of industries in economic space was poorly 'translated' into the Growth Centre theory with industrial firms that were clustered in geographical space although they could well belong to different clusters in economic space (for a review of the Growth Centre theory, see Moseley, 1974; Richardson, 1979, or Stohr and Todtling, 1979). A more recent concept originating from France which we will describe later on in more detail the filière seems to be more in line with the original ideas of Perroux. Policies based on this concept have already shown to be more valuable than conventional growth centre policies

(Groenewegen 1987). However, given the results of studies of various disciplines, we will argue that the filière concept has to be 'expanded' (Kamann, 1988a) in order to provide appropriate answers to the following questions:

- what is a network;
- how and why do actors participate;
- how do we visualise the relations that make up a network;
- how does diffusion take place?

4.2 Definitions: networks and actors

According to Perroux (1955/1970), firms operate on a plain in an abstract economic space, where they meet other actors. Actors refer first of all to individual entrepreneurs or firms in their role of competitor, supplier, customer, producer of potential substitutes or possible newcomer (Porter, 1980, Devine, 1979; Sherer, 1983; Kamann, 1985a, 1986a), but this concept may also include subsidiaries, profit centres or business units belonging to a corporate network. In the tradition of Perroux we assume a 'natural' tendency among actors to dominate a relationship and network as much as possible (Hermanssen, 1972). Or, as Johannisson (1987, p. 54) states, an actor exploits his environment; the plain where he meets the other actors is an arena. Authors in the tradition of Lewin's (1951) field theory, like Melin (1983), would even state that the position of an actor in the network the field is determined by both external and internal forces. Examples of external forces are demand, public policies, new technologies, labour relations, and other participants. Internal forces or drives are, for instance, the internal organisation and are assumed to be non-rational. Such a model however tends to underestimate the influence from the individual actor on his environment in his attempts to externalize the internal problems (resulting in negative external effects) and to internalize his external problems (by means of e.g. internal labour markets, information systems, capital supply). It is noteworthy that this approach amalgamates Mitchell's (1973) three categories of networks:

(1) information network; vehicle for the diffusion of innovations;

(2) 'exchange' network; characterized by mutual dependency;

(3) normative network; common interest versus dependency.

The first two categories fit in with Cook and Emerson's (1978, p. 725) statement that between actors, "sets of two or more connected exchange relations" are established, shaping the network. The obvious question is here, how many relationships have to exist before we can use the term 'network'. Tichy (1981) refers to three levels of networks. The first is a cluster of relations, for example coalitions or cliques. The second level is an organisational network, while the third level is an inter-organisational network, referring to interacting organisations. While this is a valuable distinction from a theoretical perspective, it does not give a practical rule for demarcating the three categories. Since Mitchell's third category is often an implicit social phenomenon that even is not recognized as a network (Hellgren and Stjernberg, 1987, p.89), further problems in the visualisation arise. Johannisson (1987) therefore makes a subsdivision into the formal and the informal structure. The formal structure is defensive, instrumental and exists apart from streams of activities. It consists of regulations, contracts and rules. Informal structures fulfill the social needs and are dynamic. "Informal structure represents the sedimentary organizing capacity of a collectivity" (Johannisson, 1987, p. 4). Leaders in the informal network translate external changes and are as such comparable with those who have 'orientation activities' with their broad socio-economic external scanning (Goddard, 1973). Having studied a great many studies that tried to visualise networks, we prefer to use a different distinction: manifest and latent network relations. Manifest network relations are flows of goods, services, information and contacts. They are the manifest signs of some kind of relationship and can be registered and measured. The strategic value of these flows and the dependency for the actors involved, are however very rarely visualized. Therefore, we would name this the latent relations. They contain some of Johannisson's elements, but focus on the strategic value of a relation related to the issue of dominance and performance. Manifest relations are the materialized dimension of latent relations.

An important part of latent relations in a network are the social relations. The essence of the social network is that results of one relation depend on the other relations (Johannisson, 1987, p. 9). The existence of a social system also means that the individual actors only have a limited freedom in their decision-making: a bounded freedom of choice. This idea is also reflected in discussions on the social paradigm and its resulting technological paradigm, determining technological trajectories (Dosi, 1982; Nelson and Winter, 1977). In an urban context, it is reflected in the discussion of social dominance in industrial centres, e.g. Detroit, Houston, Washington (Storper, 1986). This freedom of behaviour of the individual participants of the network is a direct result of the degree of dominance an actor has obtained. Thus inside the network, individual actors try to increase their freedom of decision, given the social constraints that apply in the network concerned.

4.3 The manifest network relations

4.3.1 The contents of relations

Networks based on physical exchange

The first and most wellknown type of network is based on flows of goods and services between actors. Some authors (Vernon, 1966) use this to describe the spatial distribution of labour and production. Others emphasize their role in the diffusion of innovations in networks (Perroux, 1955; Pavitt, 1984; Camagni, 1985; Davelaar and Nijkamp, 1986, or the various contributions to the Venice Conference on Innovative Diffusion, 1986). The idea, common to most contributions is that a product is invented by an actor in one branch of industry and applied in products and processes of a different branch of industry. After this, it will spread over other sub-sectors. Chips and robots are well-known examples. Depending on the origin of an innovation, three types of network relations are usually found. First of all, the supplier dominated relation has to be mentioned, for instance the chemical network described by Baranson (1978). The second type is the user dominated relation like for instance in the semisconductor industry (von Hippel, 1977). The third type of relation is the research dominated network, for instance the Swedish laser network (Laage Hellman, 1987). The phenomenon of users sectors - e.g., the food sector explains why some sectors with a relatively low R&D expenditure still are able to raise productivity by using new technologies. These innovations originate in other sectors (Mueller and Culbertsen, 1986).

The question of the origin of an innovation is related to the question who gave the impetus to the innovation or the invention that preceded (see Håkansson, 1983; Laage Hellman, 1987; Pavitt, 1984; cf. also the discussion on technology push or demand pull described by Mowery and Rosenberg, 1979). This is not necessarily the actual producer. Laage-Hellman (1987) for instance found that in more and more situations, supplier and user undertook the R&D together. This supports the statement (Dosi, 1983; H2kansson, 1987) that innovations cannot exclusively be explained by either the technology push or demand pull theory. Innovations are the result of the joint efforts of various actors.

Capital

Minority stakes in firms prove to affect the behaviour of the management involved when it is more than 17 percent because of "The mere threat of aquiring control that gives partial control over manager's behaviour" (Neun and Santerre, 1986, p. 207). Complete control is realised after 52 percent has been reached (ibid.). This means, that not only majority shareholders have to be taken into account, but also minority shareholders and their interests and policies. From the various attempts to visualise financial participations (see Vlieg, 1977) we may derive the conclusions that these networks are very complicated, interwoven and because of the obvious reluctance of firms to provide this information hard to map. The strategic value a latent feature - of capital participation - is very rarely empirically observed. It can be inferred however from case studies. We will describe some of these effects in Section 4.5.

Information: personal networks

The exchange of information can take place via printed matter, by hiring persons (from universities or competitors: i.e. the human embodied information) and finally via personal contacts (face-to-face or by telecommunication).

Printed matter - mainly journals - are an important source for diffusion and innovation. Bramer (1986) found that among the small- and medium-sized establishments journals played an important role as stimulus for innovations and also in the more advanced stages of a development process. This is confirmed by studies of Docter and Stokman (1987, 1988).

The role of exchange of human embodied information and knowledge seems to show culturally determined differences. For instance in the Netherlands exchange in terms of job-transfers between universities and the market sector does not play the same role as in the U.S.A. Exchange between firms that operate on the same plain seems more common. This is in a number of cases a matter of transfers within the same conglomerate; this could of course also indicate the existence of kartels where strategic information is shared by its members. Given the problem that very few firms will admit they buy out personnel to acquire information, empirical evidence is scarce here.

The third category consists of personal contacts. "Personal contacts are at the heart of interaction between organisations" (Cunningham and Turnbull, 1982, p. 314). Through personal contacts, policy decisions are prepared, formulated, attuned and negotiated. Much 'strategic information'- about competitors, suppliers and buyers is obtained through personal information. These - face-to-face - contacts are the source of a multiple set of other types of contacts. For example to confirm something in writing by mail, telex or fax; the telephone call and the expedition of the agreed upon good. Although modern communication systems substitute some of the face-to-face contacts - especially the more routinized - they fulfill also better their supplementarity. We will discuss here two forms of personal contacts. The first is based on face-to-face contacts that occur in interlocking systems. The second is based on contact systems that use both face-to-face contacts and telecommunication contacts.

(A): Interlocking systems

A growing number of authors (Helmers et.al., 1975; Uitham et. al., 1977, Pennings, 1980) studied personal contacts between actors in interlocking systems and 'meeting places' of employees of government agencies, private firms and semi-state bodies. In an interlocking situation, someone is member of the Board of Directors of at least two companies (Scott, 1985, p. 1). Interlocks are "intrinsically meaningful as a channel of communication" where the total network "constitutes a web of communication through which general business information and opinion can be transmitted" (Mills, 1956; quoted by Scott, 1985, p. 1). This 'opinion' coincides with what we term "paradigm fixation".

Studies on interlocking systems draw up a hierarchy in the firms that are part of it. The more interlocking relations they are involved in, the higher their position in the hierarchy. We find five different theoretical models, which differ in their selection of explanatory aspects and variables. The first model emphasizes the role of financial institutions; the finance-capital model (Stokman e.a., 1985). A second variant assumes that the economy is "structured into competing groups of co-ordinated companies, each group being subject to a specific locus" (Scott, 1985, p. 8). This co-ordination and control model focusses on the role of banks, and family or general holdings (Ziegler, et. al., 1985). The third variant is the resource dependent model and focusses on the way large companies regulate their mutual dependence in their attempt to gain access to valuable resources. Central theme is the way "to establish links ... to regulate their interdependence" (Scott, 1985, p. 9). In general, very few large and systematic networks are found here; "dyads, triads, linked together through a sparse nexus of random interlocks" (ibid. p. 10). The fourth variant is the managerial model, by many nicknamed the gin and sigar model. It assumes that senior executives of large companies have almost complete autonomy, independent from share holders and capital suppliers. Although sometimes their role as socio-economic scanning device is admitted, in this perception persons in interlocking systems - members of the Board of Directors - are figure heads or token women, just for show and public relations. They are selected to increase the prestige of the company, not its ability to solve problems. Finally, as a fifth variant, we mention the class cohesion model. In this approach, "directors are recruted from an upper class and that the patterns of interlocks express and contribute to the cohesion of this class" (Scott, 1985, p 11; Ziegler, et. al., 1985).

All five models - and the studies related to them - have in common that they do not deal with the actual contents of the information, only with its assumed strategic value. Studies that empirically tried to measure this were unable to demonstrate this however. The contents are assumed to be be conspiracies or the opposite, twaddle. In the next subssection on contact systems we describe a model that looks at the actual contents of information and its role and application.

(B): Contact systems

The Swedish Lundsschool and the British Newcastle-school have established a long and rich tradition in research on contact systems. The underlying theoretical assumption in the work of both schools is, that the hierarchical organisational structure is linked to the hierarchical urban system (Wärneryd, 1968). This assumption coincides with Vernon's (1966) observation of the distribution of production in multinational corporations. In this spatial variant of the product life cycle theory , the division of activities in the organisation follows a Tayloristic approach: head-office functions separated from basic research and production; pilot plants at reasonable distance from head-office and laboratory; standardized mass production in peripheral, rural areas (Andersson and Johansson, 1984; Kamann, 1985a; Storper, 1985;). Decreasing levels of knowledge required for activities imply a more rural orientation and increasing distance from the focus of activities (Kamann, 1985b).

Wärneryd also drew from Christaller's (1933) Central Place theory, including Pred's (1966) amendmends. Localities higher up in the national or international hierarchy provide more specialized services. These range from hamburgers to loopsholes in the tax-laws: personal services, business services and public services; the urbanisation effects. Elsewhere (Kamann, 1988a) we used a Generalized Least Squares model (LISREL) to test both the product life cycle theory and the central place theory on their empirical validity. We found they are not valid for all occupational categories but agricultural, administrative and clerical, nor for all areas. This blurs the real processes, especially when using aggregated data. It supported the assumption that network segments, related to particular activities and/or subssectors, have their own foci with industrial agglomeration effects and that these do not necessarily coincide with the urban area highest in the national hierarchy. The implication of this is that actors that require frequent face-to-face contacts with actors of the financial world, accountants, marketing agencies, consultants or politicians prefer to be located in large urban areas or the metropolitan area. Wärneryd assumes that activities in the organisation differ in their mix and types of internal and external contacts, with different actors involved. Some of the external actors are supposed to be exclusively located in metropolitan areas. These assumptions were supported by various studies (Thorngren, 1972; Goddard, 1973; Törnqvist, 1970; Andersson and Johansson, 1984). Using relocation costs and opportunity costs of missed information, an organisation could optimize its location. A first problem in this rather academic approach is that persons have a mix of activities and therefore may have different priorities and preferences for certain localities. In an organisation with numerous actors the usual social processes will take place, where coalitions or cliques and power rather than individual ratio will decide where the total group will locate.

4.3.2 Techniques to visualise manifest networks

Physical exchange: industrial clusters and filieres

Input/output analysis (Leontief, 1954; Oosterhaven, 1981) is the best known technique to visualise flows of goods and services between actors. A major draw-back of this technique is its usually high level of aggregation. Of course, this is more related to the data used than to the principles of the analysis. Czamanski (1974, 1976) in his various analyses of industrial clusters, used two types of linkages between industries: forward linkages, symbolized for industry A as X_{AB} representing the flow of goods from A forwards towards B, and backward linkages, symbolized as X_{BA}. The forward linkage from A coincides with the backward linkage from B. The relative importance of a link is expressed by four coefficients, A_{AB}, A_{BA}, B_{AB}, B_{BA} each expressing the percentage of a particular link between two industries as a percentage of all supplies to, respectively, from a particular industry. The coefficient e_{AB} stands for the degree of dependence, where

$$e_{AB} = \text{MAX}\ [A_{AB}, A_{BA}, B_{AB}, B_{BA}] \qquad (4.1)$$

where each of the four elements is derived using

$$A_{AB} = X_{AB}/\sum_{B} X_{AB} \qquad A_{BA} = X_{BA}/\sum_{A} X_{BA}$$

$$B_{AB} = X_{AB}/\sum_{A} X_{AB} \qquad B_{BA} = X_{BA}/\sum_{B} X_{BA} \tag{4.2}$$

For $e_{AB} > 0.2$, a link between A and B is significant.

The spatial attraction between industries is included in the following way. The urban attraction is expressed as follows:

$$E_{ir} = b_i P_r + E_{ir}, \tag{4.3}$$

where b_i = urban attraction on industry i
E_{ir} = employment of industry i in region r
P_r = population of region r

Elimination of the urban attraction when focussing on the spatial attraction of industry i on j gives:

$$\varepsilon_{ir} = \alpha_{ij} + \beta_{ij} E_{jr} \tag{4.4}$$

where E_{jr} = employment in industry j in region r
β_{ij} = spatial attraction between industries i and j

A new variable ϕ_{ij} is introduced because of the asymmetric relation between i and j:

$$\phi_{ij} = \beta_{ij}/\bar{E}_j + \beta_{ji}/\bar{E}_i \tag{4.5}$$

where $\bar{E}_i, \bar{E}_j$ = average employment in a region in industry i,j.

Using these formules, Czamanski found 216 spatial industrial clusters in the U.S. economy of 1963.

The major problem with this clustering approach is the cut-off point, which is often rather arbitrarily determined. In theory, one could include all exchange relations between actors. However, the - usual implicit - assumption is that the relations should be 'of importance' to the actor. However, for actor A a customer X buying 15 percent of his production may be insignificant. Customer Y however may require very specific products that stimulate A to improve its range of products while Y also provides the required assistence in meeting A's objectives. Still, Y may only buy 10 percent of A's production. Although the volume of a flow may give some indication of its strategic importance, this is only valid under rather strict assumptions of homogeneous goods, homogeneous information and homogeneous actors.

A different, more descriptive approach to visualise exchange flows is found in the filière approach: " a submode of production organization, constituted by the economically and technically interrelated operations placed between the point of availability of the raw material and that of the finished product" (Boulianne, 1982). Burink and Groenewegen (1983) give a broader definition, referring to a class of sectors, related in a chain between raw materials and capital goods at one end, and distribution and service at the other end. Machin and Wright (1985) emphasize in good Perrouvian tradition the role of nationalized corporations. Although conceptually an improvement on input/ output analysis, a number of weaknesses remain in the filirre concept. First of all, operationalisation still leads to serious measurement problems. Again, a subjective cutsoff point for relations has to be decided on. A second weakness is the static character of a filière. It reflects relations at a certain point of time, and does not indicate growing or shrinking flows, nor does it indicate the rise or fall of actors. As we know, actors are also faced with the threats of substitutes and newcomers. Products have their life cycles, technologies change and exogenous 'facts' may change, like demand and environmental demands. None of these aspects are included. A third weakness, common to all approaches in this category is that filirres - as they are visualized do not include the other manifest flows, let alone latent relations between actors. Therefore, the notion of expanded filières (Kamann, 1988a) may be developed, incorporating at least information flows and capital relations. Even expanded filirres still have the above mentioned disadvantage that they do not show the strategic relevance of its various components.

Pavitt's (1984) taxonomy could be incorporated to trace diffusion processes within filières. Since a filirre should be drawn up at a single establishment level, it also is a very useful instrument for translating actors in economic space into locations in geographical space.

Capital

Analogous to the filière approach, we find that in dynamic networks, the actual configuration of actors and their relations - in case of e.g., telecommunication industries even the names involved have changed already by the time the ink of the text describing the network has dried. An added problem for the researcher is that minority interests are difficult to trace.

Contact systems: interlocking systems and personal contacts

A great number of studies are available on interlocking systems. In some cases, the spatial dimension is implicitly present, e.g., the distribution of nationalized companies, located in the capital and companies owned by the local authorities (e.g., Ziegler et. al., 1985, pp. 73-79). Chiesi (1985, p. 209) explicitly studied spatial aspects of interlocks. He found Milan, Rome, Turin and Genova as centres for specific activities which also supports our focus concept. Bearden and Mintz (1985, p. 241) found that "... the American interlock system was divided into regional groupings within which corporations maintained denser connections to local concerns than to distant companies". These local concentrations should however be seen as "local pockets of densely interlocked corporations into a larger network, rather than as autonomous centres" (ibid.).

The contents of information contacts are studied by Goddard (1973) using a multivariate analysis. As a result, he divides contacts in three categories:

(1): orientation contacts, required for long term planning and scanning of the socio-economic environment; new developments inside our outside the actor's sub-sector should be noticed;

(2): planning contacts, related to "processes and related information flows concerned principally with the development of specific alternatives that have been identified through higher level orientation processes" (Goddard, 1973, p. 27). Both applied R&D and activities of Strategic Business units belong in this category. Telecommunication is frequently used. These activities can be performed in second order cities with good access to larger urban areas or metropolotan areas;

(3): the bulk of pre-programmed activities require routine contacts, related to progress reports, routine matters and fixed decision rules. Telecommunication is frequently used. These activities can be performed at more distant places, as long as good and reliable telecommunication facilities at reasonable prices are available.

4.4 The selection process

4.4.1 The potential network and the final segment selected

Given the potential number of actors to have contacts with, an actor selects a certain network segment. This term coincides with the term microposition (Johanson and Mattson, 1984) and net (Hägg and Johanson, 1982) and stands for the actual configuration of relations between actors. In such a segment, there are some actors an actor very rarely meets, while he may have intensive contacts with others. Because of this, the terms direct, indirect, weak and intense contacts are used.

4.4.2 Direct versus indirect; weak versus intense

When an actor has no direct contacts with particular actors since he assumes that other persons who - are included in his list of contacts - do meet these actors, this is termed indirect contacts. "I know someone, and they know someone, but I don't know who they know. The power of the network is that the participants all know it exists. we all know that we know lots of people in the Valley ... the rate of rumor-passing in Silicon Valey is simple phenomenal. Reputations, successes, people leaving a firm, new products. the mill grinds out these rumors at a prodigious rate" (Rogers and Larsen, 1984, p. 80; quoted by Hamfelt and Lindberg, 1987, p. 179).

Infrequent contacts of weak relations are 'kept in storage' for moments they become of importance: public agents in charge of e.g. building permission, grants, or subsidies. They are of no importance for the regular operation of the firm (Håkansson, 1987, p. 216). Relations with frequent and intensive contacts are termed intensive.

4.4.3 The role of personality in the actual choice

The choice of the actual network segment whether contacts will be direct or indirect, weak or intensive is a strategic choice. This choice is in spite of its importance very rarely made under full information on the potential network. The resulting incompleteness is caused by a geographical limitation of the actor's scope and his subssector fixation. This aspect seems to be open for improvement with government aid as part of an industrial policy (see Schenk and Kamann, 1987).

This stage of the selection process is largely determined by the personality of the actor, his business routines and goals (Kamann, 1986a, 1988a). When we leave the holistic concept of the firm and add the behaviour of more actors within the firm, we in fact add actors in different functions, roles and tasks, where each actor has his own personality and groups of actors show subs cultures and coalition behaviour; "individuals are multidimensional and have many roles in the total network" (Hamfelt and Lindberg, 1987, p. 180). This implies that the proper mix of actors is of great importance to the organisation. It also means that a single actor firm with restricted time available, will find it hard to have proper network contacts he actually should have. A larger multiple actor organisation can specialise and distribute the various contacts. Although this requires an internal network that also consumes time, special gatekeepers act as interface with the environment and provide relevant external information, translated into the jargon of the internal network (Allen, 1977; Tordoir, 1984). As a result of this, activities that draw their information from the internal network are less dependent on location in the source area of their relevant information - the focus - as long as the gatekeepers are in the right spot.

4.4.4 Motives for participation and types of partnership

A part of the network participation is directed towards collecting information about customers, suppliers, the competitors and his products, prices, R&D programmes and policies, new markets and changes in market shares of existing markets. Keeping or increasing the market share still is one of the most important strategies of actors. Here, in particular the information contents are important. However, in his attempts to increase his share of the market or simply to meet the challenge posed by competitors or newcomers, an actor is likely to run into obstacles or may well find some bottlenecks when implementing a proper and adequate strategy. The bottlenecks can be in research, development, finance, production, distributing, marketing, organisation and so on. It is realised by actors, that technological developments these days very rarely take place in a single firm. It is an "interplay between different organizations where independent activities are taking place simultaneously in different parts of the network" (Laage-Hellman, 1987, p. 31). This implies that firms realise that in a number of cases it is better to co-develop product and process innovations with suppliers, buyers or even competitors. "Different units have different resources and skills which are complementary in nature" (ibid. p. 37; see also Williamson, 1975). The value of the network in this case is the combination of 'resources' and 'skills' which as such is unobtainable for each of the actors involved. This positive value remains positive, even when the individual actor has to increase his dependency on other actors. Active network participation and distribution of activities enable increased specialisation of each of the actors, while increasing the need for interaction because of the required coordination. "The establishment of development relationships may in other words be a pre-condition for increasing specialization of the inshouse development process" (Axelsson, 1987, p. 131). Actors start partnerships depending on the bottleneck they find on their way and the type of specialisation they prefer (Laage-Hellman, 1987; Gold, 1986; Sundin, 1986). We find in practice that 'partnership' stands for a whole variety of forms.

Exchange of production between two or more competitors enable all actors involved to obtain scale effects. This may be required to compete with foreign producers.

Another form is the joint venture. This can be applied (de Jong, 1987) for:

(1): the development of products and the penetration of markets;

(2): the expansion on markets;

(3): consolidation of markets by horizontal integration;

(4): retreat from markets.

A third form of cooperation takes place using long term contracts between suppliers and one of their important buyers. They should guarantee timely and reliable deliveries, and enable suppliers to invest in innovations rquired by their buyer. One step further is the joint production planning between suppliers and buyers to enable zero-stock-inventories and just-in-time deliveries. The final step on this scale is the joint planning of suppliers and their industrial buyers where the latter give active support in R&D and capital.

4.5 The latent network relations and their characteristics

In manifest network relations three categories of latent features may be distinguished: (see also Håkansson, 1987, p. 84):

- multiple dependencies
- instability or fluidity
- paradigm fixation

The two forms of paradigm fixation, viz. the technological trajectory (Dosi, 1983; 1983) and the social network aspect (Storper, 1986; Kamann, 1988a), have already been discussed.

4.5.1 Multiple dependencies

Various activities between network participants are linked together. In transaction chains of actors various types of dependency between actors may be distinguised:

Technical dependency, where products and services fit technically together and result in inter-industry standards. A very rigid fit between products means an improvement of the product may upset the fit with products further upstream or downstream and therefore requires close co-operation between actors (the "Japanese model"; Korpel and Schenk, 1986). Actors that operate in a less rigid chain will find it easier to leave their network and enter new networks when they want to change their products.

Knowledge dependency means that the supplier has to know the requirements of those who use their products, while users have to know what they actually can do with their input materials, machines, hardware and sofware. Continuity dependence occurs, when a supplier sells a large share (e.g., more than 10 or 30 percent) of his output to a single buyer. The actual percentage is a function of the power of the actors involved, the profit margins and the profitability of other activities of the firm. Again, there is a reverse side; when producers are dependent on a single supplier for a particular product or service they require. Firms will try to prevent this situation by applying dual sourcing: at least two firms supply the same product. However, in a large number of cases, increased specialisation has led to single sourcing. The question is, whether it is cheaper, safer, better, faster, possible to take-over the actual supplier to ensure the continuity, make the product in-house or buy the product (Williamson, 1975, 1979).

Social dependency is the result of normal social group behaviour, where participants are likely to co-operate with other participants before establishing contacts with actors outside the network. Axelsson (1987, p. 159) uses the term soft distance to indicate the socioscultural distance between actors or entire networks: attitudes, values, norms, culture (Ronen, 1987). The term hard distance indicates the distance in kilometers.

Logistic and administrative dependency is of increasing importance in a time of improved information technology and the strategic role information handling plays these days. It means that suppliers and buyers have to use the same system to be able to communicate. Lack of standardisation between the various systems means in practice that small suppliers have to use their buyer's system and cannot easily switch to other buyers, with different systems.

Innovative dependency is in particular of interest for this paper, reason why we already discussed the question who gives the impetus to an innovation.

Financial dependency may have rather important effects. First of all, profits generated by the firm concerned will tend to be paid as dividends instead of reinvestments in new technologies and products. Secondly, the shareholding actor will obtain strategic information about new developments in techniques, products and markets and will know the actors strategy. A third point is that the actor involved may be forced to use licences, products or services supplied by its shareholder. Finally, existing actors in the firms network may be replaced by actors of the shareholders network; likewise the prices of supplies to existing network participants may be increased to weaken their competitive position on the market for final products. Complete take-overs tend to lead to centralisation of production activities in a particular location and centralisation of overhead activities in another location. The actor involved runs the risk of being converted into a standardized production branch plant. Even when it remains relatively independent as business unit, corporate planning, investment decisions and tax management are transferred to the parent company.

The various types of dependency - usually in a user dominant relationship - may result in a dual production organization (Berger and Piore, 1980). Based on the dual segmentation theory this theory assumes a primary and a secondary sector. Large and technically sophisticated corporations are part of the primary sector or core of the economy. They operate on stable and safe segments of markets applying modern, capital intensive Fordist mass production techniques. In the secondary sector or peripheral sector, relatively small firms operate with flexible technologies, catering for fluctuating and riskfull markets. Companies in the core sector externalize uncertainty and labour costs to the peripheral firms (Piore and Sabel, 1984; Stohr, 1985). In secondary firms, wages are low, prospects are poor, working conditions are bad. Allan Scott (1985, p. 17) for instance finds that many secondary firms in Orange County employ Mexican and Asian workers since these "cannot perceive or are able to demand ... recognition of rights".

Although it increases local linkages compared to the situation with branch-plants, it dominates local firms and segments the labour market. It is a new variant of Myrdal's (1954) cumulative causation.

4.5.2 Instability

Networks are not stable, they are volatile. We may divide the discussion about the dynamics of networks into three levels: the micro level, the meso level and the macro level.

The micro level

Individual actors grow when they are successful. They then evolve from a single establishment along the normal development path of functional-divisional to matrix organisations, with at present a movement to reduce decision lines resulting in semi-dependent Business Units (Chandler, 1973;

Galbraith and Nathanson, 1978; Wissema, 1987). Each step in this evolution results in a redistribution in space of functions, tasks, incomes earned and decision power. In the process of growth, takesovers and mergers are used, leading to horizontal, vertical and diagonal integration (Sherer, 1980; see James, 1985, for the various 'warsgames' involved). As a result of these take-overs, economic concentration increases, as does the role of internal networks compared to dependency on external networks.

The meso level: dynamics of the merger cycle

De Jong's (1976) merger cycle gives a theoretical concept to explain and predict the dynamics in merger behaviour. It shows that the reasons for merging activities or desintegration of activities differ and depend on the situation of the product concerned in the market. In the early stages of the product life cycle, backward integration may take place to secure appropriate deliveries of specific goods or to meet security risks (cf. Pavitt, 1984). The next bottleneck occurs when large scale production requires much capital. Inventors may join into partnerships and this early oligoploy (Dosi, 1983) may be successfull in raising capital. Others sell their company to large corporations. In the next stage, the market expands resulting in general scale effects which enables further specialisation on niches or segments of the market. Both horizontal and vertical desintegration occurs. When saturation of the market sets in, we will find actors try to consolidate their position. "... producers drop out because of losses [caused by overcapacity and reduced prices] or mergers occur in order to withstand and eliminate the heavy competition" (de Jong, 1979, p. 117). Again, we find oligopolies rising. Finally, when the market is about to collapse, sensible actors diversify, using the profits they earned in the final stages of the oligopolistic markets to purchase new firms and inventions that just started their cycle.

The concept of the life cycle has many weaknesses (Kamann, 1985a, 1986a; Storper, 1985; Schenk and Kamann, 1987). It has very little value ex-ante as a predicitive instrument and is ex-post only valid for mass-produced goods that reach the stage of mass production. Still, as a concept, it is the most attractive available when explaining mergers.

While the merger theory relates to mono-product firms, many actors in fact are multi-product firms, with products in various stages of their life cycles. This is one reason why actors these days do not merge all their activities, but only those related to a specific product or product-group. The remaining actors who deal with customized small-batch products are not included in the theory. Except that the technology used may be subject to a life cycle behaviour, which makes every step on the technological trajectory (Dosi, 1982; 1983) subject to a cycle.

The macro level: general trends

External to all these processes is a general trend that shortens life-cycles of products and techniques used. Together with a shift in the perceived core of innovative economic power, they force companies to resorganise, to adapt. The consequence for the organisation of production is that along two separate 'scales' new positions are developed. On one scale the two polarities: spatial product life cycle behaviour with head-offices and standardized branch plants as one pole and semi-dependent strategic business units. The poles on the other scale are mass-produced goods versus small-batch customized goods in flexible production. The two poles are in fact an organisational paradox. This paradox was solved by large corporations with the introduction of the dual production organisation.

Another macro trend is the increased interest for local initiatives, related to the organising power of actors in an area (Boekema and Verhoef, 1986). It results in a renewed interest in small firms of a more or less atomistic type. Some of these are organised in networks to overcome deficiencies in their selection environment (Stohr, 1985; Sundin, 1986; Piore and Sabel, 1983, 1984; Kamann, 1988a). Sometimes, they only have in common that they operate in the same location or area where they organise joint training facilities, financial institutions; in some cases competition is transformed into partnership.

4.6 Summary of the spatial aspects

(i) Spatial differentiation in the selection environment causes differences in opportunities for actors. This is a first cause for the dynamics of rising and dying actors.

(ii) New products and technologies give rise to new networks, possibly with different actors involved. Location of such a new network benefits the area concerned, while the location of the products substituted by the new network product will show decay.

(iii) Whether participants of a rising network actually receive an equal share of the benefits or not depends on the power relationship: equal partners or the dual production organisation.

(iv) Further points of importance are whether all activities-managerial, research, production, sales of the actors involved are located in one area or spread over the globe. Relevant is here, whether actors are part of a corporate network organised in the style of the spatial product life cycle or with business units.

(v) Take-overs of actors in a region by actors that operate within the same region will result in job losses because of rationalisation of production and overhead, but may increase the region's competiveness with actors from other regions. External control varying from minority shareholding to complete takesovers may in its worst form result in closure of other local actors in the network concerned and loss of jobs and capital. In cases where a large soft distance between headsoffice and local actor is involved, this may have long term consequences for investments in new activities in the area.

(vi) Areas with exclusively branch plant activities or with secondary firms operating in a dual production organisation are caught in a cumulative causation spiral. Dominance of the social networks by the branch plant managers and core actors may even be a serious obstacle for new activities.

(vii) The existence of a focus centre of activities that require many face-to-face contacts is of importance for an area. Lack of such a centre, or a distance to such foci that is prohibitive, reduces the potential for further development of existing activities and requires something completely new. These completely new activities however prove to be unpredictable in their locations of origin and further growth.

(viii) The concept of the extended filière, together with Pavitt's taxonomy is an instrument to visualise manifest network relations in economic space and is when translated into geographical space possible useful for regional industrial policies. Further research should be devoted to the development of models and techniques to measure latent network relations, since these determine the actual distribution of strategic information and incomes over the network participants.

5. Conclusions

The spatial differentiation in the selection environment is an important aspect of the origin of new ideas and innovations. It also plays an important role in the opportunities actors have to be informed about new developments of relevance to them. Whether they can actually use new developments is to a large extent determined by their network freedom and participation. Network cooperation between actors creates a synergetic surplus. Those actors who succeed to dominate other actors or even entire network segments will consume that synergetic surplus at the cost of others.

Areas where dominant actors are located will therefore through these network relations dominate areas where the dominated actors are situated. Distribution of power over the participants of a network and ability to monopolize strategic information in a network are of importance for the diffusion of innovations and for the related distribution of incomes generated.

Unfortunately, so far only a few analytical models have been devised that seek to cover the above mentioned conflicting patterns of spatial development.

References

Allen, T. (1977), Managing the Flow of Technology, M.I.T. Press, Boston.

Allen, P., Sanglier, M., Boon, F., Deneubourg, J.L. and de Palma, A. (1981), Models of Urban Settlements and Structure of Dynamic Self Organizing Systems, U.S. Dept. of Transportation, Washington D.C.

Altmann, N. and Bechtle, G. (1971), Betriebliche Herrschaftsstruktur Industrielle Gesellschaft.

Andersson, Å.E. and Johansson, B. (1984), Industrial dynamics, product cycles and employment, Working Paper of I.I.A.S.A., Laxenburg.

Axelsson, B. (1987), Supplier Management and Technological Development, in: H. Håkansson (ed.) 1987a, pp. 128-176.

Bearden, J. and Mintz, B. (1985), Regionality and Integration in the American Interlock Network, in: Stokman e.a. (eds.) 1985, pp. 234-249.

Berger, S. and Piore, M.J. (1980), Dualism and Discontinuity in Industrial Society, Cambridge.

Baranson, J. (1978), Technology and the Multinationals, : Lexington Press, Massachusetts, Toronto

Blommestein, H.J. and Nijkamp, P., Adoption and Diffusion of Innovation and Evolution of Spatial Systems.

Bluestone, B. and Harrison, B. (1982), The Industrialization of America, Basic Books, New York.

Boekema, F.W.M. and Verhoef, L.H.J. (1986), Lokale initiatieven: naar een nieuwe conceptie van regionale ontwikkelingen in theorie en praktijk, PhD thesis, Catholic University of Brabant, Tilburg.

Bollinger, L., Hope, K. and Utterback, J.M. (1983), A review of literature and hypothesis on new technology-based firms, Research Policy 12 1-14.

Boulianne, L.M. (1982), Technological change: firms and region, A case study, in: Maillat, D. (ed.) 1982, Technology, A key Factor in Regional; Development, Georgi, Saint Saphorin, pp 39-67.

Brotchie, J., Newton, P., Hall, P. and Nijkamp, P. (1985), The Future of Urban Form, Croom Helm, London.

Bramer, L. (1986), Bedrijf, vakblad, innovatie, mimeo, Geographical Institute, Groningen.

Bruno, A.V. and Tyebjee, T.T. (1982), The Environment for Entrepreneurship, Encyclopedia of Entrepreneurship, Prentice-Hall, Englewood Cliffs, pp. 288-307.

Burink, F. and Groenewegen, J. (1983), Industriebeleid à la Française, Economisch Statistisch Bulletin, pp. 622-627.

Camagni, R.P. (1985), Spatial Diffusion of Pervasive Process Innovation, Papers of the Regional Science Association, Vol. 58, pp. 83-96.

Cameron, G.C. and Johnson, K.M. (1969), Comprehensive Urban Renewal and Industrial Relocation - the Glasgow Case, in: Cullingworth, J.B. and Orr, S.C. (eds.), Regional and Urban Studies, pp. 242-271, George Allen and Unwin, London.

Chandler, A.D. (1973), Strategy and Structure: Chapter in the History of the American Industrial Enterprises, Cambridge (Mass).

Ciciotti, E. (1984), The incubation hypothesis revisited: the case of the metropolitan area of Milan, Paper presented at the 24th European Congress of the R.S.A., Milan.

Chiesi, A.M. (1985), Property, Capital and Network Structure in Italy, in: Stokman e.a. (eds.), 1985, pp. 199-214.

Christaller, W. (1933), Die zentralen Orte in Süddeutschland, Gustav Fischer, Jena.

Conference on Innovation Diffusion (1986), Venice, 17-22 March 1986.

Cook, K.S. and Emerson, R.M. (1978), Power, Equity and Commitment in Exchange Networks, American Sociological Review, Vol. 43, October, pp. 721-739.

Cooke, P. (1985), Regional innovation policy: problems and strategies in Britain and France, Environment and Planning C: Government and Policy, 3 253-267.

Cross, M. (1981), New Firm Formation and Regional Development, Gower, Aldershot.

Cunningham, M.T. and Turnbull, P.W. (1982), Inter-Organizational Personal Contact Patterns, in: H. Håkansson (ed.) 1982.

Czamanski, S. (1974), Study of clustering of industries, Halifax: Institute of Public Affairs, Dalhouse University.

Czamanski, S. (1976), Study of spatial industrial complexes, Halifax: Institute of Public Affairs, Dalhouse University.

Davelaar, E.J. and Nijkamp, P. (1986, 1987), The Urban Incubation Hypothesis: Old Wine in New Bottles?, Research Memorandum 1986-39, Free University, Amsterdam; 1987, Gesellschaft, Wirtschaft, Raum, (Fischer, M.M. and Sauberer, M.eds), AMR-Info, Vienna, pp. 198-213.

Davelaar, E.J., and Nijkamp, P. (1988), Spatial Dispersion of Technological Innovation, Journal of Regional Science, (forthcoming).

Dendrinos, D.A., and Mullally, H. (1983), Empirical Evidence of Volterra-Lotka Dynamics in US Metropolitan Areas, in: Griffith, D.A. and Cea, T. (eds.), Evolving Geographical Structures, Martinus Nijhoff, Dordrecht, pp. 170-195.

Dendrinos, D.A. and Mullally, H. (1984), Urban Evolution, Oxford University Press, Cambridge.

Devine, P.J. (1979), An Introduction to Industrial Economics, Allen and Unwin, London.

Doeringer, P. and Piore, M.J. (1971), Internal Labour Markets and Manpower Analysis, Lexington, Mass.

Docter, H.J. and Stokman, C.T.M. (1987), Diffusie van innovaties: met kennis meer kans, Economisch Instituut voor het Midden- en Kleinbedrijf, Zoetermeer.

Docter, H.J. and Stokman, C.T.M. (1988), Informatie en innovatie, Economisch Statistisch Bulletin, pp. 363-365.

Dosi, G. (1982, 1983), Technological paradigms and technological trajectories, in: Freeman, C. (ed.) 1983, Long Waves in the World Economy, Butterworth & Co. pp. 78-101; meer beleidsgericht in Research Policy, 1982, II.3.

Fagg, J.J. (1980), A Re-examination of the Incubator Hypothesis, Urban Studies, vol.17, pp. 35-44.

Flynn, N. and Taylor, A. (1986), Inside the rust belt: an analysis of the decline of the West Midlands economy. 2: corporate strategies and economic change, Environment and Planning A, Vol. 18, pp. 999-1028.

Galbraith, J.R. and Nathanson, D.A. (1978), Strategy Implementation: the Role of Structure and Process, St. Paul, Minesota, West.

Goddard, J. (1973), Office Employment, Urban Development and Regional Policy, Office Location and Regional Development, An Foras Forbartha, Dublin, pp. 21-36.

Gold, B. (1981), Technological diffusion in industry: research needs and shortcomings, Journal of Industrial Economics, 24.3, pp. 247-269.

Gold, B. (1986), Technological Change and Vertical Integration, Managerial and Decision Economics, Vol. 7, 169-176

Groenewegen, J. (1987), L'Etat développeur: de Franse weg, in: Schenk, E.J.J. (ed.) Industriebeleid in beweging, Wolters-Noordhoff, Groningen, pp. 97-120.

Haan, M.A. de (1986), Micro electronische bedrijvigheid in Zuid-Oost Nederland; een onderzoek naar economische en technologische ruimtelijke aspekten van een speerpuntindustrie, Studierapport 32, Rijksplanologische Dienst, Den Haag.

Hägg, I. and Johanson, J. (eds.)(1982), Företag i nätverk - ny syn på konkurrenskraft (Firms in networks - a nea view on competitive power), Stockholm.

Håkansson, H. (ed.)(1982), International Marketing and Purchasing of Industrial Goods, An Interaction Approach, John Wiley, Chichester.

Håkansson, H. (ed.)(1987a), Industrial Technological Development: A Network Approach, Croom Helm, London.

Håkansson, H. (1987b), Introduction in Håkansson 1987a; Product Development in Networks, id., pp. 84-127; Strategic Implications, id. pp. 210-219.

Hamfelt, C. and Lindberg, A.K. (1987), Technological Development and the Individual's Contact Network, in: Håkansson, 1987a, pp. 210-219.

Hansen, N. (1980): Dualism, capital-labor ratios and the regions of the U.S.: A Comment, Journal of Regional Science 20, pp. 401-403.

Hekman, J.S. (1980): The product cycle and New England textiles, Quarterly Journal of Economics 95, pp. 697-717.

Hellgren, B. and Stjernberg, T. (1987), Networks, An Analytical Tool for Understanding Complex Decision Processes, in: Johannisson (ed.) pp. 88-102.

Helmers, H.M., Mokken, R.J., Plijter, R.C., Stokman, F.N. and Anthonisse, J.M. (1975), Graven naar Macht: Op Zoek naar de Kern van de Nederlandse Economie, Van Gennep, Amsterdam.

Hendriks, A.J. (1986), Local Entrepreneurial Initiatives and central Government, in: Nijkamp, P. (1986), pp. 280-288

Hepworth, M.E. (1987) Geography and the information economy, Ph.D. thesis, Department of Geography, Toronto.

Hermansen, T. (1972), Development poles and development centres in national and regional development, in: Kuklinski, A.R. (ed.), Growth poles and growth centres in regional planning, Mouton, Paris/The Haguen.

Hippel, E. von (1977), The Dominant Role of the Users in Semiconductor and Electronic Subassembly Process Innovations, IEEE Transactions and Engineering Management, Vol. EM-24, No.2, pp. 60-71.

Hoogteijling, E., Gunning, J.H. and Nijkamp, P. (1986), Spatial aspects of innovation and employment: Dutch results, in: Nijkamp, P. (ed.), Technological Change, Employment and Spatial Dynamics, pp. 221-243, Springer, Berlin.

Hoover, E.M. and Vernon, R. (1959), Anatomy of a Metropolis, Cambridge University Press, Cambridge.

James, B.G. (1985), Business Wargames, Penguin, Harmonsworth.

Johannisson, B. (ed.)(1987), Organizing: The Network Metaphor, International Studies of Management and Organization, Vol. XVII, No. 1.

Johannisson, B. (1987), Beyond Process and Structure: Social Exchange Networks, in: Johannisson (ed.) pp. 3-23; Anarchists and Organizers: Entrepreneurs in a Network Perspective, idem, pp. 49-63.

Johanson, J. and Mattson, L.-G. (1984), Marketing Investments and Market Investments in Industrial Networks, Paper presented at the International Research Seminar on Industrial Marketing, Stockholm School of Economics.

Johansson, B. and Nijkamp, P. (1986), An Analysis of Episodes in Urban Event Histories, in: Van den Berg et al. (eds), Spatial Cycles, Gower, Aldershot, pp. 208-231.

Johnson, G. and Scholes, K. (1984), Exploring Corporate Strategy, London.

Jong, H.W. de (1976), Theory and Evidence concerning Mergers: an International Comparison, in: Jaquemin, A.P. and de Jong, H.W. (eds.) Markets, Corporate Behaviour and the State, Martinus Nijhoff, Den Haag/Boston.

Jong, H.W. de (1987) Joint ventures, E.S.B., 6 mei 1987, pp. 419-425

Jong, H.W. de (1988), De concentratiebeweging in de Amerikaanse/Europese economie; verklaring en resultaat, Economisch Statistische Berichten, pp. 203-207; 224-229; 248-253.

Jong, M.W. de, and Lambooy, J. (1985), Innovativiteit en stedelijke milieus, in: Molle, W.T.M. (ed.), Innovatie en Regio, Staatsuitegeverij, The Hague.

Kamann, D.J.F. (1978), Economics Impact Study of a proposed industrial site at Keenaghan, Carrickson-Shannon, An Foras Forbartha, Dublin.

Kamann, D.J.F. (1985a), Technologie en ruimte: steeds nieuwe paradigma's, in: Brouwer, M.T. and Ter Hart, H.W. (eds.) Ondernemen in Nederland, Kluwer, Deventer, pp. 171-193.

Kamann, D.J.F. (1985b), Network Dynamics of Economic Activities, paper presented at the Dutch-German seminar on regional underemployment, Karlsruhe, forthcoming in: Funck, R. and Oosterhaven, J., Karlsruher Papers on Policy Research.

Kamann, D.J.F. (1986a), Industrial Organisation, Innovation and Employment, in: Nijkamp, P. (ed.) 1986, Technological Change, Employment and Spatial Dynamics, Springer, Berlin, pp. 131-154.

Kamann, D.J.F. (1986b), Network Dynamics, Lecture at the Summer Institute of the RSA, Umeå (Sweden), June.

Kamann, D.J.F. (1988a), Spatial Differentiation in the Social Impact of Technology, Avebury (Gower), Aldershot.

Kamann, D.J.F. (1988b), Externe organisatie: een inleiding (Introduction to Industrial organisation), Groningen: Group for Research on Industrial Policies.

Kamann, D.J.F., Zwart, P.S. and Strijker, D. (1988), Offers en inzet voor een nieuwe koers, rapport in opdracht van de COR van de Coöperatieve Melkproductenbedrijven "Noord-Nederland", Group for Research on Industrial Policies, Groningen.

Kamata, S. (1985), Japan in the passing lane, Counterpoint/Unwin, London.

Knickerbocker, (1973), Oligopolistic Reaction and Multinational Enterprise, Ph.D. thesis, Harvard University Press, Boston.

Laage-Hellman, J. (1987), Process Innovation through Technical Cooperation, in Håkansson 1987a, pp. 26-83.

Laurier, J., De Graaf, P. and Läpple, D., Nieuwe patronen in de regio (RUGEON), R.O.V./Rijksuniversiteit Leiden.

Leone, R.A., and Struyk, R. (1976), The incubator hypothesis: Evidence from five SMSAs, Urban Studies 13, pp. 325-331.

Leontief, W. et. al. (1953), Studies in the Structure of American Economy, London.

Levitt, R.L. (ed.)(1985), Research Parks and Other Ventures: the University/ Real Estate Connection, The Urban Land Istitute, Washington, D.C.

Levy, J.M. (1983), Economic and social factors influencing future industrial land use in the U.S., International Workshop, Canada.

Lewin, K. (1951), Field Research in Social Science, Harper and Brothers, New York.

Machin, H. and Wright, V. (1985), Economic policy and policy-making under the Mitterand Presidency 1981-1984, New York.

Malecki, E.J. (1984), High technology growth policies and prospects for local economic development, Journal of the American Planning Assocation 50, pp. 262-269.

Malecki, E.J. (1985), Industrial Location and Corporate Organization in High Technology Industries, Economic Geography, vol. 61, pp. 345-369.

Malecki, E.J. and Nijkamp, P. (1988), Technology and Regional Development: Some Thoughts on Policy, Environment & Planning; Government & Policy, (forthcoming).

Melin, L. (1983), Structure, Strategy and Organization a Case of Decline, Proceedings of the Workshop on Strategic Management under Limited Growth and Decline, Report 83-1, EIASM Institute, Brussels.

Menzel, H. (1967), Planning the Consequences of Unplanned Action in Scientific Communication, in: Reuck, A. and Knight, J. (eds.) Communication in Science, J&A Churchill, London.

Mitchell, J.C. (1973), Networks, Norms and Institutions, in: Boissevain, J. and Mitchell, J.C. (eds.), Network Analysis: Studies in Human Interaction, Mouton, The Hague, pp. 15-35.

Mickler, O. e.a. (1977), Technik, Arbeitsorganisation und Arbeit. Eine empirische Untersuchung in der automatisierte Produktion, Frankfurt

Mills, C.W. (1956), The Power Elite, Oxford University Press, New York.

Molle W.T.M. (1983a), Industrial Location and Physical Development in the European Community, the FLEUR Model, Gower, Aldershot.

Molle W.T.M. (1983b), Technological Change and Regional Development in Europe, Papers of the Regional Science Association, Vol. 52, pp. 23-38.

Mowery, D. and Rosenberg, N. (1979), The influence of market demand upon innovation: a critical review of some recent empirical studies, Research Policy, 8.

Moseley, M.J. (1974), Growth Centres in Spatial Planning, Pergamon Press, Oxford.

Mueller, W.F. and Culbertson, J.D. (1986), Inter-industry technology Flows in the US Food-processing Industries, Managerial and Decision Economics, Vol. 7, pp. 163-168

Myrdal, G. (1957), Economic Theory and Underdeveloped Regions, Duckworth, London.

Nelson, R.R. and Winter, S.G. (1977), In search of a useful theory of innovation, Research Policy, Vol. 6, pp. 36-76.

Neun, S.P. and Santerre, R.E. (1986), Dominant Stockownership and Profitability, Managerial and Decision Economics, Vol. 7, Nr. 3, pp. 207-210.

Nijkamp, P. (1985), Metropolitan Dynamics: The Survival of the Fittest, Revue d'Economie Régionale et Urbaine, no. 5, pp. 204-212.

Nijkamp, P. (ed.)(1986), Technological Change, Employment and Spatial Dynamics, Springer, Berlin.

Nijkamp, P. (1987a), Information or Oracles in Regional Development, in: Giaoutzi, M. and Nijkamp, P. (eds.), Informatics and Regional Development, Gower, Aldershot, pp. 1-21.

Nijkamp, P. (1987b), Long-Term Economic Fluctuation: A Spatial View, Socio-economic Planning Sciences, vol. 21, no.3, pp. 189-197.

Nijkamp, P., and Schubert, U. (1985), Structural Change in Urban Systems, Sistemi Urbani, no. 2, pp. 155-176.

Nijkamp, P., Poot, J. and Rouwendal, J. (1988), R&D Policy in Space and Time, Paper European Congress Regional Science Assocation, Stockholm.

Norton, R.D., (1979), City Life Cycles and American Urban Policy, Academic Press, New York.

Oakey, R.P. (1981), High Technology Industry Location, Gower, Aldershot.

Oakey, R.P. (1983): New Technology, government policy, and regional manufacturing employment, Area 15, 61-65.

Oomkes e.a. (1985), De Zuivelindustrie, Wetenschapswinkel voor Economie RUG, Groningen.

Oosterhaven, J. (1981), Interregional Input Output Analysis and Dutch Regional Policies Problems, Gower, Aldershot.

Paelinck, J.H.P. (1965), La Théorie du Développement régional polarisé, Cahiers de l'ISEA, Série L, No. 15, pp. 10-11.

Pavitt, K. (1984), Sectoral Patterns of technical change: towards a taxonomy and a theory, Research Policy, No. 13, pp. 343-373.

Pennings, J. (1980), Interlocking Directorates, Jossey-Bass, San Francisco.

Perroux, F. (1955), Note sur la notion pôle de croissance, Economie Appliqué, Série D, Vol. 8.

Piore, M.J. and Sabel, Ch. (1983), Italian Small Business Development: lessons for U.S. Industrial Policy, in: Tyson, L. and Zysman, J. (eds.), American Industry in International Competition, Ithaca, pp. 391-421.

Piore, M.J. and Sabel, Ch. (1984), The Second Industrial Divide, Basic Books, New York.

Porter, M.E. (1980), Competitive Strategy, Collier-MacMillan, London.

Porter, M.E. (1983), The Technological Dimension of Cooperative Strategy, Research on Technological Innovation, Management and Policy, Vol. 1, JAI Press, Greenwich, Conn.

Pred, A.R. (1966), The Spatial Dynamics of U.S. Urban-Industrial Growth, 1800-1914, MIT Press, Cambridge, Mass.

Pred, A.R. (1977), City-Systems in Advanced Economies, John Wiley, New York.

Premus, R. (1982), Location of High Technology Firms and Regional Economic Development, OECD, Paris.

Premus, R. (1985), The High-Tech Market for University Research Parks, in: Levitt, R.L. (ed.), pp. 24-33.

Rees, J. (1979), Technological Change and Regional Shifts in American Manufacturing, Professional Geographer 31, pp. 45-54.

Richardson, H.W. (1979), Aggregate Efficiency and Interregional Equity, in: Folmer, H. and Oosterhaven, J. (eds.), Spatial Inequalities and Regional Development, Martinus Nijhoff, Boston.

Rima, A., and Van Wissen, L.J.G. (1988), A Dynamic Model of Household Relocation, North-Holland Publ. Co., Amsterdam.

Rogers, E.M. and Larssen, J.K. (1987), Silicon Valley Fever, Basic Books, New York.

Ronen, S. (1987), Comparative and Multinational Management, John Wiley & Sons, New York.

Rothwell, R. (1982), The Role of Technology in Industrial Change, Regional Studies 16, pp. 361-369.

Schenk, E.J.J. and Kamann, D.J.F. (1987), Naar een ruimtelijk georienteerd specifiek technologiebeleid, in: Schenk, E.J.J. (ed.) Industriebeleid in beweging, Wolters-Noordhoff, Groningen, pp. 221-260.

Scherer, F.M. (1980), Industrial Market Structure and Economic Performance, Rand McNally, Chicago.

Scott, J. (1985), Theoretical Framework and Research Design, in: Stokman e.a. (eds.) 1985, pp. 1-9.

Scott, A.J. (1985), High Technology Industry and Territorial Development: The Rise of the Orange County Complex, 1955-1984, Working Paper Series, nr. 85, Department of Geography, University of California, Los Angeles.

Scott, A.J. and M. Storper (1987), High Technology Industry and Regional

Development: A Theoretical Critique and Reconstruction, International Social Science Journal, UNESCO, Paris.

Shapero, A. (1971), An Action Program for Entrepeneurship, Report prepared for the Ozarks Regional Commission (Multi-Disciplinary Research Inc., Austin, TX).

Stohr, W. (1985), The Spatial Division of Labour and Entrepreneurial Strategies; Regional Development Strategies and the Spatial Division of Labour; International Economic Restructuring and the Territorial Community, UNIDO/IS571; resp. pp 37-56 and 221-237.

Stohr, B.W. and Tödtling, F. (1979), Spatial Equity: Some AntisTheses to Current Regional Development Doctrine, in: Folmer, H. and Oosterhaven, J. (eds.), Spatial Inequalities and Regional Development, Martinus Nijhoff, Boston.

Stokman, F.N., Ziegler, R. and Scott, J. (eds)(1985), Networks of Corporate Power, Polity Press, Cambridge.

Stokman, F.N., Wasseur, F.W. and Elsas, D. (1985), The Dutch Network: Types of Interlocks and Network Structure, in: Stokman e.a. (eds.) 1985, pp. 112-130.

Storper, M. (1985), Essentialism in economic geography: oligopoly and the product cycle, Paper, University of California, Los Angeles.

Storper, M. (1986), Technology and New Regional Growth Centres: the Economics of Discontinuous Spatial Development, in: Nijkamp, P. (ed.) 1986, pp. 46-75.

Suarez-Villa, L., The Evolution of Regional Economics, 1988 (forthcoming).

Sundin, E. (1986), The Entrepreneurial Setting of Regional Economic Renewal, Paper presented at the European Advanced Summer Institute of the RSA, Umea, Sweden, June 1986

Thomas, M.D. (1987), Schumpeterian Perspectives on Entrepreneurship in Economic Development: a Commentary, Geoforum, vol. 18, no. 2, pp. 173-186.

Thorngren, B. (1970), How Do Contact Systems Affect Regional Development?, Environment and Planning, Vol 2, pp. 409-427.

Thorngren, B. (1972), KOMM'71 Preliminary Summary, mimeo.

Thorngren, B. (1973), Communication Studies for Government Office Dispersal in Sweden, Office Location and Regional Development, An Foras Forbartha, Dublin, pp. 47-58.

Thwaites, A.T. (1982), Some evidence of regional variations in the introduction and differentiation of industrial products within British industry, Regional Studies 16, pp. 371-381.

Tichy, N.M. (1981), Networks in Organizations, in: Nystrom, P. and Starbuck, W. (eds.), Handbook of Organizational Design, Oxford University Press, Oxford, pp. 225-249.

Tordoir, P. (1984), De relaties tussen technologische innovatie en agglomeratie, EGI paper No. XXVIII, University of Amsterdam, Economic-

Geographical Institute, Amsterdam.

Törnqvist, G. (1970), Contact Systems and Regional Development, Lund Studies in Geography, Series B, No. 35, Gleerup, Lund.

Uitham, C., De Vries, B. and Zijlstra, G.J. (1978), Het Nederlandse Kernenergiebeleid, Tijdschrift voor Politieke Economie, No.1, pp. 106-150.

Utterback, J.M. and Abernathy, W.J. (1975), A Dynamic Model of Process and Product Innovation, Omega, Vol. 3, No. 6, pp. 639-656.

Vernon, R.F. (1966), International investment and international trade in the product cycle, Quarterly Journal of Economics, Vol. 80, pp. 190-207.

Vlieg, F. (1977), De West-Duitse industrie achter kernenergie, mimeo, University of Groningen, Groningen.

Wärneryd, O. (1968), Interdependence in Urban Systems, Regionkonsult, Gothenburg.

Wettman, R.W. (1983): Technological development in a regional context: The case of Germany, Paper OECD Workshop on 'Technology and Regional Policy', Paris.

Wever, E. (1984), Nieuwe Bedrijven in Nederland, Van Gorcum, Assen.

Williamson, O.E. (1975), Markets and Hierarchies: Analysis and Antitrust Implications, Free Press, New York.

Williamson, O.E. (1979), Transaction-Cost Economics: The Governance of Contractual Relations, The Journal of Law and Economics, 22, pp. 233-261.

Wissema, J.G. (1987), Unit Management, Van Gorcum, Assen.

Ziegler, R., Bender, D. and Biehler, H. (1985), Industry and banking in the german Corporate Network, in: Stokman e.a.(eds.), pp.91-111.

Ziegler, R., Reissner, G. and Bender, D. (1985), Austria Incorporated, in: Stokman e.a. (eds.), pp. 73-90.

11 R & D technology transfer in economic development: the role of regional technological capability

E. J. MALECKI

1. Introduction

The process of economic development is in large part an outcome of the mechanisms by which technology is acquired and utilized. Technology originates primarily through research and development (R&D), the deliberate effort to discover and learn about new or improved products and processes (Malecki, forthcoming). Within nations, R&D resources exhibit a marked geographical concentration, and are thought to be a central cause of regional development differences. (See the chapter in this volume by Thwaites and Alderman.) An R&D effort relies critically on a pool of technical workers, whose talents help them to seek out solutions from a variety of sources. The "technological capability" embodied in this subset of the labor force - whether employed in private or government-controlled enterprises - also provides a firm, region and nation with the skills necessary to evaluate new technologies as they emerge in other places. If technology is not developed internally then it must be acquired or transferred from elsewhere. As new techniques and products are learned, acquired, evaluated and improved upon, a firm or region comes to know about best-practice technology and to be prepared for new and substitute products.

The transfer of technology takes place via investment in branch production plants by transnational firms, through licensing, and by way of other agreements between firms in different countries. In all of these cases the control of the technology remains at its origin. There are other forms of learning of new technology, including learning by using, improving, and monitoring equipment and processes obtained outright (Bell, 1984). The commonality among all of these forms of technology acquisition is that they entail dependence of a firm, region, or country on external sources of technology, and can perpetuate the dependence of some regions and nations on others, particularly if a region's own technological capacity is weak (Fransman and King, 1984; Molero, 1983).

Regions are not isolated, especially in the context of science and technology. Even nations are unable to contain knowledge and the fruits of

R&D within their borders. "Technological knowledge and capability are international rather than national" (Nelson, 1984, p. 75). They certainly are not regional or local, but there are important connections between the local economy and international and national dynamics (Nijkamp and Schubert, 1985; Rees, 1987). Concerns over capability and competitiveness have been raised frequently in recent years, primarily in an international context (Cohen and Zysman, 1987; Guile and Brooks, 1987; Spence and Hazard, 1987; Teece, 1987). Regional fortunes are dependent on the activities of the firms operating within them, on the level and types of government research activities, on the strength of their regional institutions, such as universities, and on the nature of the regional culture regarding education, innovation, and entrepreneurship. These constitute the elusive "regional factor" which has proven difficult to define exactly (Thomas, 1987). However, unlike scientific knowledge and technology, skills and entrepreneurship are more local phenomena, not as readily transferable from place to place. They depend on local economic structure as well as on regional attitudes and culture (Brugger and Stuckey, 1987; Sweeney, 1987).

The transfer of technical knowledge and capability from one place to another is an aspect of regional change to which little attention has been paid in the regional literature (Malecki, 1983), but is has links to the large body of research on the diffusion of innovations (Brown, 1981). As with innovation diffusion, technology transfer has a cost; however, technology transfer may also have restrictions on how it can be utilized and learned by its recipient (Giaoutzi, 1985; Stewart, 1981).

This chapter reviews the topics of R&D and technology transfer from a development perspective. Examples are drawn from both the developed and the underdeveloped worlds, because it is clear that there is much to be learned from the experience of Third World nations as some succeed in developing a technological capability, while others do not.[1] The notion of **technological capability** provides an operational, if complex, concept for assessing the development situation of regions and nations.

2. Technology Transfer

Technology transfer is defined in many ways, but generally as "the conveyance of either a man-made tangible good/process or intangible know-how from those who possess it to those who do not" (Frame, 1983, pp. 72-73).

---

[1] For a useful review, see Fransman (1986a), and the reviews and case studies in Fransman and King (1984), Fransman (1986b), and McIntyre and Papp (1986).

Although "technology is not usually produced directly for sale", its transfer involves the transfer of production capacity to a different firm or country (OECD, 1981). Any technology transfer process has six elements: (1) a transfer item (product, process, know-how), (2) a technology donor, who may possess ownership rights to the transfer item, and who may or may not be willing to make the technology accessible to others, (3) a technology recipient, (4) a transfer mechanism, (5) a rate of diffusion of technology, and (6) the absorptive capacity of the recipient (Frame, 1983).

It is a form of innovation diffusion where the adopters are organizations rather than individuals, and where substantial knowledge (absorptive capacity) may be required in order to use an innovation effectively. The amount of knowledge necessary may be quite small, as in the case of turn-key plants, or it may involve the recipient substantially defining the technology. Innovation diffusion among firms was somewhat underplayed in urban and regional research for many years, and this weakened its ties to the larger literature in economics and business (Camagni, 1985; Malecki, 1977, 1983; see also the chapter by Orishimo in this volume). In some respects, the topic of technology transfer itself has only recently entered the literature on the economics of technological change. Mansfield (1968) and Rogers (1983) failed to cover the issue of technology transfer at all. Rosegger (1980) and Stoneman (1983), however, make a close link between international diffusion of innovation and technology transfer, defining the latter as the spread of innovations among industries and among economies.

The term technology transfer is usually used to describe international transactions, but it applies to domestic situations as well (Frame, 1983). Technology transfer occurs when firms purchase products from each other, license each other's technology, trade patents, or examine competitors' products. Such commercial transactions entail the purchase of tangible goods and intangible know-how. Technology transfer is also used to refer to the international transfer of a new production process within a multinational firm (Prasad, 1986). Many nations have programs to facilitate the transfer of technology developed in their government R&D laboratories to private sector firms (Chaudhuri, 1986; Hough, 1975).

Technology transfer may be thought of as two distinct processes (Headrick, 1988, p. 8). One is the relocation, from one area to another, of equipment and methods, along with the experts to operate them. The other is the diffusion from one society to another of the knowledge, skills, and attitudes related to a particular device or process. In Headrick's view, the geographic relocation indeed is distinct from the cultural diffusion of technology.

Transfer agents include exporters (salesmen, foreign aid officials, engineers on contract) -diffusion agents in Brown's (1981) terminology- who specialize in the geographic relocation of technology. Their expertise depends on an absence of native experts. Importers of technology, on the other hand, such as students, purchasing agents, and spies, are more likely to seek its cultural diffusion and to spread a particular device or process as a means of developing human capital and complementary technologies in their home countries (Headrick, 1988, p. 8-10). Graham (1988) has shown this in the case of the Brazilian firm Engesa's simultaneous acquisition and internal development of technologies for its military vehicle products, and Bae and Lee (1986) and Kim et al. (1987) show much the same process in Korean industry.

The success of technology transfer depends more on the ability and willingness of the importing society to accept the technology than on the support of the exporter. Technology has increased over the past century, and technology transfer has become more difficult from the point of view of the recipient. Where preindustrial craftsmen could formerly understand, copy, and improve technologies, this now requires knowledge of science and engineering, not to mention literacy. As a consequence, underdeveloped countries now must turn to the advanced countries for their capital equipment, thereby losing the benefits - multiplier effects, backward linkages, externalities - of the capital goods industries (Headrick, 1988, p. 11). Katz (1982a, 1982b) underscores the difficulties confronting Third World countries in attempting to create competitive local industries. Firms must be able to recognize certain technologies as outdated and inadequate, and be able to identify technologies which need to be acquired from foreign sources rather than be developed (with greater delay) domestically. Katz (1982b) suggests that the state and its S&T bodies can act as a "talent scout" in the field of world-wide technology and as a centralized buyer of technology in world markets. The latter is increasingly common via various countertrade arrangements whereby technology is purchased subject to the seller taking specified products in return (Alexandrides and Bowers, 1987). Local or domestic technological capability is indispensable if transferred technology is to be altered, modified and adapted to local conditions. Local learning by doing and informal skills are more important than R&D in mastering imported technology (Fransman and King, 1984; Segal, 1987). Cohen and Zysman (1987) reiterate this point in their contrast between strong American R&D and strong Japanese production and organizational skills.

Technology transfer without development of technological capability has been especially common as developing countries attempted to increase their production capacities in a minimal amount of time. Headrick (1988, p. 13), using his distinction identified above, explains:

> "Many factors favor the geographic relocation of technology over its cultural diffusion. Exporters of Western technology have an interest in selling machines, spare parts, and expertise. For importers, geographic relocation, while costlier, is faster and entails fewer risks than cultural diffusion. And governments need foreign equipment and experts, but shy away from the political dangers involved in cultural or social changes. Thus risk aversion leads to technological dependency. The cultural diffusion of technology, in contrast, is a much more difficult task. It takes a willingness to accept changes, a strong political cohesiveness, and a common vision of the future. Western societies, facing less culture shock, have readily imported industrial technologies. Among non-Western societies, only Japan had the requisite cultural and social base. Others have had to undergo political revolutions first" (p. 13).

In the Third World, scientific and technological activities are less well developed, especially with regard to their linkage and integration with production activities (Fransman and King, 1984; Silveira, 1985). The separation in developing countries between R&D and production highlights the fact that neither fully accounts for technological capability (Bhalla and Fluitman, 1985; Segal, 1987; Sharif, 1986). R&D is essential to the improvement and innovativeness of technologies over the longer run, and it makes it possible to obtain or adopt technologies which are nearer to the international state-of-the-art. Production skills allow for continual innovation and significant inputs to the design of new products and processes.

The question of whether transferred technology is "inappropriate" for less developed countries arises frequently (Bhalla, 1982; Clark, 1985; Dicken, 1986, pp. 361-370; Stewart, 1981, 1987). Perhaps more importantly for Third World nations, however, technology transfer affirms their dependence on nations and firms from which they obtain technology. Control over the pace and form of technology remains where R&D and production process technology are ongoing (Buttel et al., 1985; Kaplinsky, 1984). Suarez-Villa (1987) maintains that international diffusion of process technologies and know-how takes place mainly in the very late or mature phases of the product life cycle when production skills are relatively more important.

A standard of competitiveness, **technological mastery,** is seen as a goal, whether by firms or by nations. It requires a series of steps: (1)

production engineering, or the operation of existing plants; (2) project execution, or the establishment of new production capacity; (3) capital-goods manufacture, embodying technological knowledge in physical equipment and facilities; and (4) R&D, the specialized activity to generate new technological knowledge (Dahlman and Westphal, 1982). These broaden considerably the R&D-focused technological indicators on which most regional science research has dwelt.

Increasingly, too, corporate R&D is seen to be an inadequate indicator of technological capability for industrial firms. It has never been quite accurate to confine a firm's technologically-oriented activities to R&D alone (Gold, 1979). Contracted research at universities or other firms, acquisition or equity participation in other firms, licensing, joint ventures in research and production - either with competitors or with suppliers - are now commonplace (Friar and Horwitch, 1986; Graham, 1986). Global strategies are deemed imperative in order to compete (Hout et al., 1982; Porter, 1983; Spence and Hazard, 1988). Especially in Europe, where national firms are constrained to relatively small domestic markets, coordinated R&D and technology in information technologies (the ESPRIT program) and in biotechnology are taking place (Nueno and Oosterveld, 1986; Sharp and Shearman, 1987). Japanese firms have led the way in demonstrating the competitiveness provided by cooperation with suppliers and an emphasis on organization and manufacturing skills. It may be that production capability is more critical to competitiveness than is R&D (Cohen and Zysman, 1987).

Traditionally, corporate R&D is concentrated primarily in product development. Knowledge transfer and feedback occur throughout the process of technological change and in all of its stages (Kline and Rosenberg, 1986). However, firms must be able, once they have created or acquired technology, to nourish and sustain it in production. Especially for process technology, learning by using and by doing are essential components of the technological activities of firms (Dosi, 1984). Firms have three ways of adding to their stock of knowledge: through learning by doing and using, through R&D (or learning by studying), and through imitating others (diffusion between firms) (Dutton and Thomas, 1985). Technological capability, although often measured by R&D effort or R&D personnel, includes a wider array of technological assets, and draws upon a wider variety of information sources, including customers and suppliers as well as in-house capabilities.

Within large firms with differentiated facilities, product development and refinement customarily occur within product-line labs and new-product plants, rather than in central research facilities (Malecki, 1987). Interaction or "coupling" with marketing and with manufacturing become critical as product design, production engineering, and market acceptance all must be assessed and incorporated into decision-making. These tasks recur intermittently throughout a product's "life", and perhaps more within firms which specialize and innovate as part of their routine activity (Sabel et al., 1987). More typically in the product cycle pattern, product development peaks early in the commercial life of a product, whereupon process innovation begins to take precedence as the firm prepares for large-volume production. The information sources likewise vary across the stages (de Meyer, 1985).

The product cycle model, especially its regional variant, has undergone much criticism in recent years (De Bresson and Lampel, 1985; Steiner, 1987; Taylor, 1986). Basically, the criticisms address two elements. First, products are rarely constant and unchanging, even when in mass production and somewhat standardized. Improvements are made, often in back-and-forth competition between rival firms; production techniques are enhanced. Second, not all products fit the pattern of achieving mass production where low cost and high volume are the norm. Many products, especially custom and small-batch products, never attain large volumes (De Bresson and Lampel, 1985; Kim and Lee, 1987). Particularly if they emphasize unique processes, their location need not be in agglomerations, and a rigid hierarchy of regions based on innovativeness need not be defined (Steiner, 1987).

The geographical distribution of product and process R&D is dependent on the locational decisions of firms for their R&D facilities, for the work of product development and its overlaps with other corporate functions are highly reliant on professional and technical workers (Malecki, 1980, 1987). The locational concentration of R&D in every country favors established regions (especially national capitals) and reinforces the regional economic growth of those areas where universities, industrial R&D, and national government R&D facilities and contracts are plentiful.[2] These concentrations of R&D provide an important, but not a comprehensive, indicator of regional technological capability.

3. Best-Practice and Regional Competitiveness

Technological capability is not fixed or permanent, since technology

---

[2] See Breheny and McQuaid (1987) for references to studies of this type.

and the abilities of competitors are constantly changing (Dosi, 1984; Nelson and Winter, 1982). It relies on firms, on their activities, and how close they are to the state-of-the-art (**best-practice**) at any point in time. "Best practice" technology typically refers to process technology (machinery, equipment, management practices) which ranges from that of the most efficient (best-practice) producers to the most inefficient (worst-practice) (Le Heron, 1973; Malecki, 1983, pp. 93-94). The spectrum from best-practice to worst-practice technology is largely a function of the age or vintage of capital equipment employed. Newer vintages will incorporate or embody newer concepts, techniques, and knowledge which tend to give an advantage to firms - and regions - where newer technology is employed (Varaiya and Wiseman, 1981). Firms and countries which are not near the current "technology frontier" both in science and in production find it increasingly difficult to keep up with changes in other places (Cohen and Zysman, 1987; Katz, 1982a; Spence and Hazard, 1988).

The level of technology in a region, however, goes beyond considerations of machinery and plant design. It encompasses the stock of knowledge within a firm or region concerning both what is being made and how things are made. Technically progressive firms obtain knowledge from customers and suppliers as well as generate it internally. For a region, links to outside information networks and a high level of technical skill define a level of best-practice that in many respects is cultural and which goes well beyond what is encompassed by R&D (Brugger and Stuckey, 1987; Sweeney, 1987; see also the chapter by Kamann and Nijkamp in this volume). Most of all, then, technological capability relies on people in technological roles within organizations. Not all people will be able to serve as "gatekeepers" but their role is pivotal. For regions, the gatekeeper function also is critical, if informal. (The chapter by Kamann and Nijkamp expands on some of these issues.) Sweeney (1987) suggests that "extension agents", much like those who have helped to diffuse agricultural technology around the world, can serve this purpose.

Sweeney (1987) believes that technology transfer depends critically on the level of information that is "in the air" in a locale (Andersson, 1985; Rogers, 1982). This quantity of information hinges on the firm size mix in the region: small firms much more willingly share information with each other, largerly because this open flow has mutual benefits. Large firms are more likely to have to devote much of their information-oriented time to intra-firm communications, such as between any facility and the firm's headquarters or between facilities. This is particularly true of R&D, for which interaction both with headquarters and with production plants and sales offices is crucial.

Johansson and Westin (1987) suggest that a region where imports are diverse and state-of-the-art are most likely to be innovative themselves.

> The critical element in the process is the 'innovation response to import flows.' If the import node and its surrounding region of smaller nodes have a sufficiently rich production structure embedded in an active R&D environment, with access to a creative and knowledge-oriented labor force, then this together with the import of new products constitute a prerequisite for successful innovations. The latter may range from new products in the form of imitations and modifications of imported products adapted to local demand to the creation of products which are complementary to developing import flows. When this type of response is successful, it brings about the birth of local product cycles (Johansson and Westin, 1987, p. 15).

The diffusion of process innovation and the resulting distribution of best-practice (and older) technologies is a decisive determinant of regional economic competitiveness. To a large extent, however, it remains somewhat a mystery as to why investment in new technologies varies from region to region. The simple and attractive finding, that the largest firms adopt earliest, has numerous proponents and supporters (Ettlie and Rubenstein, 1987; Ewers, 1986; Rees, Briggs, and Oakey, 1984; Thwaites, 1982). Large firms also do most of the R&D in any industry, bolstering the relationship between R&D and technological capability at the level of the firm. For regions, the implication is often that R&D leads to general economic competitiveness (Meyer-Krahmer, 1985).

The view that innovation is a continual process thrusts some complications into this argument. Rosenberg (1982) and others have noted that innovations - especially process innovations - do not remain constant during their diffusion. Rather, in large part because most process innovations are the product innovations developed by supplier firms, improvements and modifications alter the characteristics of the innovation over time. Adoption may lag because of the expectation, based on past experience, that newer and better innovations will come along in the near future.

A number of other factors contribute to delays in adoption, and again, regional variations are present. Perhaps most important are access to information and channels of supply for the innovation-factors which vary markedly from place to place (Brown, 1981; Sweeney, 1987). Information may be withheld or proprietary to an innovator, or it may be slow to accumulate as firms adopt and gain experience with new technology. The slowness of diffusion is often noted, and is part of the reason why equilibrium rarely characterizes economic change (Nelson, 1981; Soete and Turner, 1984). These institutional factors provide elements of a more realistic model of the diffusion of technological innovation, even if they are much more complex than what we have relied on in the past.

Doz (1986) depicts well the dilemma facing large multinational firms. They must, on the one hand, strive to coordinate and control their far-flung operations and, on the other hand, they face considerable pressure to decentralize activities - including R&D - to their various host countries. The dilemma is not an easy one. As a result, firms may perform R&D outside a handful of advanced countries, but the amounts tend to be small, the focus tends to be on minor adaptions, and it often is separated from local R&D (Teece, 1981).

4. Large Firms, Technology Transfer Mechanisms and Regional Technological Capability

The acquisition and sale (licensing) of technology provide an alternative to internal R&D for the purchasing firms, and furnish a return on prior R&D investments and a source of revenue for further R&D by the firm selling its technology. From the point of view of a purchasing firm, licensing results in faster commercial development and market entry or greater market share than costly internal R&D would permit. It also can provide specialized knowledge and skills, and even stimulate a licensee's technological capability by making it more aware of the advantages of outside developments (McDonald and Leahey, 1985). The advantages of licensed technology depend heavily on how current the technology is, and on whether the licensee is permitted to retain the rights to any improvements made.

Licensing of production to another firm is an indirect means of technology transfer. In order to gain access to a foreign market and where direct foreign investment is not feasible or attractive, for example, a firm might choose to "license" a domestic firm to be the sole authorized producer of a product within that country.[3] Restrictive agreements prohibit the domestic firm from some activites, such as exporting, improving upon the product, or setting prices (Odle, 1979; Prasad, 1981). Thus, from the point of view of the multinational firm, licensing is not a mechanism for

[3] Much licensing has nothing to do with technology, but only represents use of a trademark and results in immediate market recognition (e.g. clothing with Mickey Mouse or other characters). Whether such a creative symbol or image is "technology" is open to question.

exploiting technolocial assets; technology transfer is a result, not a motive, for the licensor (Adam et al., 1988).

In recent years, however, there has been a significant change in the use of licensing as a vehicle for the sale and transfer of technology as a primary motive. The stimulus for this change is the capability of an increasing number of countries to acquire technology this way and their unwillingness to permit unrestricted or unnecessary foreign direct investment. Japan, Western European and Eastern European countries have been successful in using licensing, and the more advanced less developed countries prefer licensing. It has become a widely available option because of the increased competition among suppliers of technology and the resulting need to sell existing technology to be able to finance future R&D programs (Adam et al., 1988).

The firms most likely to be involved in licensing, in fact, are large firms and those firms which have the highest R&D intensities (the ratio of R&D expenditure to annual sales) (Adam et al., 1988; McDonald and Leahey, 1985). The short product life cycles in R&D-intensive industries might in fact be the motivation for firms in those sectors. This advantage may be illusory, however, especially in such industries, since it is costly to maintain worldwide patent protection. Kumar (1987) believes that direct investment is most likely among large firms whose products embody more than technical knowledge alone, but also a high degree of product differentiation and the capabilities, or complementary assets, involved therein. Licensing alone is more common in capital goods, where the knowledge is embodied almost fully in the products.

The larger issue within which licensing rests is that of appropriability - the ability of a firm to retain the profits from innovation. Teece (1986) provides a valuable typology of the "regimes of appropriability," the environmental factors, other than firm and market structure, that govern an innovator's ability to capture the profits generated by an innovation. (The chapter in this volume by Giaoutzi puts appropriability in a regional context). The most important dimensions of such a regime are the nature of the technology and the efficacy of legal mechanisms of protection. Patents, copyrights, and trade secrets (e.g. recipes and chemical formulas) are legal means of keeping a technology out of competitors' hands, but these do not guarantee that imitation cannot take place. Most common is the evolutionary transformation of products and processes over time. In the early, "preparadigmatic" stage of a technology's development, an innovation can become worthless quickly as a new design leapfrogs it. As standardization of design takes place, an industry **paradigm** emerges, and the firm which gains most will tend to be the one which has a full array of complementary assets in order to not only exploit the innovative design but also to manufacture it competitively and bring it successfully to market.

The most important means of retaining control of an innovation are the **complementary assets** involved in the commercialization of a technology. These include competititve manufacturing, suppliers, distribution, service, a successful tradename (e.g. Nutrasweet), and complementary technologies (Chaudhuri, 1986; Cohen and Zysman, 1987; Porter, 1987; Teece, 1986, p. 289). Traditional neoclassical trade theory refers exclusively to regimes of tight appropriability and zero transactions cost, where it does not matter whether an innovating firm has an in-house manufacturing capability, domestic or foreign. It can simply engage in arms-length contracting (such as licensing and co-production) for the sale of the output of the activity (R&D) in which it has a comparative advantage. In a regime of weak appropriability, manufacturing may be necessary if an innovator is to appropriate the rents from an innovation. This is especially true where the requisite manufacturing assets are specialized to the innovation. If an innovator's manufacturing costs are higher than those of imitators, for example, the bulk of the profits will go to the imitators. This is a concern in the United States at present, where "hollow corporations" have lost manufacturing capability relative to those in Japan and elsewhere in Asia (Business Week, 1986; Cohen and Zysman, 1987).

The spectrum of complementary assets encompass a range of capabilities which support and sustain the development and enhancement of technology. R&D capability is but one part of the range. The capabilities across the entire production process need not be present within a single firm. In fact, the strength of industrial districts lies in the interfirm sharing of ideas and capabilities, as demonstrated in the chapters by Garofoli and by Maillat in this volume. The interconnectedness of sectors is a key aspect of the diffusion of product innovations from supplier sectors to user sectors (Scherer, 1982).

5. Entrepreneurship and Technological Capability

Entrepreneurship is a form of direct technology transfer, where knowledge is taken from one firm to a newly founded one by an individual. Coffey and Polèse (1985) have placed entrepreneurship at the center of the process of local economic development, clearly in reaction to the well-documented consequences of "external control" (Watts, 1981). This is summed up well by Sweeney (1985, p. 97):

> Areas dominated by large firms tend to have low entrepreneurial vitality. Such firms have internalised their information resources and networks ... Areas dominated by branch plants of large firms also have low entrepreneurial vitality, their networks are with their distant parent company. Sub-contracting, an important source of start-up for new companies, is minimised by the large integrated firm and even more by branch plants. An important means of developing local networks for technology and information transfer is thus lost.

Allen et al. (1983) bolster this argument in their findings about technology transfer to firms in Ireland, Spain, and Mexico. They found that information in all three countries flowed mainly through informal channels - that is, through interpersonal contact. As Sweeney suggests, foreign subsidiaries obtained the greatest proportion of their technology from their parent firms. In fact, such plants had several channels of technology blocked to them, such as government-sponsored research, trade fairs, and industry associations, which are quite important to domestic firms. Firms in the same industry are also much more likely to be a source of technological knowledge to domestic firms. Allen et al. conclude that domestic firms in many ways have easier access to foreign technology than do the subsidiaries of multinational firms.

Orientation toward national and international markets is what distinguishes among types of small firms. Brusco (1986) suggests three types: the traditional artisan common to many rural areas; dependent subcontractors who produce part and components to be included in products of larger firms which sell nationally and internationally; and small firms in industrial districts. The technological sophistication of the latter group of firms must be quite great, often for a large number of clients, which indicates that their level of quality and technical standards are high. In industrial districts, an additional type of firm is found: producers of machinery necessary for the production of the district's specialized commodity. In industrial districts, "the local communities have accumulated managerial, technical and commercial competence and capacity... This knowledge is not characteristic of one particular group but... is spread to all social strata" (Brusco, 1986, p. 198). Both Russo's (1985) case study of the ceramic tile industry and Sabel et al.'s (1987) example of German textile-machinery firms illustrate that the interaction between machinery firms and their users generates a continual flow of information concerning improvements and modifications. In other words, information based on "learning by using" flows back and forth between the two industries (Kline and Rosenberg, 1986; Rosenberg, 1982). Indeed, Lorenzoni and Ornati (1988) suggest that firms in such "constellations" are more willing to seek information from outside sources, such as consultants, universities, and other firms.

The "environmental texture" of such districts - suppliers and infrastructures particular to the production requirements of a group of firms - supports the needs of new and small firms and to some extent replaces the need for internalized complementary assets (see Maillat's chapter in this volume). Economic theory is limited in explaining the industrial district phenomenon (Santarelli, 1987), and Garofoli has shown that there are different degrees of innovativeness in such districts (see his chapter in this book). However, recent work by Shearman and Burrell (1987) links the notion of industry cycle with different levels of interaction, mobility, and barriers to entry.

To replace branch plant economies, Coffey and Polèse propose four stages of local development centered on the generation of new firms:

(1) the emergence of local entrepreneurship;
(2) the growth and expansion of local enterprises (to replace jobs lost by closed branch plants);
(3) the maintenance of local enterprises under local control;
(4) the attainment of an autonomous local control structure and of a local business service sector.

By focusing on innovation as an essential element in entrepreneurship, Martin (1986) disagrees with Coffey and Polèse (1985) on the potential of backward regions to spawn entrepreneurs. Martin believes that only densely populated regions, in the vicinity of large urban centers, will produce sufficient entrepreneurship. In the context of small, peripheral regions, innovative development is "too good to be true" (Martin, 1986, p. 17; Brugger, 1986, p. 42; Friedmann, 1986). The human capital of such regions is too limited and too mobile: i.e., many of the talented people will simply leave. Unless a local economy meets some fairly large threshold size, its base of potential entrepreneurs - and the likelihood that they as a group will be able to come up with successive rounds of innovations as the product cycle progresses - will be inadequate to compete with other regions (Martin, 1986).

In what regions and under what sorts of conditions, then, can we expect to find higher levels of entrepreneurship? As Martin and Friedmann suggest, large urban regions have certain advantages over rural areas. In discussing "creative regions", Andersson (1985) maintains that communication is a key to understanding the fundamental role of metropolitan regions in the creative process. His description of contact networks and communication is reminiscent of Silicon Valley as described by Rogers and Larsen (1984). Andersson (1985, p. 19) proposes that creativity as a social phenomenon primarily develops in regions characterized by: (1) high levels of competence; (2) many field of academic and cultural activity; (3) excellent possibilities for internal **and** external communications (emphasis in the original); (4) widely shared perceptions of unsatisfied needs; and (5) a general situation of **structural instability** facilitating a synergistic development (Stöhr, 1986). Andersson's conditions for regional creativity can be translated into more conventional policy variables, but in generall, they focus on three main elements: (1) the presence of professional and technical labor (competence); (2) urban agglomeration, or a threshold size of place, where cultural activity and communication will be heightened; and (3) conditions that promote synergy or instability. Oakey (1984) places even greater stress on agglomeration economies as a necessary condition for regional high technology development. Pottier (1985) believes that advanced technology is possible only in large urban regions where this agglomeration of attributes is concentrated. The importance of instability is reinforced by Cappellin's (1985) research on regional productivity differences in the EEC. He found that differing regional capabilities to change and to mobilize and reallocate resources within and among sectors were far more important than the regional sectoral mix. In essence, then, entrepreneurship follows from regional technological capability, and is enhanced by communication of new ideas and knowledge among technical people.

Local information channels and knowledge are fundamental to entrepreneurs. Large firms try to maximize information-intensive contacts by centralizing in large urban centers. Around these cluster service activites for whom large firms are the dominant market. However, localities which are dominated by large firms are not centers of entrepreneurial vitality, partly because there is a lack of openness in the flow of information from the large to the small. Large bureaucracies prefer to talk to large bureaucracies (Sweeney, 1985, pp. 92-94). Pottier (1988) also scrutinizes the problems that traditionally have resulted from large firms, and he notes that they have begun to take advantage of local traditions, know-how, and innovation through acquisition of small firms.

"Entrepreneurial vitality is very much a local phenomenon ... Prosperity and economic growth of regions and localities are strongly associated with the strength and vitality of the small firm sector in the region or locality" (Sweeney, 1985, p. 94; Beesley and Hamilton, 1986). It must be stressed that new firms are rarely attracted from elsewhere; entrepreneurship either is a standard endeavor in a region or it is not. Nearly all (84%) of firms are founded within commuting distance of where the founder previously worked (Cooper, 1985).

Diversity within the small firm sector - not a dependence on a narrow range of technologies or markets - reinforces this vitality. "In other words, localities having high entrepreneurial vitality have characteristics similar to those of technically progressive firms, openness in giving as well as taking information, continuing effort in their own search for information and good internal communication flows. The vitality, once initiated, becomes self-reinforcing and sustaining" (Sweeney, 1985, p. 95-96). It also is a key element in successful exchange of research information from R&D to users (Segal, 1987). In such a contact system, users also become important sources of innovation, both of product designs and of production processes (Von Hippel, 1988).

A locality with high entrepreneurial vitality has a high birth rate and a high death rate of small firms. However, other hallmarks of such a region may be harder to identify empirically: diversity and quality in the information flows, responsiveness in meeting the needs of entrepreneurs, and openness in provision of information (Sweeney, 1987). This "information rich environment" is the principal characteristic of "creative regions" (Andersson, 1985: Johansson, 1987). It also is a characteristic of technology transfer. Information flow, long the hallmark of innovation diffusion studies (Brown, 1981; Morrill et al., 1988), remains a key part of technological learning within organizations (Winter, 1987) and within regions (Harrington, 1987).

The contact networks of technical workers are a principal source of technical knowledge. Firms rely on the heterogenous contacts and networks of individuals in order to acquire information from other organizations (Kok and Pellenbarg, 1987). The information "gatekeeper" role is a well-known one; only recently has the diversity of contact of technical workers been studied (Hamfelt and Lindberg, 1987; see also the chapter by Kamann and Nijkamp in this volume). Increasingly, it is seen to be vital for firms to recruit personnel with certain contacts and personal networks in place, to encourage contact to be formed and maintained, and to circulate people in order to more fully develop their networks of contacts to the advantage of the firm (Håkanson, 1987; Shapero, 1985). The common contact networks of rival firms even results in informal trading of know-how between competitors (Von Hippel, 1988).

Cultural characteristics also constrain or facilitate information flow (Keeble and Wever, 1986). Cultural influences on entrepreneurship include the entrepreneurial culture of a locality - the examples of others, an already strong small firm sector, the tradition of being self-employed (Illeris, 1986; Sweeney, 1985, p. 91). Regions with high rates of new firm formation (NFF) have many 'good' entrepreneurs and many in-migrants with high educations and salaries. Rural areas are burdened by cultural/social factors, such as no entrepreneurial class, a social structure dominated by church or a single industry (e.g. mining), and high rates of out-migration. Education levels appear strongly related to a greater range of information sources and a wider set of customers and markets (Miller et al., 1987).

The tendency for some regions to have a greater proportion of successful, more rapidly growing small firms may be even more important than geographical differences in new firm formation. Again, social and occupational influences appear to be most significant in reinforcing existing spatial contrasts. In a recent Dutch study, successful entrepreneurs (as opposed to those of failed or failing firms) were simply more prepared for starting a business: they had more starting capital, were more oriented to markets outside the local region, had a larger number of clients, and were more likely to have worked for a large firm previously (Wever, 1986). Keeble and Kelly (1986) also found that virtually all of the founders of new computer electronics firms had a higher education degree. Overall, there is a strong tendency for entrepreneurship to be strongest precisely in those regions which need it least, suggesting that to rely on new and small firms will not eliminate regional economic differentials.

The "entrepreneurial climate" of a place is probably the greatest single variable influencing entrepreneurship as an ongoing part of regional economic change. This climate relies almost entirely on a well-connected network of investors, especially informal investors, previous entrepreneurs, and an aura of nonroutine, innovative activity (Gruenstein, 1984; Miller and Coté, 1987; Senker, 1985; Shapero, 1984; Sweeney, 1987). Without such a climate, a region faces both capital shortages and reduced entrepreneurial activity. The entrepreneurial climate of a region is influenced by the power elite and government policy in a community (Schell, 1983; 1984). Most important to the "entrepreneurial climate" of a region, however, are entrepreneurial "role models". These can promote and encourage local entrepreneurship and make seed and early financing available to new entrepreneurs (Schell, 1983, 1984; Gruenstein, 1984). Financing for start-up firms can come from many sources, including personal funds, family and friends, local bankers, and outside lenders, but it tends to come from informal (noninstitutional) sources that operate almost entirely via a network of personal, and local contacts (Gaston and Bell, 1987; Wetzel, 1986). In addition to their role in financing new firms, previous entrepreneurs share a variety of routine business knowledge that is necessary to operate a business, on matters such as regulations, taxes, accounting, suppliers, customers, and distribution, as Maillat stresses in his contribution to this volume.

> "Entrepreneurship alone does not lead to economic development. The kinds of enterprises that could have led the colonies toward economic development required other elements that were in short supply in the colonial world. To create modern industries and businesses, entrepreneurs also needed information about foreign machines, technical processes, and business practices, information which was not forthcoming from the educational system. In other words, they had to be importers or creators of technological culture" (Headrick, 1988, p. 352).

6. Conclusion

Pavitt (1986) suggests that much of technical change is cumulative - within firms at least. First, firms accustomed to organized R&D on old technologies may be more likely to learn how to attract and use scientists and engineers in the new technologies. Second, any new development often depends on upstream or downstream improvements in more conventional technologies. Third, there is evidence that effective assimilation of electronics-based technology depends on the more general competence in management (Cohen and Zysman, 1987). Technological accumulation along known technological "trajectories" (Dosi, 1984) is much more likely than is leapfrogging of new countries and regions.

In many respects, technology transfer is more complex than is the linear model of R&D and innovation. Formal mechanisms, such as licenses, joint ventures, and direct investment, require problems of distance (geographical and hierarchical) to be overcome. Inputs and outputs must be coordinated, and assignments of responsibility among the parties must be clarified. Informal technology transfer takes place largely interpersonally. Even if between different organizations, it may be an effective way to obtain information through gatekeepers and their personal relationships across organizations. Informal information flow, both of technology and of routine business knowledge, facilitates simple technology transfer between firms in industrial districts and entrepreneurial regions. Incremental improvements, which do not necessarily result from technology transfer, may be so if they are a response to information from elsewhere. Even internal improvements of product designs or production processes frequently arise from knowledge about new practices used by other firms. The awareness of best-practice products and processes requires considerable knowledge, often as a result of formal R&D in large firms, but also from more spontaneous mechanisms in small firms. It seems useful to emphasize the range of technological activities employed by firms in a technologically dynamic economic environment. These activities all influence regional economic capabilities and ultimately the economic status of regions and nations.

References

Adam, Y., Ong, C.H., and Pearson, A.W. (1988), Licensing as an alternative to foreign direct investment: an empirical investigation, Journal of Product Innovation Management 5:32-49.

Alexandrides, C.G. and Bowers, B.L. (1987), Countertrade: practices, strategies, and tactics, John Wiley, New York.

Allen, T.J., Hyman, D.B., and Pinckney, D.L. (1983), Transferring technology to the small manufacturing firm: a study of technology transfer in three countries, Research Policy 12:199-211.

Andersson, A.E. (1985), Creativity and regional development, Papers of the Regional Science Association 56:5-20.

Bae, Z. and Lee, J. (1986), Technology development patterns of small and medium sized companies in the Korean machinery industry, Technovation 4:279-296.

Beesley, M.E. and Hamilton, R.T. (1986), Births and deaths of manufacturing firms in the Scottish regions, Regional Studies 20:281-288.

Bell, M. (1984), 'Learning' and the accumulation of industrial technological capacity in developing countries, in: Fransman, M. and King, K. (eds.), Technological capability in the third world, pp. 187-209, St. Martin's, Press, New York.

Bhalla, A.S., ed. (1979), Towards global action for appropriate technology, Pergamon, Oxford.

Bhalla, A.S. and Fluitman, A.G. (1985), Science and technology indicators and socio-economic development, World Development 13:117-190.

Breheny, M.J. and McQuaid, R.W. eds. (1987), The development of high technology industries: an international survey, Croom Helm, London.

Brown, L.A. (1981), Innovation diffusion: a new perspective, Methuen, New York.

Brugger, E.A. (1986), Endogenous development: a concept between utopia and reality, in: Bassand, M. et al. (eds.), Self-reliant development in Europe, pp. 38-58, Gower, Aldershot.

Brugger, E.A. and Stuckey, B. (1987), Regional economic structure and innovative behaviour in Switzerland, Regional Studies, 21:241-254.

Brusco, S. (1986), Small firms and industrial districts: the experience of Italy, in: Keeble, D. and Wever, E. (eds.), New firms and regional development in Europe, pp. 184-202, Croom Helm, London.

Business Week. (1983), The hollow corporation, March 3.

Buttel, F., Kenney, M., and Kloppenburg, J. (1985), From green revolution to biorevolution: some observations on the changing technological bases of economic transformation in the Third World, Economic Development and Cultural Change 34:31-55.

Camagni, R.P. (1985), Spatial diffusion of pervasive process innovation, Papers of the Regional Science Association 58:83-95.

Cappellin, R. (1983), Productivity growth and technological change in a regional perspective, Giornale degli Economisti e Annali di Economia 42:459-482.

Chaudhuri, S. (1986), Technological innovation in a research laboratory in India: a case study, Research Policy 15:89-103.

Clark, N. (1985), The political economy of science and technology, Basil Blackwell, Oxford.

Coffey, W.J. and Polèse, M. (1985), Local development: conceptual bases and policy implications, Regional Studies 19:85-93.

Cohen, S.S. and Zysman, J. (1987), Manufacturing matters: the myth of the post-industrial economy. Basic Books, New York.

Cooper, A.C. (1985), The role of incubator organizations in the founding of growth-oriented firms, Journal of Business Venturing 1:75-86.

Dahlman, C. and Westphal, L. (1982), Technological effort in industrial development - a survey, in: Stewart, F. and James, J. (eds.), The economics of new technology in developing countries, pp. 105-137, Frances Pinter, London.

De Bresson, C. and Lampel, J. (1985), Beyond the life cycle: organizational and technological design. I: an alternative perspective. Journal of Product Innovation Management 2:170-187.

De Meyer, A. (1985), The link between technological life-cycle and the flow of information in an R&D department, in: Mar, B., Newell, W.T. and Saxberg, B.O. (eds.), Managing high technology, pp. 353-366, Elsevier, Amsterdam.

Desai, A.V. (1984), India's technological capability: an analysis of its achievements and limits, Research Policy 13:303-310.

Dicken, P. (1986), Global shift: industrial change in a turbulent world, Harper and Row, London.

Doz, Y. (1986), Strategic management in multinational companies, Pergamon, Oxford.

Dutton, J.M. and Thomas, A. (1985), Relating technological change and learning by doing, in: Rosenbloom, R.S. (ed.), Research on technological innovation, management and policy, Vol. 2, pp. 187-224, JAI Press, Greenwich, CT.

Ettlie, J.E. and Rubenstein, A.H. (1987), Firm size and product innovation, Journal of Product Innovation Management 4:89-108.

Ewers, H.-J. (1986), Spatial dimensions of technological developments and employment effects, in: Nijkamp, P. (ed.), Technological change, employment and spatial dynamics, pp. 157-176, Springer-Verlag, Berlin.

Frame, J.D. (1983), International business and global technology, Lexington Books, Lexington, MA.

Fransman, M. (1986a), Technology and economic development, Westview Press Boulder, CO.

Fransman, M., ed. (1986b), Machinery and economic development, St. Martin's Press, New York.

Fransman, M. and King, K., eds. (1984), Technological capability in the third world, St. Martin's Press, New York.

Friar, J. and Horwitch, M. (1986), The emergence of technology strategy: a new dimension of strategic management, in: Horwitch, M. (ed.), Technology in the modern corporation: a strategic perspective, pp. 50-85, Pergamon, New York.

Friedmann, J. (1986), Regional development in industrialised countries: endogenous or self-reliant? in: Bassand, M. et al. (eds.), Self-reliant development in Europe, pp. 203-216, Gower, Aldershot.

Gaston, R.J., and Bell, S.E. (1987), Business angels: mobilizing private sector seed capital, Economic Development Commentary 11 (1):3-5.

Giaoutzi, M. (1985), Factors affecting the capacity of technological change: the case of less developed countries, Papers of the Regional Science Association 58:73-82.

Gold, B. (1979), Productivity, technology, and capital, Lexington Books, Lexington, MA.

Graham, C.P. (1988), Technology and development: a case study of the Brazilian firm Engesa, S.A., Unpublished M.A. thesis, University of Florida.

Graham, M.B.W. (1986), Corporate research and development: the latest transformation, in: Horwitch, M. (ed.), Technology in the modern corporation: a strategic perspective, pp. 86-102, Pergamon, New York.

Gruenstein, J.M.L. (1984), Targeting high tech in the Delaware Valley, Business Review, Federal Reserve Bank of Philadelphia (May-June):3-14.

Guile, B.R. and Brooks, H., eds. (1987), Technology and global industry: companies and nations in the world economy, National Academy Press, Washington.

Håkanson, H. (1987), Industrial technological development: a network approach, Croom Helm, London.

Hamfelt, C., and Lindberg, A.K. (1987), Technological development and the individual's contact network, in: Håkanson, H. (ed.), Industrial technological development: a network approach, pp. 177-219, Croom Helm, London.

Harrington, J.W. (1987), Strategy formulation, organisational learning, and location, in: Van der Knaap, B. and Wever, E. (eds.), New technology and regional development, pp. 63-74, Croom Helm, London.

Headrick, D.R. (1988), The tentacles of progress: technology transfer in the age of imperialism, 1850-1940, Oxford University Press, New York.

Hough, G.W. (1975), Technology diffusion: federal programs and procedures,

Lomond Books, Mt. Airy, MD.

Hout, T., Porter, M.E., and Rudden, E. (1982), How global companies win out, Harvard Business Review 60, 5:98-108.

Illeris, S. (1986), New firm formation in Denmark: the importance of the cultural background, in: Keeble, D. and Wever, E. (eds.), New firms and regional development in Europe, pp. 141-150, Croom Helm, London.

Johansson, B. (1987), Information technology and the viability of spatial networks, Papers of the Regional Science Association 61:51-64.

Johansson, B. and Westin, L. (1987), Technical change, location, and trade, Papers of the Regional Science Association 62:13-25.

Kaplinsky, R. (1984), Automation: the technology and society, Longman, London.

Katz, J.M. (1982a), Technology and economic development: an overview of research findings, in: Syrquin, M. and Teitel, S. (eds.), Trade, technology, and equity in Latin America, pp. 281-315, Academic Press, London.

Katz, J.M. (1982b), Technological change and development in Latin America, in: French-Davis, R. and Tironi, E. (eds.), Latin America and the new international economic order, pp. 192-215, St. Martin's Press, New York.

Keeble, D. and Kelly, T. (1986), New firms and high-technology industry in the United Kingdom: the case of computer electronics, in: Keeble, D. and Wever, E. (eds.), New firms and regional development in Europe, pp. 184-202, Croom Helm, London.

Keeble, D. and Wever, E., eds. (1986), New firms and regional development in Europe, Croom Helm, London.

Kim, L. and Lee, H. (1987), Patterns of technological change in a rapidly developing country: a synthesis, Technovation 6:261-276.

Kim, L., Lee, J. and Lee, J. (1987), Korea's entry into the computer industry and its acquisition of technological capability, Technovation 6:277-293.

Kline, S.J. and Rosenberg, N. (1986), An overview of innovation, in: Landau, R. and Rosenberg, N. (eds.), The positive sum strategy, pp. 275-305, National Academy Press, Washington.

Kok, J.A.A.M. and Pellenbarg, P.H. (1987), Innovation decision-making in small and medium-sized firms: a behavioural approach concerning firms in the Dutch urban system, in: Van der Knaap, B. and Wever, E. (eds.), New technology and regional development, pp. 145-164, Croom Helm, London.

Kumar, N. (1987), Intangible assets, internalisation and foreign production: direct investments and licensing in Indian manufacturing, Weltwirtschaftsliches Archiv, 123:325-345.

LeHeron, R.B. (1973), Best-practice technology, technical leadership, and regional economic development, Environment and Planning 5:735-749.

Lorenzoni, G. and Ornati, O.A. (1988), Constellations of firms and new ventures, Journal of Business Venturing 3:41-57.

Malecki, E.J. (1977), Firms and innovation diffusions: examples from banking, Environment and Planning A 9:1291-1305.

Malecki, E.J. (1980), Corporate organization of R and D and the location of technological activites, Regional Studies 14:219-234.

Malecki, E.J. (1983), Technology and regional development: a survey, International Regional Science Review 8:89-125.

Malecki, E.J. (1987), The R&D location decisions of the firm and 'creative' regions, Technovation 6:205-222.

Malecki, E.J. (forthcoming), Technological innovation and paths to regional economic growth, in: Schmandt, J. and Wilson, R.W. (eds.), Growth policy in the age of high technology.

Mansfield, E. (1986), Industrial research and technological innovation, Norton, New York.

Maritn, F. (1986), L'entrepreneurship et la développement local: une evaluation, Canadian Journal of Regional Science 9:1-23.

McDonald, D.W. and Leahey, H.S. (1985), Licensing has a role in technology strategic planning, Research Management 28, 1:35-40.

McIntyre, J.R. and Papp, D.S. (1986), The political economy of international technology transfer, Quorum Books, Westport, C.T.

Meyer-Krahmer, F. (1985), Innovation behaviour and regional indigenous potential, Regional Studies 19:523-534.

Miller, H.M., Brown, E.E. and Centner, T.J. (1987), Southern Appalachian handicrafts industry: implications for regional economic development, Review of Regional Studies 17, 3:50-58.

Miller, R. and Coté, M. (1987), Growing the next silicon valley, Lexington, MA: Lexington Books.

Molero, J. (1983), Foreign technology in the Spanish economy: an analysis of the recent evolution, Research Policy 12:269-286.

Morrill, R., Gaile, G. and Thrall, G. (1988), Spatial diffusion, Sage, Newbury Park, CA.

Nelson, R.R. (1981), Research on productivity growth and productivity differences: dead ends and new departures, Journal of Economic Literature 19:1029-1064.

Nelson, R.R. (1984), High-tech policies: a five-nation comparison, American Enterprise Institute, Washington.

Nelson, R.R. and Winter, S.G. (1982), An evolutionary theory of economic change, Harvard University Press, Cambridge, MA.

Nijkamp, P. and Schubert, U. (1985), Structural change in urban systems, Sistemi Urbani 7:155-176.

Nueno, P. and Oosterveld, J.P. (1986), The status of technology strategy in Europe, in: Horwitch, M. (ed.), Technology in the modern corporation: a strategic perspective, pp. 145-166, Pergamon, New York.

OECD (1981), North/south technology transfer: the adjustments ahead, Organisation for Economic Co-operation and Development, Paris.

Oakey, R.P. (1984), High technology small firms, St. Martin's Press, New York.

Odle, M.A. (1979), Technology leasing as the latest imperialist phase: a case study of Guyana and Trinidad, Social and Economic Studies 28:189-233.

Pavitt, K. (1986), 'Chips' and 'trajectories': how does the semiconductor influence the sources and directions of technical change? in: MacLeod, R.M. (ed.), Technology and the human prospect, pp. 31-54., Frances Pinter, London.

Porter, M.E. (1983), The technological dimension of competitive strategy, in: Rosenbloom, R.S. (ed.), Research on technological innovation, management and policy, Vol. 1, pp. 1-33, JAI Press, Greenwich, CT.

Porter, M.E. (1987), Changing patterns of international competition, in: Teece, D.J. (ed.), The competitive challenge, pp. 27-57, Ballinger, Cambridge, MA.

Pottier, C. (1985), The adaptation of regional industrial structures to technical changes, Papers of the Regional Science Association 58:59-72.

Pottier, C. (1988), Local innovation and large firm strategies in Europe, in: Aydalot, P. and Keeble, D. (eds.), High technology industry and innovative environments: the European experience, pp. 99-120, Routledge, London.

Prasad, A.J. (1981), Licensing as an alternative to foreign investment for technology transfer, in: Hawkins, R.G. and Prasad, A.J. (eds.), Technology transfer and economic development [Research in international business and finance, vol. 2], pp. 193-218, JAI Press, Greenwich, CT.

Prasad, S.B. (1986), Technology transfer: the approach of a Dutch multinational, Technovation 4:3-15.

Rees, J. (1987), What happened to macroeconomics? Environment and Planning A 19:139-141.

Rees, J., Briggs, R. and Oakey, R. (1984), The adoption of new technology in the American machinery industry, Regional Studies 18:489-504.

Rogers, E.M. (1982), Information exchange and technological innovation, in: Sahal, D. (ed.), The transfer and utilization of technical knowledge, pp. 105-123, Lexington Books, Lexington, MA.

Rogers, E.M. (1983), Diffusion of innovations, third edition, Free Press, New York.

Rogers, E.M. and Larsen, J.K. (1984), Silicon valley fever, Free Press, New York.

Rosegger, G. (1980), The economics of production and innovation, Pergamon, Oxford.

Rosenberg, N. (1982), Inside the black box: technology and economics, Cambridge University Press, New York.

Russo, M. (1985), Technical change and the industrial district: the role of interfirm relations in the growth and transformation of ceramic tile production in Italy, Research Policy 14:329-343.

Sabel, C., Herrigel, G., Kazis, R. and Deeg, R. (1987), How to keep mature industries innovative, Technology Review 90, 3:27-35.

Santarelli, E. (1987), Generation and diffusion of new technologies, Rivista Internazionale di Scienze Economische e Commerciali 34:829-852.

Schell, D.W. (1983), Entrepreneurial activity: a comparison of three North Carolina communities, in: Hornaday, J.A. et al. (eds.), Frontiers of entrepreneurship research 1983, pp. 495-518, Babson College, Center for Entrepreneurial Studies, Wellesley, MA

Schell, D.W. (1984), The development of the venture capital industry in North Carolina: a new approach, in: Hornaday, J.A. et al. (eds.), Frontiers of entrepreneurship research 1984, pp. 55-72, Babson College, Center for Entrepreneurial Studies, Wellesley, MA.

Scherer, F.M. (1982), Inter-industry technology flows in the United States, Research Policy 11:227-245.

Segal, A., ed. (1987), Learning by doing: science and technology in the developing world, Westview Press, Boulder, CO.

Shapero, A. (1984), The entrepreneurial event, in: Kent, C.A. (ed.), The environment for entrepreneurship, pp. 21-40, Lexington Books, Lexington, MA.

Shapero, A. (1985), Managing professional people, Free Press, New York.

Sharif, M.N. (1986), Measurement of technology for national development, Technological Forecasting and Social Change 29:119-172.

Sharp, M. and Shearman, C. (1987), European technological collaboration, Routledge and Kegal Paul, London.

Shearman, C. and Burrell, G. (1987), The structures of industrial development, Journal of Management Studies 24:325-345.

Silveira, M.P.W., ed. (1985), Research and development: linkages to production in developing countries, Westview Press, Boulder, CO.

Soete, L. and Turner, R. (1984), Technology diffusion and the rate of technical change, Economic Journal 94:612-623.

Spence, A.M. and Hazard, H.A., eds. International competitiveness, Ballinger, Cambridge MA.

Stewart, F. (1981), International technology transfer: issues and policy options, in: Streeten, P. and Jolly, R. (eds.), Recent issues in world development, pp. 67-110, Pergamon, Oxford.

Stewart, F. (1987), The case for appropriate technology: a reply to R.S. Eckaus, Issues in Science and Technology 3, 4:101-109.

Stöhr, W. (1986), Regional innovation complexes, Papers of the Regional Science Association 59:29-44.

Stoneman, P. (1983), Economic analysis of technological change, Oxford University Press, Oxford.

Suarez-Villa, L. (1987), Entrepreneurship and the international diffusion of innovations in manufacturing: a general approach, Rivista Internazionale di Scienze Economische e Commerciali 34:369-391.

Sweeney, G.P. (1985), Innovation is entrepreneur-led, in: Sweeney, G. (ed.), Innovation policies: an international perspective, pp. 80-113, Frances Pinter, London.

Sweeney, G.P. (1987), Innovation, entrepreneurs and regional development, St. Martin's Press, New York.

Taylor, M.J. (1986), The product cycle model: a critique, Environment and Planning a 18:751-761.

Teece, D.J. (1981), Technology transfer and R&D activities of multinational firms: some theory and evidence, in: Hawkins, R.G. and Prasad, A.J. (eds.), Technology transfer and economic development [Research in international business and finance, vol. 2], pp. 39-74, JAI Press, Greenwich, CT.

Teece, D.J. (1986), Profiting from technological innovation: implications for integration, collaboration, licensing and public policy, Research Policy 15:285-305.

Teece, D.J. ed. (1987), The competitive challenge, Ballinger, Cambridge, MA.

Thomas, M.D. (1987), The innovation factor in the process of microeconomic industrial change: conceptual explorations, in: Van der Knaap, B. and Wever, E. (eds.), New technology and regional development, pp. 21-44, Croom Helm, London.

Thwaites, A.T. (1982), Some evidence of regional variations in the introduction and diffusion of industrial products and processes within British manufacturing industry, Regional Studies 16:371-381.

Varaiya, P. and Wiseman, M. (1981), Investment and employment in manufacturing in U.S. metropolitan areas 1960-1976, Regional Science and Urban Economics 11:431-469.

Von Hippel, E. (1988), The sources of innovation, Oxford University Press, New York.

Watts, H.D. (1981), The branch plant economy, Longman, London.

Wetzel, W.F. (1986), Informal risk capital: knowns and unknowns, in: Sexton, D.L. and Smilor, R.W. (eds.), The art and science of entrepreneurship, pp. 85-108, Ballinger, Cambridge, MA.

Wever, E. (1986), New firm formation in the Netherlands, in: Keeble, D. and

Wever, E. (eds.), New firms and regional development in Europe, pp. 54-74, Croom Helm, London.

Winter, S.G. (1987), Knowledge and competence as strategic assets, in: Teece, D.J. (ed.), The competitive challenge, pp. 159-184, Ballinger, Cambridge, MA.

12 SMEs, innovation and territorial development

D. MAILLAT

1. Prologue

For about fifteen years small and medium-sized enterprises (SME) have been regarded as the main source of local and regional dynamism. The large company is thus no longer the only engine of development. It was in fact the economic recession, and the accompanying changes in production organisation, which revealed the vulnerability and deficiencies of large companies and, conversely, the dynamism of SMEs.

The 1960s showed the development of research work stressing the predominant role of large companies and multinational groups in the functioning of the industrial countries' economies. The emphasis was placed on the Fordian organisation of labour, the spatial division of functions and the opposition between central and peripheral regions. Since the end of the 1970s, the increase in the number of small businesses and their substantial contribution to creating jobs, have led to a revival of the discussion of new forms of production organisation (particularly the flexible manufacturing systems) and regional development mechanisms. Thus attention was drawn to their ability to innovate, their contribution to the performance of previously little-developed regions and their role in the revitalization of certain industrial regions.

The phenomenon of the emergence of SMEs has many different causes, which point to its being permanent. But the stock of SMEs is very heterogeneous. Their role in regional development should therefore be analysed carefully. An analysis of the individual behaviour of SMEs is inadequate. In a region SMEs do not operate in isolation from one another; they belong to a network which, in order to be efficient, must have a territorial dimension, that is, the intensity of the flows must be specific to the region. This being the case, the organisation of territorial systems conditions the dynamism of SMEs just as it does that of the regions. In this chapter we shall be considering, first of all, whether or not the phenomenon of the emergence of SMEs is a permanent one. Next we shall recall the role SMEs play with regard to innovation and the revival of entrepreneurship. Then we shall show that SMEs have characteristics which make them capable of generating new regional organisation models. This will lead us on to describe the elements of a theory of local dynamism. Finally, we shall present the various types of territorial production organisation which are likely to grant the SMEs a dynamic role. We shall conclude by emphasising the importance of the milieu as a factor for integrating SMEs.

2. The Phenomenon of the Emergence of SMEs and its Permanence

While large companies have generally stopped contributing to employment since the mid 1970s, after previously playing a central role, it may be said that SMEs have developed in the opposite direction. After their decline in the 1960s they have performed very dramatically since the 1970s (Greffe, 1985; Keeble and Wever, 1986; Storey, 1986; Julien, 1987; Aydalot and Keeble, 1988). The phenomenon is sufficiently significant and well documented in each country to be acknowledged as being a general one. However, one is justified to wonder whether it is temporary or permanent. A number of reasons has been put forward to explain the re-appearance of SMEs (Keeble and Wever, 1986; Storey, 1986; Brusco, 1986; and Piore and Sabel, 1984).

Some explanations give this phenomenon a temporary character. This is the case with the "recession push theory" (Gudgin, 1984). According to this theory, the creation of new enterprises may be explained in at least two ways:

- some people who are unemployed or who were threatened with unemployment often thought the best way to find a job was to create it themselves. It may be assumed that under different circumstances they would not have done so. This theory is controversial in that there is not necessarily a high correlation between the unemployment level and the rate at which new enterprises are created. However, the theory would be relevant to a certain extent to the regions with an industrial tradition (Keeble and Wever, 1986);
- in many cases the rationalisation which large companies have to carry out or the abandonment of some of their unprofitable production lines have allowed small enterprises to infiltrate markets which had previously been closed to them. Of course, a better level of economic activity might modify this situation and cause the disappearance of the firms created in these conditions (Shutt and Whittington, 1987).

While the recession push theory explains the emergence of SMEs in temporal terms, other explanations, in contrast, are based on long-term structural considerations. We propose to mention four of them:

- the "fragmentation strategy" effect. The economic recession has obliged the large companies to adopt new strategies. In order to limit the risks inherent in the need for innovation and in market instability, and to regain a certain degree of control over the labour process, large companies (the shipbuilding, automobile and electrical engineering industries) are organising themselves differently and in particular are now making more use of subcontractors. The typical example is that of FIAT in the early 1970s, which contracted its own activities but which at the same time did more subcontracting and even encouraged some of its employees to set up their own businesses. This fragmentation strategy (decentramento produttivo) involves transferring production segments to outside the parent company, the aim being to cut costs and achieve the flexibility needed to respond to market fluctuations (Santilli, 1987). This phenomenon assigns a new role to the small business, since it becomes a functional and increasingly crucial element of the production system. The fragmentation strategy can take several forms (Shutt and Whittington, 1987):
    - "Decentralisation of production: large plants are broken up, but retained under the same ownership, by hiving off into smaller plants or by creating new subsidiary companies;
    - Devolvement: large firms cease to own units directly, but retain revenue links with them, i.e., licensing or franchising;
    - Dis-integration of production and innovation: large firms cease to own units of production or innovation, but retain control through market power or, latently, through the power to repurchase the units".
- The "income growth theory" (Brusco, 1982; Storey, 1982). For about fifteen years following the increase in real incomes, household demand has diversified considerably. Today's consumers want specialised, more sophisticated and tailor-made goods and services. Consumption is thus less and less directed towards standardised goods. On the contrary, it is becoming increasingly personalised. Consequently, the path is clear for small series and numerous small businesses which can take very specific market shares. These SME's specialise in products with low economic durability or which are constantly changing, whereas the large companies' production remains more standardised (Delattre et al., 1984). This is equally true of manufacturing industries and service industries.

- The "technological change theory" (Keeble and Wever, 1986; Aydalot and Keeble, 1988). New technologies (electronics, telecommunications, industrial materials, biotechnology, artificial intelligence) are responsible for the creation of many new products, new production processes and new markets. The extremely fast pace of technological change favours the creation of new businesses by highly-qualified entrepreneurs (technical entrepreneurship) or those with advanced scientific skills. These are what are known as new innovative firms, characterised by a high percentage of skilled labour, an ability to adjust rapidly to change, and a strong propensity to export their products (Maillat, 1988). These enterprises play a very important role in the production system, as they introduce a special production method. This special production method "is characterised, first, by a productive combination in which the scientific and technical information factor is essential and, second, by a rapidly evolving production process, since innovation, which is an integral part of this production process, constantly challenges production processes and the nature of products" (Planque and Py, 1986).
- The effect of "osmosis between manufacturing and service industries". Today it is no longer conceivable to regard service industries as totally separate from manufacturing industries. On the contrary, many service activities (professional services) are intimately linked to the development of industry, both upstream of the production process (research and development, prototypes) and at the intermediate level and downstream (sales, marketing). To retain their dynamism, manufacturing firms constantly have to turn to specialised service firms (Bailly and Maillat, 1986; Cappellin, 1987). This development of the production system is of course very favourable for the creation of new service activities, generally SMEs.

In fact, the causes which explain the phenomenon of the emergence of SMEs are multiple and usually well-established. Thus one may expect the phenomenon to be more or less permanent (Julien and Morel, 1986). But the multiplicity of the causes is responsible for the heterogeneousness of SMEs. High-tech firms coexist with low-tech firms; industrial companies with service and craft firms. These enterprises have profiles (in terms of work-force qualification, proportion of goods exported, ability to innovate, etc.) which are very different from one another and their behaviour is far from uniform.

Furthermore these new SMEs are not necessarily independent. Many remain linked to large companies and their fragmentation strategy (Shutt and Whittington, 1987). This may of course cast doubt on their ability to play an independent role in the restructuring of activities. But the doubt can be dispelled when one considers that the emergence of SMEs is part and parcel of the appearance of a new mode of production organisation.

In the context of this new production organisation, the so-called flexible specialisation (Piore and Sabel, 1984) or flexible accumulation (Scott, 1988), which is of course due to the compartmentation of demand and the possibilities for different production techniques offered by the new technologies, the process of forming new enterprises is broadly independent of large companies' strategy. In this context, the SME is characterised by its autonomy and market independence, it is less and less subject to a single giver of orders, and its ability to innovate is by no means inferior to that of large companies (Santilli, 1987).

3. SMEs and Innovation

Today there is extensive literature on SMEs' ability not only to innovate (including in the fields of new technologies) but also to acquire and retain market power (Rothwell and Zegveld, 1983; Bramanti and Senn, 1987; Saxenian, 1985). However, the SMEs' innovative ability is contested. It is often argued that SMEs have lost their pre-eminence in major innovations because technology has become more complex, thereby increasing the importance of industrial research as a major source of innovation. But as the OECD notes (1986), "even if one considers that the proportion of SMEs which deliberately engages in innovative activities as defined above does not exceed 10% to 20%, their role is nevertheless important as they help to accelerate the development of industrial applications of products which are the result of discoveries or new ideas originating in major industrial, public or university research laboratories. Their innovative role is essentially to disseminate technologies which are currently being transferred from leading sectors, particularly those which, like micro-elecronics and new materials, play a fundamental part in ongoing structural change. SMEs invent very diverse applications for these technologies, thereby helping to disseminate the technology throughout the industrial system".

Furthermore, if one discards the idea that innovation concerns not only the new technologies and if one accepts that there is no restriction on the degree of complexity of innovation (the latter can extend from the skateboard to the compact disk or from the mechanical tool to the most complex robot), then there are many innovative SMEs. Their innovations are of course more akin to the improvement or development of products or production processes than to elaborate research work (Cf. the work of GREMI in Aydalot, 1986). Their activity coincides with the broad notion of innovation which consists of introducing to a given market a new or improved production technique, good or service. The work of Pavitt et al. (1987) summarizes the situation well. After examining 4,378 significant innovations introduced in British industry since 1945, it appears that 80% of them are to be found in the manufacturing sector. Within this sector the distribution of innovations is far from uniform. They are, for example, much more numerous in the instruments sector than in the printing and food industries. As regards the size of companies, the results show that innovativeness per employee has increased for both small and large enterprises, while medium-sized firms are marking time. Consequently, the ratio of innovativeness to size may be represented by a U-curve. True, this curve is not to be found in all sectors, but it does reflect the important role of small businesses as a source of innovation.

4. SMEs and the Spirit of Enterprise

The phenomenon of the emergence of SMEs has helped to re-popularise the idea of entrepreneurship. Indeed, the SME is managed by an entrepreneur who is assumed to have the qualities of responsibility, spontaneity, imagination and a high degree of economic know-how (detecting new opportunities, developing strategies, identifying new resources) and relational know-how with people and the environment. The role attributed to the "new entrepreneur" dovetails neatly with the current neo-Schumpeterian ideology (Suarez-Villa, 1986). In fact, contrary to the fragmentation theory, most of these new SMEs are created by people who had previously worked for other SMEs and not for large companies (Keeble and Wever, 1986). Before setting up his own business the entrepreneur thus has some job experience which enabled him to acquire or develop his expertise (Maillat, 1986).

The analysis of entrepreneurship led to the identification of several types of entrepreneur (Arzeni, 1986; Santilli, 1987). It is possible to distinguish:

- technical entrepreneurship. A person (for example, an engineer) with a high level of technical skills, coming from a large company or a university and exploiting new technologies;
- "sheltered" entrepreneurship. A person operating under a franchising

contract or a sub-contracting system;

- "social" entrepreneurship. An unemployed person who sets up his own business or makes available his skills to train fledging entrepreneurs (tutorship);
- venture entrepreneurship. A person who acts as a "venture capitalist";
- the entrepreneur who was previously a wage-earner (or worker, employee, manager). A person benefitting from the facilities offered by the firm for which he works to set up his own business. This may involve hiving off;
- underground entrepreneurship. An enterprise set up in what is known as the underground or black economy;
- intrapreneurship. A person authorized to create a new activity within the company for which he is working.

The purpose of this list is basically to demonstrate that one should not stop at the idea that the only entrepreneurs are technical entrepreneurs. As Suarez-Villa (1986) notes, "clearly, a definition of entrepreneurship that focuses only on technological innovation is insufficient to consider the myriad of other innovative activities and roles that are part of the entrepreneurial function".

Thus the phenomenon of the emergence of SMEs cannot be reduced to job creation, but is a vector of new facts (the SMEs' innovative ability, the revival of the spirit of enterprise, etc.) likely to lend fresh momentum of the production system and territorial development.

5. The Regional Impact of the Emergence of SMEs

SMEs indisputably have characteristics which make them the vectors of a new strategy for the regions. Moreover, there is factual evidence to back this up, since many regions owe their dynamism to SMEs. It may therefore be assumed that changes in spatial organisation evolve parallel to their emergence (Courlet, 1987; Julien, 1987; van der Mark and P. Nijkamp, 1987). However, although globally the SME phenomenon is present in all the industrialised countries, regionally the situation is more differentiated (Gould and Keeble, 1984; Korte, 1986).

Keeble and Wever (1986) beleive three main trends are emerging in Europe:

- Europe's largest diversified cities and their surrounding metropolitan regions tend to exhibit high rates - and large volumes - of new firm formation, in both manufacturing and service industries.

- The lowest rates - and small volumes - of new manufacturing firm formation are now to be found in Europe's old specialised urban-industrial decline... While this pattern must reflect in some way inherited over-specialisation on declining manufacturing and mining industries, it is noteworthy that many much less-specialised cities in northern Europe - Dublin, Copenhagen, Manchester, Edinburgh - are also characterised by low firm formation rates. This suggests that the relative death of entrepreneurial activity in Europe's old industrial regions may be as much associated with the urban, as with the industrial character of these areas.
- High rates, though lower volumes, are characteristic of a number of previously unindustrialised, rural regions. The most striking examples here are the rural regions of southern France (the Midi), various regions in Britain (the South West, East Anglia, West Midlands and central Wales, the Scottish Highlands), north-central Ireland, Denmark's Jutland, and Adriatic central Italy (Umbria, Marche, Abruzzi)".

While the first two conclusions appear to be normal, the third highlights a new phenomenon, that of the development of a new category of regions characterised by so-called "diffuse", "spontaneous" or "rampant" industrialisation. These regions testify to a new dynamic based on the formation of SMEs or their multiplication by hiving off. Their dynamism is based on innovation, the spirit of enterprise and the effects of proximity.

6. The NEC Model and its Spread

This new industrialization model has some noteworthy characteristics (Fuà and Zacchia, 1983; Courlet, 1986). It is based:

- on the choice of specific market niches. The small enterprises which are at the base of this industrialisation select products or production lines which correspond to their size. They prefer to produce products which enhance the value of traditional skill or know-how, while adapting it. They often hold market niches left vacant by large companies (small-scale domestic appliances, small-scale engineering, etc.) or those from which they can oust the large firm (clothing, shoes). They specialise in small series (certain clothes, certain metallic supplies, spare parts) or made-to-measure production (manufacture of robots). The most successful companies are those which know how to find specialised or regional products or services (Julien,

1987);

- on integrated SME systems. The small businesses involved in this form of industrialisation belong to a localised production network. The preference is for industries in which it is possible to fragment the production process into phases or products, without using outmoded technologies. This fragmentation of activities implies a dense network of interdependences between the enterprises, and thus a genuine division of labour. In this integrated system SMEs maintain links such as financial stakes, supply contracts or temporary informal agreements (Courlet, 1986).

Although this flexible organisation model may be found in several peripheral and semi-peripheral regions of Europe and North America it is undoubtedly in Italy that it has flourished most. In that country it is the cause of a new form of economic organisation of the territory which modifies the conventional image of an opposition between the industrialisation north with its large companies and the under-developed and mainly agricultural south. The reality is now more complex, since new development zones are appearing. These are mainly the regions of North-East and Central Italy (often referred to by the acronym NEC) which contain about 37% of the national population. They are characterised by the presence of SMEs organised into "local production systems" (Garofoli, 1983; Courlet, 1986; Brusco, 1986; Santilli, 1987).

The production organisation of the NEC cannot be explained by decentralisation from large companies and/or the central regions; rather it is the result of the enhancement of the value of local resources: local entrepreneurial ability, specific skills and know-how accumulated over time, qualification of locally-trained workers, locally-accumulated financial resources, innovative technologies introduced locally, etc. (Fuà, 1983; Garofoli, 1987). This is why the local production systems of the NEC type enjoy a certain degree of independence in the design, production, management and marketing of the product. The degree of independence may go as far as generating an internally-controlled accumulation process. In this case what we have is a self-centered local development phenomenon which some authors compare with "marshallian industrial districts", i.e., spatially agglomerated production complexes together with their dependent labour markets and intercalated human communities (Scott, 1987; Bellandi, 1986).

By emphasising local resources and the interdependence of activities at the local level, the NEC model explains the behaviour of other regions where the same type of phenomena are appearing. As Scott (1988) emphasises, the lessons to be learned from the contemporary industrialisation of the Third Italy are just as important as those of Japan.

7. Towards a Theory of Local Dynamism

Enhancing the value of local potential, highlighting the interdependence at the territorial level and the role of local entrepreneurial ability result in a development model which indicates that the destiny of some regions is not necessarily dictated by their attractiveness for the branches of large companies, but that it depends on their ability to generate SMEs and to organise them into a coherent whole. In other words, local companies may be considered as a source of social and technological innovation. It is a new development model which is emerging, its emphasis being on local dynamism. The challenge is to identify for those areas the strategic key factors which provide them with a sufficient growth potential based on an active (bottom up) self-organising capability (Maillat, 1984; van der Mark and Nijkamp, 1987).

In fact, local dynamism challenges the traditional model based on the large company as the engine of regional development (Aydalot, 1984; Quévit, 1985; Santilli, 1987). In Aydalot's view (1984), for example, "the major dynamisms have been upset. It is not the same sectors, the same products, enterprises, technologies, countries and the same regions which propagate and generate development. Already in the sixties industrial growth was faster in the peripheral regions and countries. Today it is clear that regions which were for a long time agricultural and poor have embarked on a process which is not only one of catching up but which points to an avenue for a new type of development". This view may seem optimistic, but it does stress the fact that for about fifteen years, since the large companies have had to reduce their employment levels, they are no longer able to control regional development as powerfully. This responsibility now falls on the region itself, through the creation of new small enterprises.

To explain the role of SMEs in regional development, one has to switch from a theory of localisation to the mobile parts of large companies to a theory of the development and functioning of territorial production systems based on SMEs.

One might suppose that the regions which are distinguished by the dynamism of their SMEs have better entrepreneurs than the others, and that they offer a more favourable environment for starting up and supporting new enterprises (network effects).

8. Local Differences in Entrepreneurship

A discussion of local entrepreneurship often includes the idea that the region will regain its dynamism if it regains its entrepreneurs (Coffey and Polèse, 1985). Attention is thus focused on a small group of people whose characteristics are not easy to specify either (O'Farrel, 1986; Suarez-Villa, 1986). However, we noted above that job experience was crucial. In addition to this important factor of previous experience, it is necessary to remember that most SME entrepreneurs are of local origin. In general, empirical studies conducted in Europe clearly show that "most entrepreneurs start activities with which they are already familiar. In that way a relationship arises between the industrial composition of new firms and the existing firms in a region" (Wever, 1986). This means that the local rate of new firm start-ups is directly related to the economic and social fabric of the region (Keeble and Wever, 1986; Illeris, 1986; O'Farrel, 1986).

In their summary Keeble and Wever (1986) pinpoint three types of explanation: structural, socio-cultural and economic. A region's economic structure, particularly the nature of its industries and size of its firms, play a role in the possibilities for creating new enterprises. Evidently, "new enterprise creation rates in communities dominated by iron and steel or petrochemical complexes are bound to be lower than in areas specialising in clothing manufacturing, diversified electronics production or tourism". In addition, in the regions where large companies predominate, the spirit of enterprise has often been destroyed, whereas it is much more alive in regions with many SMEs. From the socio-cultural point of view, it is the occupational profile of the local population that is important. A good correlation is generally found between the proportion of skilled workers and the rate at which new enterprises are formed. The spirit of enterprise is also stronger in regions which have a local tradition of self-employment than in those with a more proletarian tradition. Finally, the rate at which new businesses are created also depends on economic considerations, particularly "on the local availability of factors of production such as premises or capital, and demand for new firm products from particular geographical markets".

At the local level the creation of enterprises is thus not only the mechanical result of a market opportunity, it depends on the characteristics of the environment.

9. Belonging to a Territorial Network

The other aspect of local dynamism based on SMEs is the fact that the latter do not function independently of each other: they belong to a territorial network. The network which they supply and from which they receive, multi-faceted services in the fields of technology, training, management, marketing and financing is essential to their proper functioning (Planque, 1987). Now this network is localised since SMEs, particularly those which belong to local entrepreneurs, have territorial base. These enterprises maintain more links (whether commercial or non-commercial) among themselves than do branches belonging to large companies with local firms.

It should be emphasised that the links which are forged in such networks are not only structural in nature - they are above all cultural. Indeed, the networks observed in today's dynamic regions existed before they underwent strong growth: links between entrepreneurs or technicians in the former industrial regions, links between former students in the high-tech regions and family or trade-union links in Italy (Crevoisier, 1988).

The territory in which local SMEs operate thus constitutes a zone of support from which they draw their dynamism. They are therefore concerned by the way it is organised. This distinguishes them from the large groups, which show little interest in the network of the area in which their branches operate.

It is consequently possible to detect the emergence of a special level of production organisation which is neither micro-economic nor macro-economic and which consists of territorialised, open industrial systems (in which proximity phenomena are important). Territorialised industrial systems thus appear as a level of production organisation that is different from that of the individual entrepreneur (micro-economy) and from that of the national production system, structured as it is into industries and sectors (macro-economy).

10. Territorialised Industrial Systems

In recent years several types of territorialised industrial systems in which networked SMEs play an important role have appeared (Scott, 1988). Together with Perrin (1988), we shall distinguish four groups of network which have, or are likely to develop, a strong territorial base:

1. Territorialised networks built up around the large local companies.The core of these networks is formed by large companies which in the course of their development maintained special links with the territory in which they were located, so much so that they shaped its industrial and urban structures (Fiat and Olivetti in the Piedmont, Peugeot in Franche-Comté, Philips in the south Netherlands, Volkswagen in the

Hannover region, etc.). In the course of time these firms have built up networks of SMEs around them, including in research and development and machine manufacturing. The behaviour of these local firms differs from that of multi-branch companies which practice a functional and spatial division of labour and which therefore do not help to create networks in the regions in which they locate branches.

2. Territorialised networks formed from SMEs. The composition of these territorialised networks is the result of historical processes which favoured the formation of SMEs with similar industrial profiles and whose territorial economic relations have been cemented by social traditions. This is the case, for example, of Prato (Brusco, 1982; Ritaine, 1987), Savoie (Courlet, 1986), Oyonnax (Raveyre and Saglio, 1984) and the Jura Arc (Maillat, 1984). This type of network is characterised by the intensity of the relations between a large number of SMEs. This leads to a very high degree of specialisation in the zone, which may go as far as the mono-industry. This type of organisation presupposes both a strong division of labour between the firms, a marked specialisation in the different phases of the production process and fierce competition at the horizontal level between firms producing the same product or involved in the same phase of the production process. These characteristics form the basis for continuous technical improvement of the local production system (Garofoli, 1986; Maillat, 1988).

To these two types of network, whose organisational forms may be described as old, two others can be added, whose emergence is more recent.

3. Territorialised networks formed by "horizontal fragmentation". The formation of these networks is part of the strategy of large companies. Traditionally these firms exploited the logic of function specialisation (R&D, production, etc.) and located the latter in the most appropriate regions (vertical dis-integration of the company). The new strategy involves delocalising to a given region not one of the firm's functions but all the functions contributing to the production of a given good (horizontal fragmentation). In other words, the strategy involves grouping together in a single location, i.e. territorialising, all the links in the production process (from upstream to downstream, i.e. from R & D, through production, control

etc. to sales) (Loinger and Peyrache, 1986). If several large companies adopt this approach on a given site (Sophia Antipolis, Toulouse, Cambridge, for example) it may result in the formation of a network which generates a local dynamism and which becomes independent of the original firms.

4. Territorialised networks formed by incubation. These networks take the form of a collective organisation in which various institutions (public and private research laboratories, enterprises, local authorities, etc.) make up an original structure which, in a more spontaneous manner (Silicon Valley, Orange County) or a more deliberate manner (science park, technology park) contribute to the emergence of many firms using new technologies. These networks are characterised by numerous inter-relationships between production firms, research laboratories and firms offering specialised services (in organisation, international marketing, venture capital, finding partners, etc.). Furthermore, the phenomenon of hiving off (from research firms or laboratories) is characteristic of their dynamism. The development of this type of network is due broadly to local initiative. Their management of innovation is highly territorialised. It gives a territorial basis to the applications of science and technology and relies on the effects of proximity to stimulate the innovative process (OECD, 1986; Storey, 1986).

Each of the above described networks forms a territorialised industrial system which is very independent, due to the high degree of territorial integration in evidence between the companies that make up the network. It should be pointed out that independence here depends on the extent to which local industry controls capital, technologies, qualifications and markets. The networks in question are also characterised by the presence of numerous SMEs which, due to the links they maintain, form a geographical clustering of activities in which relations of competition/co-operation give rise to an endogenous development dynamic. Of course there exist other "systems", also comprising SMEs, but which on account of their low degree of territorial integration are not "autonomous" and which do not govern their development, i.e., they are not capable of creating a genuine local dynamism.

11. The Milieu, the Origin of Local Dynamism Based on SMEs

As has been emphasised, the characteristic of territorialised industrial systems based on SMEs is the fact that the latter belong to a network. It is from this that they derive their dynamism. This means that local dynamism does not result from the action of a single firm but from their overall behaviour. It is in fact the notion of the milieu which best illustrates this phenomenon. Let us quote Aydalot and Keeble (1988): "the central concern of the local environment-based approach is to understand the firm in its local and regional context, and to ascertain what conditions external to the enterprise are necessary both for the creation of new firms and the adoption of innovations by existing ones. The firm, and the innovating firm, are not viewed as pre-existing in or separate from the local environment, but as being a product of it. Local milieus are regarded as the nurseries, the incubators, of innovation and innovative firms... The historical evolution and characteristics of particular areas, their social and economic organisation, their collective behaviour, the degree of consensus or conflict which characterises local society and economy, these are the major components of innovative behaviour. ...This approach implies that innovative behaviour is as much dependent on variables defined at the local and regional level as on national scale influences. Access to technological know-how, the availability of local industrial linkages and inputs, the impact of close market proximity, the existence of a pool of qualified labour - these are the innovation factors which will determine areas of greater or lesser innovative activity within national space".

The milieu may be defined as a territorialised complex comprising a production system, a culture and protagonists having a common approach to problems and opportunities. It is certain that the spirit of enterprise, organisational practices, companies' behaviour and the way technology is used are all elements which form part of the milieu. The milieu is thus composed of material and non-material elements. The material elements are organised around the territorial production system (nature of companies and their local integration, coherence of all the activities, etc.), the local labour market and the way it works (nature of jobs, types of chains of mobility, etc.) and the territorial scientific system (training system, research institutes, production and accumulation of the region's knowledge). These various elements function in symbiosis and give the milieu its specific characteristics. The latter are reinforced by the non-material elements, particularly by the technical culture. At the territorial level, the technical culture results from the knowledge and skills which are acquired from contact with production (a knack, for example) and from technical and scientific knowledge taught by regional vocational schools.

This technical culture is communicated, acquired and renewed by exchanges and contacts which require a certain proximity (Maillat and Vasserot, 1988; Crevoisier, 1988; Maillat, 1988).

The notion of the milieu thus enables us to differentiate the functional logic from the territorial logic (Crevoisier, 1988). Functional logic corresponds to the Fordian production method; it is this logic which led to the spatial fragmentation of production and to the spatial division of labour. The territorial logic accounts for the development certain regions are undergoing thanks to the innovation and dynamism generated by the milieu.

Innovation originates from putting information into contact with resources (capital, skills, spirit of enterprise, creativity). Now, the milieu corresponds to an integrated complex of resources. However, these are not always sufficient to maintain or trigger a process of innovation. That is why an innovative milieu is one which remains open towards the outside world; it must be sensitive and receptive to change (new patterns of behaviour, new services, new technologies). It is a collective process in which all the elements and all the protagonists of the milieu are involved. This explains why local dynamism cannot be associated with the presence of SMEs alone, even if they were innovative. The examples of the territorialised industrial systems we have mentioned attest to this.

In conclusion, the role of the milieu is all the more important when considering SMEs. If absolutely necessary the large company is capable of doing without the milieu, since it can internalise all the elements of the innovative process. SMEs cannot do this. It is the milieu which enables them to mobilise the information they need. The links with the elements that make up the milieu are indispensable as they provide them with the various elements of the innovative process. Furthermore it is the milieu, by virtue of its cultural dimension in particular, which is behind the local entrepreneurial ability. Considered from the point of view of local dynamism, therefore, the emergence of SMEs is only meaningful if the latter are integrated in or generated by the milieu. If this is not the case, SMEs remain dependent on the outside and the territorial development logic is unable to manifest itself.

12. Conclusions

Faced with the mass of information on the role of the SME and their regional impact, we have emphasised three factors which we believe to be crucial: the permanence of the SME phenomenon, their ability to innovate and experiment with new technologies, and their ability to revitalise the spirit of enterprise. SMEs are thus not only important because they are responsible for the creation of new jobs but also and above all because they help to transform the production system.

From the regional point of view, it is a question of knowing whether they have a similar effect in each region. The reply is in the negative; they do not have the same effect everywhere. The presence of SMEs in a region does not necessarily mean development or revitalisation. The arrival of SMEs in a region may be the result of the corporate strategy of large companies (for example, vertical dis-integration). Because SMEs depend on outside entities, in this case they do not help to generate "autonomous" local dynamism. Nor is the existence or emergence of independent or local SMEs in a region necessarily the sign of a specific regional dynamic. True, these SMEs provide jobs, but they do not provide the region with a chance to control its development. Indeed, if local dynamism based on SMEs is to manifest itself, one condition has to be met: SMEs have to belong to a territorialised network. We have seen that there are several types of territorialised networks comprising SMEs. They are all characterised by a common feature: territorial integration of companies. That being so, dynamism is no longer expressed through a single firm but by the action of them all. In their most elaborate form, these networks belong to a milieu, that is, a complex of commercial and non-commercial proximity-based relations organised around a territorialised production system. When these conditions are fulfilled, one can speak of local dynamism in which SMEs play an active role. Special conditions must therefore obtain if one is to be able to claim that territorial development depends on SMEs.

This conclusion is important for regional policy. It means that it is not sufficient to promote the establishment of SMEs in a region to stimulate local dynamism. It is preferably a milieu-oriented approach that should be adopted. Rather than providing specific assitance to companies, it would be preferable to aim to set up territorialised networks by encouraging inter-company and inter-branch relations (co-operation, partnership), with a view to developing a territorialised production system. Regional policy must also pay particular attention to the development of technical culture and specific skills. The non-material elements of the milieu form truly external entities which are indispensable to the innovativeness of companies, particularly SMEs. In fact, what particularly characterises a milieu policy are actions which aim to increase the density and permanence of material and non-material relations in networks with a territorial base.

References

Arzeni, S. (1986), Entrepreneurship Promotion in Europe, in: Ergas, H. (ed.), A European Future in High Technology?, CEPS, Paris.

Aydalot, P. (1986), La place des petites et moyennes entreprises dans le développement régional: Les leçons de l'expérience récente, The Role of Small and Medium Sized Enterprises for Regional Development, Samos, Greece.

Aydalot, P. (1986), L'aptitude des milieux locaux à promouvoir l'innovation, in: Federwish, J. and Zoller, H. (eds.), Technologies nouvelles et ruptures régionales, Economica, Paris.

Aydalot, P. and Keeble, D. (eds.) (1988), High Technology Industry and Innovative Environments, Routledge, New York-London.

Aydalot, P. (ed.) (1986), Milieux innovateurs en Europe, GREMI, Paris.

Bailly, A. and Maillat, D. (1988), Le secteur tertiaire en question, ERESA-Economica, Geneva.

Bellandi, M. (1986), The Marshallian Industrial District, Dipartimento di Scienze Economiche, Università degli studi di Firenze, Studi e Discussioni 42.

Bramanti, A. and Senn, L. (1987), Innovations, ruptures et stratégies de développement dans le cas des régions à tissu industriel diversifié, Les politiques d'innovation technologique au niveau local, Paris.

Brusco, S. (1986), Small Firms and Industrial Districts: The Experience of Italy, in: Keeble, D. and Wever, E. (eds.), New Firms and Regional Development in Europe, pp. 184-202, Croom Helm, London.

Cappellin, R. (1987), The Diffusion of Producer Services in a Metropolitan System, Les stratégies régionales d'innovation et la compétitivité des entreprises, Neuchatel, IRER.

Cappellin, R. (1988), Transaction Costs and Urban Agglomeration, Revue d'économie régionale et urbaine 2, pp. 261-278.

Castells, M. (1984), Towards the Informational City? High Technology, Economic Change and Spatial Structure, Institute of Urban and Regional Development U.O.C. Berkeley Working Paper 430.

Coffey, W. and Polese, M. (1985), Local Development, Conceptual Bases and Policy Implications, Regional Studies 2, pp. 85-94.

Courlet, C. (1987), Local Industrial Policies in favour of Small and Medium-Sized Firms: The French Case, Leuven.

Courlet, C. (1986), Stratégies de P.M.E. et développement local en Savoie, Economie et Humanisme 289, pp. 38-45.

Crevoisier, O. (1988), Functional Logic and Territorial Logic and how they inter-relate in the Region, Regional Science Association, Summer Institute, Arco, Italy.

Delattre, M. and Eymard-Duvernay, F. (1984), Le progrès des P.M.E. dans la crise: signe d'un relâchement du tissu industriel, Critiques de l'économie politique 26, pp. 119-133.

Fischer, M. and Nijkamp, P. (1988), The Role of Small Firms for Regional Revitalization, The Annals of Regional Science, 22, pp. 28-42.

Fua, G. and Zacchia, C. (1983), Industrializzazione senza fratture, Il Mulino, Bologna.

Garofoli, G. (1986), Le développement périphérique en Italie, Economie et Humanisme 289, pp. 30-37.

Garofoli, G. (1983), Le aree-sistema in Italia, Politica e Economia 14 (11).

Giaoutzi, M., Nijkamp, P. and Storey, D. (1988), Small and Medium Size Enterprises and Regional Development, Routledge, London.

Gould, A. and Keeble, D. (1984), New Firms and Rural Industrialization in East Anglia, Regional Studies 3(18), pp. 189-201.

Greffe, X. (ed.) (1984), Les P.M.E. créent-elles des emplois?, Economica, Paris.

Hitchens, D.M. and O'Farrel, P.N. (1987), The Comparative Performance of Small Manufacturing Firms in Northern Ireland and South East England, Regional Studies 6(21), pp. 543-553.

Illeris, S. (1986), New Firm Creation in Denmark: The Importance of the Cultural Background, in: Keeble, D. and Wever, E. (eds.), New Firms and Regional Development in Europe, p. 141-150, Croom Helm, London.

Julien, P.A. (1987), Le développement micro-régional, la PME et la théorie micro-économique, Canadian Journal of Regional Science 2, pp. 133-147.

Julien, P.A. and Morel, B. (1986), La belle entreprise: la revanche des PME en France et au Québec, Boréal, Montreal.

Keeble, D. and Wever, E. (eds.) (1986), New Firms and Regional Development in Europe, Croom Helm, London.

Korte, W. (1986), Small and Medium-Sized Establishments in Western Europe, in: Keeble, D. and Wever, E. (eds.), New Firms and Regional Development in Europe, p. 35-53, Croom Helm, London.

Leo, P.Y. (1986), Les milieux régionaux de P.M.I.: Une approche statistique et régionalisée des choix stratégiques des P.M.I. 39-82, Centre d'économie régionale, Université d'Aix-Marseille, Aix-en-Provence.

Loinger, G. and Peyrache, V. (1986), Le concept de grappes de technologies appliqué à l'espace économique régional, in: A.S.R.L.F. (ed.), Technologies nouvelles et développement régional, pp. 451-471, GREMI, Paris.

Maillat, D. (1984), Les conditions d'une stratégie du développement par le bas: le cas de la région horlogère suisse, Revue d'économie régionale et urbaine 2, pp. 257-274.

Maillat, D. and Vasserot, J.-Y. (1988), Economic and Territorial Conditions for Indigenous Revival in Europe's Industrial Regions, in: Aydalot, P. and

Keeble, D. (eds.), <u>High Technology Industry and Innovative Environments</u>, p. 163-183, Orutledge, New York, London.

Maillat, D. (1988), The Role of Innovative SMEs and the Revival of Traditionally Industrial Regions, in: Giaoutzi, M., Nijkamp, P. and Storey, D. (eds.), <u>Small and Medium Size Firms and Regional Development</u>, Croom Helm, London.

Maillat, D. (1984), Mobility Channels: an Instrument for Analyzing and Regulating the Local Labour Market, <u>International Labour Review</u> 123(3), pp. 349-362.

Nijkamp, P. and Van der Mark, R. (1987), <u>The Potential Role of High Tech Firms in Backward Areas</u>, Les politiques d'innovation technologique au niveau local, Paris.

O'Farrel, P. (1986), Entrepreneurship and Regional Development: some Conceptual Issues, <u>Regional Studies</u> 16(20), pp. 565-574.

Pavitt, K., Robson, M. and Townsend, J. (1987), The Size Distribution of innovating Firms in the U.K.: 1945-1983, <u>Journal of Industrial Economics</u> 35(3), pp. 297-316.

Perrin, J.-C. (1988), New Technologies, Local Synergies and Regional Policies in Europe, in: Aydalot, P. and Keeble, D. (eds.), <u>High Technology Industry and Innovative Environments</u>, pp. 139-162, Routledge, New York, London.

Piore, M.J. and Sabel, C.F. (1984), <u>The Second Industrial Divide</u>, Basic Books, New York.

Planque, B. and Py, B. (1986), <u>La dynamique de l'insertion des P.M.E. innovatrices dans leur environnement</u>, Les P.M.E. innovatrices et leur environnement local et économique, Aix-en-Provence.

Quevit, M. (1986), <u>Le pari de l'industrialisation rurale</u>, ERESA - Economica, Paris.

Raveyre, M. -F. and Saglio, J. (1984), Les systèmes industriels localisés: éléments pour une analyse sociologique des ensembles de P.M.E. industriels, <u>Sociologie du Travail</u> 2, pp. 157-176.

Ritaine, E. (1987), Prato ou l'exaspération de la diffusion industrielle, <u>Sociologie du Travail</u> 2, pp. 139-156.

Rothwell, R. and Zegveld, W. (1983), <u>Innovation and the Small and Medium sized Firms</u>, Frances Pinter, London.

Santilli, G. (1987), La Troisième Italie: Le rôle moteur des P.M.E. dans la mutation de l'économie italienne, <u>Problèmes Economiques</u> 2041, pp. 8-14.

Saxenian, A. (1983), The Genesis of Silicon Valley, in: Hall, P. and Markusen, A. (eds), <u>Silicon Landscapes</u>, pp. 20-34, Allen and Unwin, Boston.

Scott, A.J. (1987), Flexible Production Systems and the Regional Development: the Rise of new Industrial Spaces, <u>International Journal of Urban and Regional Research</u>, pp. 171-186.

Shutt, J. and Whittington, R. (1987), Fragmentation Strategies and the Rise of Small Units, Regional Studies (21), pp. 13-23.

Storey, D.J. (ed.) (1985), Small Firms in Regional Economic Development, Cambridge University Press, Cambridge.

Storey, D.J. (1986), The Role of SMEs in Job Creation in the European Community, Role of Small and Medium Size Industries in Regional Development, Samos, Greece.

Suarez-Villa, L. (1986), Innovation, Entrepreneurship, and the Role of Small and Medium Size Industries, Role of Small and Medium Size Industries in Regional Development, Samos, Greece.

Swyngedouw, E.A. and Anderson, S.D. (1987), Le schéma spatial de la production de haute technologie en France, Revue d'économie régionale et urbaine 2, pp. 321-349.

Wadley, D. (1986), Restructuration régionale, O.C.D.E., Paris.

13 Technological change as a driving force behind structural dynamics

M. GIAOUTZI

1. Introduction

The period from the late 60s to the early 80s was marked by considerable structural changes in many parts of the world. In this period many formerly thriving industrial regions fell into stagnation and decay characterized by severe problems of deindustrialization and job loss.

In many countries **regional disparities** were also exacerbated because at the same time a limited number of areas experienced significant growth in output and employment due to the rise of **new technology** based industries and services. These changes were often associated with the emergence of the microelectronics technology, with much of this growth taking place in new and **small business firms** or in a limited number of **multinational enterprises** operating in this field.

Two types of dichotomies can now be identified in this context. The first one takes place at the level of **industrial organization** in which the requirements of advanced technologies induce an increasing trend towards **large scale** companies (e.g., telecommunications, aircraft, etc.) due to their highly oligopolistic-competitive framework. This goes together with a contrasting tendency in which the requirements of modern technology induce the emergence of a wide variety of advanced **small** and **medium sized** firms (mainly in the business service sector), which play the role of the hinterland for large-scale firms and create the seedbed conditions for a flexibility and diversity in local-regional-national industrial structure. Thus the first type of dichotomy appears to release two antagonistic forces, -viz. a **large-scale internationalization versus a small-scale localization-** which result from the above mentioned drastic changes in the technological and economic conditions of our society (see Nijkamp and Malecki, 1988 and Giaoutzi et al., 1988).

The second dichotomy refers to the increasing discrepancy between **core and periphery** in which contrasting spatial power relations can clearly be identified among various regional, technological, financial and economic systems all over the world (see e.g. the emergence of economic metropolises such as Tokyo, Milano, etc.). The above relations do not only take place in an international core-periphery context but also and with the same intensity in regions within a national economy.

In the light of the above trends the aim of this paper will be to explore the implications of technological change for **spatial-economic reorganization** by using an analysis framework based on a systemic approach.

2. Main Trends in High Technology and Regional Development

It has been widely accepted that technological change forms the driving force behind structural dynamics (see Nijkamp, 1986) and hence has had a positive effect on the growth of industrial economies. But this high rate of technical change has not been obsered to the same extent in all countries, nor have in peripheral areas the same level of technological sophistication been reached as those existing in most central industrialized societies.

Thus unequal technological development is likely to produce uneven growth, where there can be winners and potentially absolute losers (Goddard et al., 1985). The above observation has had severe implications for regional development and it has given rise to a number of research endeavours by trying to identify the spatial-economic potential imposed by the new logic of advanced technologies.

Among the most important lines of analytical research dealing with the above problem can be mentioned:

a) the product and spatial life cycle model;
b) the seedbed/incuabator hypothesis;
c) the locational factors approach;
d) the neo-Fordist approach.

These four lines of inquiry will now be discussed briefly.

The **product life cycle model** claims that each industrial sector evolves through a cycle of invariant stages, which are defined as innovation growth and maturity (Norton and Rees, 1979 and Vernon, 1979). Increasing outputs in each industry lead to the reorganization of production through changes in the industry's technology, beginning with skilled labour-intensive methods in the innovation phases, evolving through a stage of rapid growth and resulting into deskilled, capital-intensive mass-production methods, usually in large branch plants.

Sicne the product life cycle is defined by a sequence of techniques, each with quasi-fixed factor coefficients, it also describes the temporal pattern of change in the industry's locational requirements.

As the market grows, routinization and mechanization of production alter the industry's labour demand in favour of increasing lower skills.

Scale increases also make possible the standardization of external linkages, thereby reducing unit spatial interaction costs. As a consequence of these developments, the industry's spatial structure shifts progressively from centralization and agglomeration -often described in product cycle theory in terms of **seedbed** phenomena- to decentralization and geographical dispersal.

Very much in line with the above approach is the more recent work of Freeman (1984).

The **seedbed/incubator hypothesis** is mainly concerned with the geographical conditions under which new sectors of production arise on the landscape and are nurtured through their initial phases of growth. An older version of the hypothesis suggested that the centers of large cities are the best incubators for fledging sectors, for it is here that these sectors are said to find the manifold agglomeration economies they need in order to survive the pressures of economic competition (Struyk and James, 1975).

A more recent version of the hypothesis proposes that the locations of the seedbeds themselves may be periodically adjustable and that new seedbeds are now starting to spring up in specific favourable parts of the spatial structure, where they function as 'new territorial innovation complexes' (Storper, 1985). In both versions the assumed spinoffs from existing firms complete the dynamic attitudes of the seedbed phenomenon (Nijkamp and Davelaar, 1987).

A great deal of work along these lines has been done by Nijkamp and Davelaar (1989), Ciciotti (1984), Cooper (1985) and Cooper et al., 1985).

It is clear the the location factors which predominate in 'high tech' industries are broader than those common in previous generations of regional development efforts. This has led to the **locational factors approach**, which is trying to identify the factors that are likely to define the underlying attractiveness of particular regions for high technology. A wide variety of such factors is usually mentioned in the literature, such as venture capital availability, presence of experienced entrepreneurs, technically skilled labour force, accessibility of suppliers, accessibility of customers or new markets, favourable governmental policies, proximity of universities, availability of land or facilities, accessibility of suporting services, attractive living conditions (Nijkamp and Malecki, 1988). The above factors are also found in work by Aydalot (1986), Boullianne and Maillat (1983), Brochie et al. (1985), Camagni and Rabelottie (1986), Ciciotti (1984), Giaoutzi and Nijkamp (1987), Giaoutzi et al. (1988), Goddard et al. (1985), Hieronymi et al. (1983), Kamann (1988), Maillat (1982), Oakey (1982) and Stöhr (1986).

Finally, the so-called **Neo-Fordist approach** mainly deals with the logic of industrial organization based on the theoretical assumption that each broad epoch of growth in capitalism tends to be associated with and based upon a particular cluster of dominant production sectors. The epoch of growth associated with each of these clusters is typified by a regime of accumulation, and each regime of accumulation is in turn sustained by a set of macro-social and political arrangements that can be called a mode of social production.

The new industrial cluster based on high technology industry has hence seemed to introduce a **new regime of accumulation** and a corresponding mode of **social regulation**. Its emergence, like that of the major clusters that preceded it -and the service industry that accompanies it-, has also been associated with a series of major and largely unforeseen geographical shifts. These shifts are the outcome of the inextricable interwinding of the specific features of high tech industry as a unique physical phenomenon within broader trans-sectoral tendencies in the organization of production and the division of labour in capitalism.

Thus according to this approach the historical geography of high-tech industry presents us with a new set of detailed contingent expressions of capitalist development and hence with a new set of theoretical questions and tasks (see Scott and Storper, 1986). Work along these lines can be found in Aglietta (1976), Gillespie (1983), Liepietz (1986), Scott and Storper (1986), Storper (1985) and Swijngedouw (1987).

3. The Characteristics of Technology and Industrial Environment

In the course of the efforts to understand the types of dichotomies emerging at the level of both **industrial organization** (large-scale vs. small-scale firms) and the level of **spatioeconomic organization** (viz. core-periphery) of the world regions, there has been a renewed interest in the economics of innovation. This has led to a considerable progress in the empirical description and theoretical conceptualization of the **sources**, **characteristics**, **directions** and **effects** of technological change. Referring to the work of Dosi, Pavitt and Soete (see Dosi, 1985 and Dosi et al., 1986b) we summarize here some of the major findings which are directly related to the characteristics of technology and industrial environment.

Technological change can in general be characterized by:

- varying degrees of **appropriability** of innovations;
- **uncertainty** on technical and commercial outcome of innovative efforts;
- **opportunity of achieving technical advances** for various efforts;

- **cumulativeness in the patterns of innovation** and in the capabilities to innovate;
- **tacitness** of the knowledge and expertise on which innovative activities are based.

In this respect particular search and learning processes draw on technology-specific 'knowledge bases' which are related -in different combinations- to both freely available **information opportunities** and more **'local'** and **tacit** skills, experiences and heuristics embodied in people and organizations (see Dosi et al., 1986a).

Technologies develop along relatively ordered paths which are shaped by the technical properties, the heuristics and the cumulative expertise embodied in each 'technological paradigm' (Nelson and Winter, 1987). In this context Winter (1986) also defines different 'technological regimes' according to whether the 'knowledge base' supporting innovative research is primarily 'universal' and thus, external to individual firms or alternatively 'local' and firm-specific.

In line with the above discussion it appears that **'diversity'** is a fundamental characteristic of industrial environments which undergo technical change.

Diversity has been nurtured by certain characteristics of individual firms:

- different technological capability to innovate;
- different degrees of success in developing-imitating-adopting product and process innovations;
- varying cost structures.

These forms of diversity have been defined as technological **asymmetries**, in other words unequivocal technological gaps between firms which can be ranked as 'better' and 'worse' according to their costs of production and their product characteristics. However, in a static context technological differences between firms contribute to determine economic performance, while in a dynamic context the evolution of industries is shaped by the relative pace of innovation and imitation of individual firms (Dosi et al., 1986b).

Diversity relates also to differences among firms as to their search procedures as well as the characteristics of their techniques of production and their products. This set of sources of diversity has been identified in the same analytical context as **technological variety**; in other words, all these technological differences which do not correspond to the above mentioned technological asymmetries but rather to a set of different but equally suitable elements.

Another source of diversity stems from the behavioural differences of firms as to their strategic decisions (e.g., investment, pricing, R&D, etc.) which will be identified as **behavioural diversity** (see Dosi et al., 1986b).

The feedback processes linking the evolution of the above mentioned forms of diversity to the fundamental characteristics of each technology are to a large extent conditioning the range of the two types of dichotomies displayed in both industrial organization structures and spatioeconomic structures of the world regions.

More precisely, a set of important characteristics of each technology is critical for both the locational requirements and the type of industrial organization required for the creation, successful commercialization and adoption of each technology. Examples of such factors are:

- the sources of basic knowledge (which require information accessibility, availability or access to R&D, skill availability, etc.);
- degrees of appropriability and tacitness of innovation (which require knowledge and ability to evaluate technologies, bargaining ability, lack of institutional constraints, etc.);
- complexity of research (which require financial means to support it, access to R&D, skill availability, critical size of firm, information accessibility, etc.);
- production and products (conditioned by the range of the market, competitiveness, etc.);
- existence and role of various forms of economies of scale (determined by the location of scale of firm);
- cumulativeness of technological learning (conditioned by the scale of firm, type of product, technological environment, etc.).

The **degrees and forms of diversity** between the economic agents which include the levels and distribution of their technological capabilities and the variety of their search processes and behavioural rules, together with the **endogenous evolution of incentives, constraints and selection mechanisms** (which include the evolution of relative profitabilities of different technologies, firm size, cash flow, market shares, etc.) are also critical to the spatio-functional requirements of technologies and are thus clearly conditioning the polarization exhibited in the above two types of dichotomies under discussion.

4. A Systemic Framework toward Understanding the Impact of High-Tech on Regional Development

After having discussed the characteristics of technology and industrial environment, we will now make an effort to identify the factors at work conditioning the spatial configuration of the world regions and thus try to understand the underlying mechanisms conditioning the observed tendencies towards the ever increasing discrepancies between core and periphery.

Each spatial system functioning in the context of technological change can be described by a number of factors classified both as exogenous and endogenous to the system.

We here draw a parallel with the systemic so-called Petit model (see Figure 1; Dermatt and Taylor, 1982), which springs off from an open system, in which organizational structure of an actor is treated as being dependent upon the external constraints and contingencies confronting the firm - in our case the spatial entity system. As an open system involving input, transformation and output, the model treats the organization as a set of systems within systems, none of which can survive in isolation, but all of which can gain negative entropy by moving towards states of increased order and organization.

At the center of the organization we find the 'Industrial Structures' of the subsystems, which produce the goods and services characterizing a particular organization. This component of the organization is generally considered to require the highest degree of stability in order to perform its task efficiently with the technology that it contains and, therefore, requires the greatest protection possible from the activities of the exogenous environment. To achieve an approximation of a closed system each unit/firm of the industrial structure is buffered by the managerial subsystem within each firm, which coordinates the operation within and between firms, resolves conflicts, allocates tasks and formulates policy.

The industrial structure's managerial subsystems are in turn surrounded by the **Institutional Level** of the organization which is trying to deal with all these elements of the 'selection environment' that impinge upon the organization as a whole. Beyond and surrounding these systems which constitute the spatial system, is the exogenous system.

The approach described above is however not 'reductionistic', since it provides a perspective whereby order can be recognized in all subsystems within and around the spatial subsystem.

Clearly the different positions of subsystems that make up the organization reflect variable levels of **control** and **autonomy with each**

Figure 1 Petit's Model on Firm-Environment Interaction Structural Contingency Model

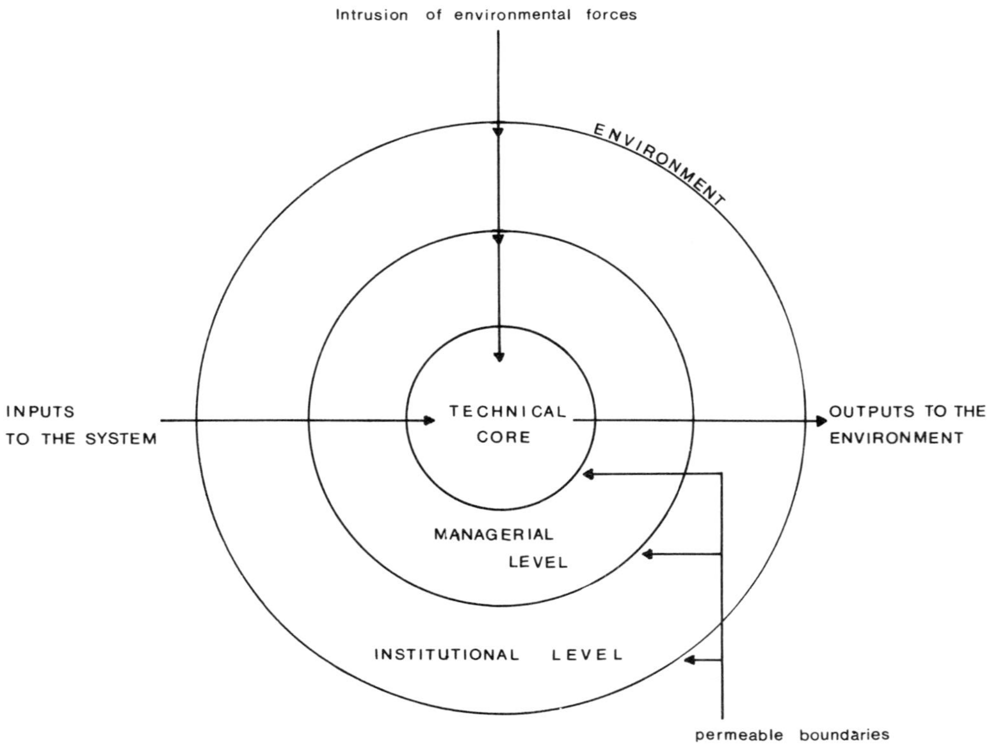

subsystem having degrees of power over others which may vary over time.

The boundaries of an organization also poses an element of imprecision, being ultimately dynamic. Furthermore, parallelling the input-transformation-output sequence in the organization is a feedback mechanism whereby subsystems relate to one another and the exogenous environment intrudes into the organization. The feedback mechanisms converts the organization into a patterned set of events able to achieve a reasonable level of **self-organization** (Dermott and Taylor, 1982).

These characteristics act to prevent the intrusion of the environment into the organization from a completely determining structure. Indeed, the forces of exogenous environment can only intrude into various levels of the organizational system to the extent that the boundaries between layers are permeable. Given the relationship between different levels or subsystems within the organization and the intrusion of environmental forces into it, organizational systems will adapt their behaviour, structure and interrelationships to the internal and external pressure exerted upon them.

The mechanisms facilitating this adaptive behaviour can be characterized by means of the processes of **differentiation** and **integration.** Differentiation refers to the variety of organization appearances of a firm, whilst integration refers to the inter-firm linkage patterns. Integration can be distinguished into horizontal and vertical desintegration (see Nijkamp, 1982 and Swyngedouw, 1987).

Horizontal differentiation is thought to lead to the establishment of entities to confront homogeneous sections of the environment.

Vertical differentiation is thought to establish the appropriate levels and spans of control in the bureaucratic hierarchy of an organization.

Integration is considered to be essential for the co-ordination of differently structured entrepreneurial entities, removing internal tensions and sustaining the performance of the organization.

It is also quite conceivable that **similar structure states** could be achieved from different previous conditions and via different routes (the so-called **equifinality** property of open systems. A corollary of this proposition is that under similar external circumstances similar organization forms may be created, a possibility which is central to the structural contingency paradigm by Petit. The explanatory value of the above model lies evidently in the adequate conceptualization and specification of the set of key variables.

To assess the benefits obtained from the introduction of the structural contingency concepts into the geography of core-periphery relations it is necessary to elaborate the issues relating to exogenous environment and technological innovation in more detail. The specification of the **'selection environment'** confronting and surrounding the 'industrial structure' of a spatial subsystem is perhaps the essence of the structural contingency model.

In a broad sense, 'selection environment' can be defined as everything outside the focal organization, i.e., 'industrial structure'. In the terminology of various authors this is the general or societal environment, an all-embracing notion which has also been referred to as macro-environment (see e.g., Dermott and Taylor, 1982). In this respect the following sets can be identified:

1. **cultural** - values and norms of a particular society;
2. **technological** - the depth of technological and scientific knowledge in society;
3. **educational** - the general literacy level of a population together with aspects of specialization and professionalism;
4. **political** - the political climate in a society and the structure of power;
5. **legal** - running from constitutional considerations to the impact of specific laws;
6. **natural resources** - the natural endowment of renewable and non-renewable resources;
7. **demographic** - the nature of human resources;
8. **sociological** - class structure and social institutions;
9. **economic** - the general economic framework of society together with the level of investment;
10. **infrastructural** - the availability in networks (including information);
11. **scientific** - the existence of universities and research institutes.

However, it is extremely difficult, if not impossible, to put this broad definition of a 'selection environment' in operation, since any such operationalization would be of limited empirical value.

Of greater importance from a measurement point of view are the aspects of **'domain'** and the **'task environment'** of the organization. The **'domain'** is the group of organizations with which a focal organization could interact, the points within the selection environment with which that organization might potentially have contact, given the activity or range of activities it undertakes (Dermott and Taylor, 1982). To establish this domain, however, requires clearly domain consensus. Not only does an organization stake out a sphere of potential operations and impact amongst other organizations as a result of its charter, but it also needs to be recognized as significant within that sperhe by other organizations.

This narrowing down of the conceptualization of environment from the selection environment to the organizational domain has been taken one step further in the structural contingency framework with the development of the concept of the **'task environment'**, a concept which is potentially more capable of being put in operation.

The **task environment** was defined by Thompson (quoted in Dermott and Taylor, 1982) as the set of organizations with which a focal organization actually establishes exchange and relationships. Such exchanges can include materials, goods of information and can be considered as fundamental to the survival of the organization and its decision-making.

Dill and Duncan, quoted in Dermott and Taylor (1982), categorized the organizations comprising this external task environment as:

1. customers;
2. suppliers;
3. competitors;
4. regulators (including government and unions) and
5. technology.

The three components of environment selection -domain and task environment- represent arbitrary points along a continuum depicting the directives or indirectives with which environmental forces impinge upon the operations of a focal organization.

The relationship is by no means perfect, however, as the impact of a given set of forces or synergies upon an organization will be influenced by the amount of relevant information available to it and the way it is mentioned. This environmental continuum also relates to the **scale** at which the environment is considered within the focal organization. The selection environment for example, is likely to be delineated by national and cultural considerations, encompassing all potentially relevant systems within a given environment.

The domain may be relevant in considering the spatial entity, region or nation within which an industrial structure is located, while the task environment is defined by the actual contacts of the focus organization at any point in time.

The above systemic expression of forces at work can by no means be considered as an integrated operational approach for structural dynamics, since the level of abstraction accompanying it -despite its theoretical consistency- would rather mislead the empirical pioneer. Nevertheless, this framework leaves prospects for further elaboration by incorporating in the model both the characteristics of technology and industrial environment as such as well concrete empirical arguments found by other researchers, in particular when the abstract notions turn into specific sets of variables.

In conclusion, despite the lack of specificity, the systemic model based on the Petit paradigm offers a wide variety of interesting analytical departures for technological research at the firm level. In particular, the linkage between behavioural explanations and overall consistency with entrepreneurial functioning in an open environment is worth noting here.

References

Aglietta, M. (1976), A Theory of Capitalist Regulation, New Left Books, London.

Aydalot, P. (ed.) (1986), Milieux Innovateurs en Europe, GREMI, Paris.

Boulianne, L.M. and Maillat, D. (1983), Technologie, Enterprises et Region, Georgi, Saint-Saphorin.

Brochie, J., Newton, P., Hall, P. and Nijkamp, P. (1985), The Future of Urban Form, Groom Helm, London.

Camagni, R. and Rabelotti, R. (1986), Innovation and Territory: The Milan High-Tech and Innovation Field, in: Aydalot, Ph (ed.), Milieux Innovateurs en Europe, pp. 101-128, GREMI, Paris.

Ciciotti, E. (1984), The Incubator Hypothesis Revisited: The Case of the Metropolitan Area of Milan, paper presented at the 24th European congress of the R.S.A., Milan.

Cooper, A.C. (1985), The Role of Incubator Organizations in the Founding of Growth-oriented Firms, Journal of Business Venturing, 1, pp. 75-86.

Cooper, A.C., Dunkelberg, W.C. and Furuta, R.S. (1985), Incubator Organizations: Background and Founding Characteristics, in: Hornaday, J.A. et al. (eds.), Frontiers of Entrepreneurship Research, pp. 61-79, Babson College, Center for Entrepreneurial Studies, Wellesley MA.

Dermott, P.M. and Taylor, M. (1982), Industrial Organization and Location, Cambridge.

Dosi, G. (1985), The Microeconomic Sources and Effects of Innovation: An Assessment of some Recent Findings, SPRU, University of Sussex, Brighton.

Dosi, G., Pavitt, K. and Soete, L. (1986a), The Economics of Technical Change and International Trade, Wheatsheat, Brighton.

Dosi, G., Orsenigo, L. and Silverberg, G. (1986b), Innovation Diversity and Diffusion: A Self Organization Model, paper presented at the Conference on Innovation Diffusion, Venice.

Freeman, C. (1984), Keynes or Kontradiev; How can we get back to Full Employment, in: Marstrand, P. (ed.), New Technology and the Future of Work and Skills, pp. 2-16, Frances Pinter, London.

Freeman, C. (1986), The Economies of Industrial Innovation, MIT Press, 2nd edition.

Giaoutzi, M. and Nijkamp, P. (eds.) (1988), Informatics and Regional Development, Gower, London.

Giaoutzi, M., Nijkamp, P. and Storey, D. (eds.) (1988), Small and Medium Size Enterprises and Regional Development, Routledge, London.

Gillespie, A. (1983), Technological Change and Regional Development, Pion Ltd, London.

Goddard, J.B., Thwaites, A.T. and Gibbs, D.C. (1985), Technological Change and Regional Development, Research Paper, CURDS, University of Newcastle.

Hieronymi, O., Gabus, A., Hattemer, H. and Sallin, E. (1983), La Diffusion de Nouvelles Technologies en Suisse, Goergi, Saint-Saphorin.

Kamann, D.J.F. (1988), Spatial Differentiation in the Impact of Technology on Society, Avebury, Aldershot, UK.

Liepietz, A. (1986), New Tendencies in the International Division of Labour: Regimes of Accumulation and Modes of Regulation, in: Scott, A.J. and Storper, M. (eds.), Production, Work, Territory: The Geographical Anatomy of Industrial Capitalism, pp. 16-39, Allen and Unwin, Boston.

Maillat, D. (1982), Technology: A Key Factor for Regional Development, Georgi, Saint-Sopherin.

Nelson, R. and Winter, S. (1977), In Search of a Useful Theory of Innovation, Research Policy, vol. 8, pp. 1-22.

Nijkamp, P. (ed.) (1986), Technological Change, Employment and Spatial Dynamics, Springer Verlag, Berlin.

Nijkamp, P. (1987), The Strategic Role of High Tech Firms in Peripheral and Border Areas, Research Memorandum, Dept. of Economics, Free University of Amsterdam.

Nijkamp, P. and Davelaar E.J. (1987), The Urban Incubator Hypothesis: Old Wine in New Bottles?

Nijkamp, P. and Davelaar, E.J. (1989), Spatial Dispersion of Technological Innovation: The Incubator Hypothesis, Journal of Regional Science, forthcoming.

Nijkamp, P. and Malecki, E.J. (1988), Technology and Regional Development: Some Thoughts on Policy, Environment and Planning C, vol. 4, pp. 232-248.

Norton, R.D. and Rees, J. (1979), The Product Cycle and the Spatial Decentralization of American Manufacturing, Regional Studies, no. 13, pp. 141-151.

Oakey, R.P. (1982), High Technology Industry and Industrial Location, Gower, London.

Scott, A.J. and Storper, M. (eds.) (1986), Production, Work, Territory: The Geographical Anatomy of Industrial Capitalism, Allen and Unwin, Boston.

Stöhr, W.B. (1986), Territorial Innovation Complexes, in: Aydalot, Ph. (ed.), Milieux Innovateurs en Europe, pp. 29-56, GREMI, Paris.

Storper, M. (1985), Oligopoly and the Product Cycle: Essentialism in Economic Geography, Economic Geography, no. 61, 260-282.

Struyk, R. and James, E. (1975), Intrametropolitan Industrial Location, Lexington Books, Lexington, Mass.

Swyngedouw, E. 91987), Planning!!.... What Planning?, in: Angenout, L. (ed.), Planning without a Passport, pp. 121-132, Geographical Studies,

Amsterdam.

Vernon, P. (1979), A Product Life Cycle Hypothesis in a New International Environment, Oxford Bulletin of Economics and Statistics, vol. 14, pp. 98-116.

Winter, (1986), The generation and utlization of technology: a cross industry analysis, paper presented at the conference on 'Innovation Diffusion', Venice, January.

14 Technology diffusion: an overview of theories and a process-space model

I. ORISHIMO AND G. INOUE

1. Introduction

Generally diffusion is defined as the social process, by which an innovation spreads through a social system over time (Hayward, 1979; Mansfield et al., 1971). It is not necessarily defined in terms of a spatial context but in terms of the process, referring to the way the innovation spreads from firm to firm in an industry or from industry to industry during the period concerned (Malecki, 1983; Davies, 1979).

Nowadays, immaterial factors such as information rich environment, creativity (Andersson, 1985) and entrepreneurship - rather than material factors - attract the attention of many researchers as an important driving force for individual firms as well as for regional development (Thwaites and Oakey, 1985). Also, information technology and knowledge engineering are the most rapidly developing fields of modern technology, and such fields are drastically influencing the economy, in particular since technological innovation diffuses generally via information technology (Johansson, 1987).

As a result, scientific research has to face many unknown factors such as the change of industrial structure and the fear for the unemployment as a result of the rapid development of information technology. These unknown factors related mostly to spatial problems such as interregional as well as international productivity competition through the advancement of R & D activities. Because of the above mentioned background, the topic of technology diffusion - not only in terms of process but also of space - has to be seen as one of the core problems of the main stream of the present day's characteristics of the economy.

From the viewpoints mentioned above, we wish to approach the analytical problem posed above by decomposing our research into two parts, one being a survey of recent main streams on the theories of technology diffusion, and the other one consisting of a formulation of a synthesised process-space model of technology diffusion and an empirical test of the model, followed by a reflection on the model.

2. Theories of Technology Diffusion

In this section, we will survey the research on technology diffusion by paying attention to the following topics:

(a) disciplines;

(b) methodologies;

(c) objects of analyses, and

(d) time horizons.

2.1 Disciplines

There are many disciplines which are addressing themselves to technology diffusion. These are economics, regional science, management science, geography, history, and psychology.

Recent research has generally been more or less affected by various ideas of 'classical' researchers such as Marx (1867), Schumpeter (1942), Hägerstrand (1967). In the sequel I wish to refer to some major researchers in specific disciplines who follow these classical researchers by focusing first on various contributions in the economics field, and then on other fields.

2.1.1. Economics

Several decades ago already, Solow (1957) developed his theory on innovations based on a production fucntion approach, and many researchers have followed - on the basis of the neo-classical production theory or growth theory - his approach. Major contributions in this research tradition were provided by Shen (1973), Allen (1982), Soete and Turner (1984), Carlberg (1981), and Norton and Bass (1987).

Second, in addition to the above mentioned growth theory, there is a group of contributions which are more or less based on the evolutionary ideas of Schumpeter (1942). Especially Schmookler (1966) and Nelson and Winter (1982) further developed and applied these evolutionary theories. Moreover, Kamien and Schwartz (1982), and Dasgupta and Stiglitz (1980) have - along similar lines - paid explicit attention to the relationship between innovation diffusion and changes in market structures.

Third, from the viewpoint of industrial economics, Rosegger (1976, 1978), Gold (1964, 1981) and Hayward (1979) have developed their research on innovation diffusion with a strong empirical orientation. Their works indicates that the formal model of technology diffusion represents only one side of the problem.

Fourth, focusing on the international technology transfer, Nabseth and

Ray (1974), and Hill (1985) have made interesting contributions to innovation research, which show us the diffusion process by country.

Fifth, from the viewpoint of labour economics, Freeman (1979, 1982), and Freeman et al. (1982) have provided path finding research on innovation and labour markets. Freeman accomplishes his work mainly on the basis of a historical descriptive approach.

Finally, it is worthwhile to call attention for earlier research by Rosenberg (1976), who developed important historiographical analyses covering several hundred years.

2.1.2. Other disciplines

In the field of geography, the influence of Hägerstrand's idea of information fields is more or less recognised in most of the research. On the other hand, there are many common or similar theories such as dynamic systems analysis in the area of economics. Pred (1985) and Pedersen (1970) are building upon Hägerstrand's research through the concept of the rank-size rule and the system of cities approach. Brown (1981) contributed very much to the diffusion research through his broad survey of diffusion theories. There are also other geographers such as Smith (1976), Rees (1979), Morill and Mannien (1975), Garst (1973), Curry (1981) and Olsson (1974), who have provided creative contributions to this field.

In the case of regional science, there are interesting surveys by Thomas (1975) and Malecki (1981, 1983) to whom this paper owes very much. Furthermore, real-world questionnaire survey work done by Thwaites and Oakey (1985), research on the long-term perspective based on Kondratiev long waves by Rothwell (1982), and the logistical system's approach by Johansson (1987) should be mentioned here.

In the case of management science, there are also many common methodologies such as Lotka-Volterra models. Among the researchers, Norton and Bass (1987), Kalish and Lilien (1983), Easingwood et al. (1983), Schmittelein and Mahajan (1982), Rosegger (1975), Burns and Stalker (1961), and Utterback and Abernathy (1975) are to be mentioned in this context.

In the field of sociology, there are among others Rogers (1962) and Hayward (1979). In case of political or administrative science, Roessner (1980) and Pachico and Ashby (1983) should be mentioned. And finally, in the field of energy systems analysis, Marchetti's work (1980) is noteworthy.

2.2 Methodologies

Despite the existence of different disciplines in the field of technology research, there are several common important methodologies for the research in this field. First, there is dynamic systems theory - usually based on Volterra-Lotka type of differential equation systems - which may be regarded as one of the prominent theories which are often used in innovation research. Important contributions which use analogous methodologies are provided among others by Camagni (1985), Davies (1979), Easingwood et al. (1983), Jeuland (1979), Mahajan and Peterson (1978), Metcalfe (1981), Sharif and Ramanathan (1982), Sonis (1982), and Souder and Quaddus (1982).

Secondly, especially for the macroscopic approaches to technology research, statistical analysis has become a meaningful tool used by various researchers such as Batholomew (1976), Davies (1978), Globerman (1975), Griliches (1957), Karmeshu and Pattria (1980), Mansfield (1968), Olson (1982), Romeo (1977) and Schmittelein and Mahajan (1982). Finally, Allen (1982) addresses himself especially to a stochastic approach.

2.3 Objects of analysis

It is noteworthy that in the wide spectrum of research undertaken in the various fields mentioned above, the objects of analysis are also very pluriform. Though the concept of innovation is usually not rigorously defined, the object of innovation research is often differentiated by product, process and mixed innovations. From the behavioural point of view of economic actors, the characteristics of managers or individuals are then important objects in the different fields of scientific analysis. In addition, this classification of objects of analysis, the focus of innovation research on either intra- or inter-firm, either intra- or inter-industry, and either intra- or inter-regional perspectives should be mentioned.

2.4 Time horizons

There are important distinctions regarding the time horizon of the analysis concerned (see Figure 1).

From the viewpoint of long term and long distance analysis of technology diffusion, the concept of information field developed by Hägerstrand is not necessarily sufficient for the analysis. The concept of culture seems to be more important from this viewpoint. If we address ourselves to a long run perspective regarding regional development, as is done for instance in the research by Andersson (1986) or Oishi (1985), it is very difficult to come up with quantitative results based on numerical data.

Long-term economic development depends more on immaterial location factors such as the regional distribution of knowledge heritage, creativeness, and Schumpeterian entrepreneurship than on material location

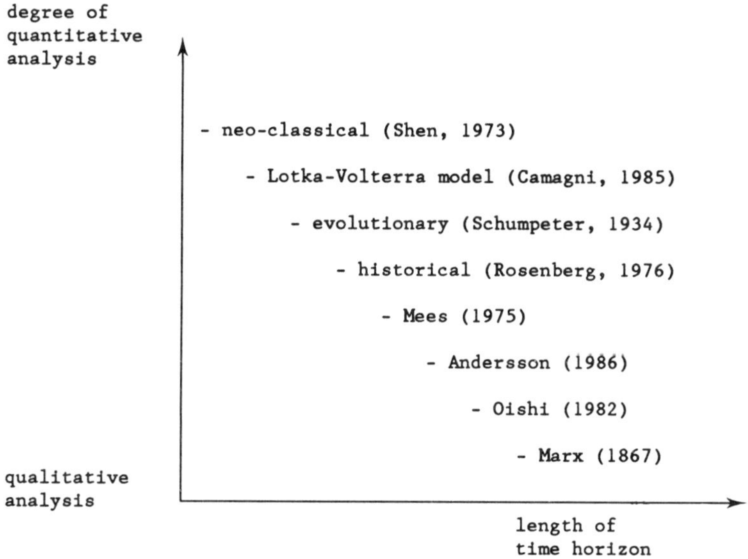

Figure 1 Relation between mode of analysis and length of time horizon

factors in the traditional location theory. What we have to consider is the fact that economic development is a kind of technological diffusion, which is contained in the cultural diffusion. Keynes (1936) suggested already that economic property is excessively dependent on a political and social atmosphere which is congenial to the average businessman, and on a local cultrue conducive to the formation of new small firms by technological entrepreneurs.

Also Sonis (1982), who is addressing himself to rather short term technology diffusion problems, stressed the importance of an 'active environment of adoption unit', where an active environment should contain both the physical environment and the information environment and cultural heritage. The latter refers to a spatial distribution of various kinds of research institutes, education systems, and other cultural centers. Thus an information-rich environment contains the education and training system in a region, the quality and intensiveness of the information flows between firms and entrepreneurs and their friends, the cycle time between the need for and the delivery of information for decision-making, and the technical culture and progressiveness which determine the innovative potential of a region (Sweeney, 1987).

In other words, our objectives and our cultural heritage are the mother of the innovation in a region (Schmookler, 1966). Our objectives are closely related to our contact patterns with a heterogeneous culture, which might be named as a line of cultural discontinuity. It means the possibility of contacts with a heterogeneous culture in various kinds of context, such as traditional versus non-traditional or rural versus urban cultures, etc. For instance, in the case of Greece, the age of Socrates is said to be the age of contradiction between traditional culture and non-traditional culture (Armstrong, 1947). Also in the case of oriental regions such as China and Japan, we are experiencing several times in our history similar lines of discontinuity of culture.

Regrettably, the researchers referred to in the previous sections, have not given due attention to this cultural background. Some notions in this context will be introduced in the next section.

3. A Process-Space Model

After the general survey in the first part, we will now focus our attention on a process-space model of technology diffusion.

Although in the general regional science literature the relationship between the transportation network and the location of individual firms (and the resulting regional development) has intensively been studied, the equally important relationship between the information network and the locational choice of individual firms (and the regional development consequences) has not yet received much attention. Therefore, it is interesting to analyse the structure of a regional economy, which is influenced by both the transportation and the information network.

In this section we will address ourselves to the analysis of the effects of the information network on entrepreneurial location and regional development, along the idea of Hägerstrand's information field. Although many difficulties will be faced in measuring information networks and related indicators of innovation (as a result of the high-technology oriented or R & D activities), such an endeavour would be useful for policy formulation on a balanced regional development with respect to communication and the information networks (e.g., tariff policy or telephone systems, etc.).

3.1 A single center model

Hägerstrand's concept of information field was based on a single point center of information transfer, i.e.,

$$F = Cd^{-n} \qquad (1)$$

where F is the number of migratory units per km^2, d the distance from the origin of the center of the migration field, c a normalisation constant which indicates the value of F when d = 1, and n a distance decay exponent which expresses the steepness of the migration slope (Hägerstrand, 1967, p. 187). The migration slope is supposed to represent the frequency slope of communication impulses.

Usually, this diffusion model for a given innovation is reduced as follows. Let m_t be the number of firms which from the total population of k firms, have adopted the innovation at hand at time t. Then we may define a parameter for the adoption rate as follows:

$$(t) = \frac{m_{t+1} - m_t}{k - m_t} \qquad (2)$$

Next, we may assume the following explanatory model for this adoption rate:

$$(t) = f_t\left(\frac{m_t}{k}, \pi , S \right) \qquad (3)$$

Where π denotes the profitability of implementing the innovation relative to that of alternative investments, and S the size of the investment outlays required to install the innovation as a proportion of the average total assets of firms in the industry (see Davies, 1979).

Based on the hypothesis of equation (3), the following popular formulation is obtained:

$$\frac{dm_t}{d_t} \frac{1}{k-m_t} = A + \beta \frac{m_t}{k} \qquad (4)$$

$$m_t/k = \{ 1+\exp(-\beta-\tau t) \}^{-1} \qquad (5)$$

We may now consider diffusion model (3) in a spatial context. Let s denote the location of the firm; then π and S should be a function of the location s. Considering the concept of an information field, we may now formulate the system of the spatial technology diffusion. Then, instead of (1), we obtain equation (6) which is illustrated in Figure 2.b.

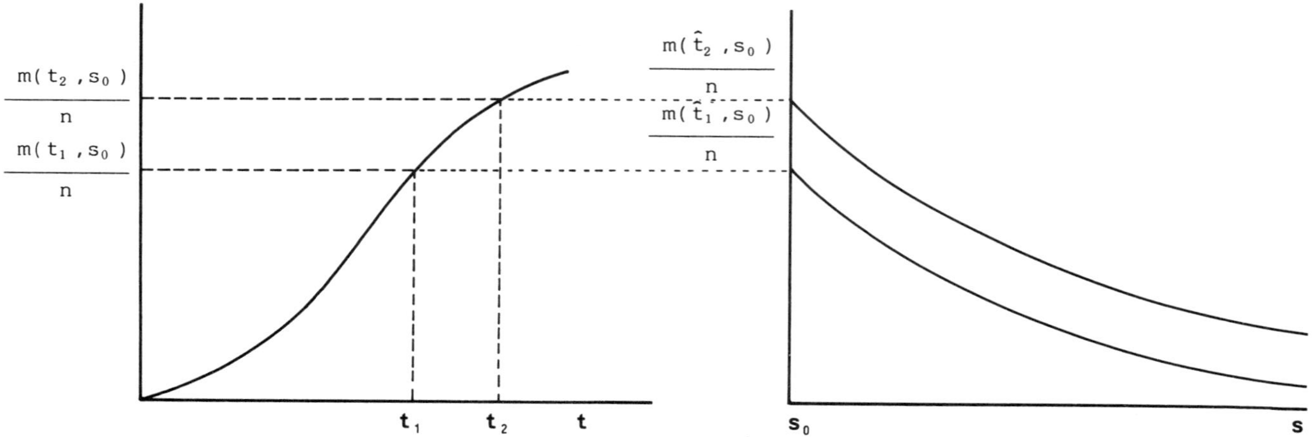

a. Process Diffusion Curve

b. Spatial Diffusion Curve

Fig. 2 Process-Spatial Relation

$$\frac{m_t(s)}{k} = e^{-\alpha(t)s} \qquad (6)$$

If we aggregate the spatial diffusion curve (6) at a certain point of time, we get the following result:

$$\int_0^\infty e^{-\{\alpha(t)s\}}ds = [\frac{1}{\alpha}e^{\{-\alpha(t)s\}}]_0^\infty = \frac{1}{\alpha(t)} \qquad a>0 \qquad (7)$$

From equations (4) and (7) we may derive:

$$\hat{m}^t/_h = \int_0^\infty e^{-\alpha(t)s} \; ds = \frac{1}{\alpha(t)} \qquad (8)$$

$$= \frac{1}{1+\exp(-\beta-\tau\hat{t})}$$

with:

$$\alpha(t) = 1+ \exp(-\beta-\tau\hat{t}) \qquad (9)$$

The parameter in equation (6), α (t), is explained by the parameters of equation (5), β and τ. Thus in a process diffusion in time (Fig. 2.a), taking any time unit t_i, we can draw a spatial diffusion curve in each time unit t_i in a cross-sectional manner, as shown in Fig.2.b. In other words, each point of the curve shown in Fig.2.a. can be projected on the corresponding curve in Fig.2.b. In a three-dimensional space which is composed of time (t), space (s) and innovation acceptance rate $\frac{m(t,s)}{k}$, we can illustrate the surface of the diffusion process in time.

3.2 A multiple center model

In the case of multiple centers, instead of the case of a single center, the expression of the information field is written as follows. Suppose the number of centers is equal to h. The distance fro i-th location to the j-th center is denoted by s_{ij}, whilst the parameter of the information field with respect to the i-th center is denoted by a_i (a_i>0). Then the aggregated information field is written as equation (10) (see Fig.3).

$$\sum_{j=1}^{h} A_j \; e^{-\alpha_i(t)s_{ij}} \qquad (10)$$

Figure 3 Multiple Centers' Case

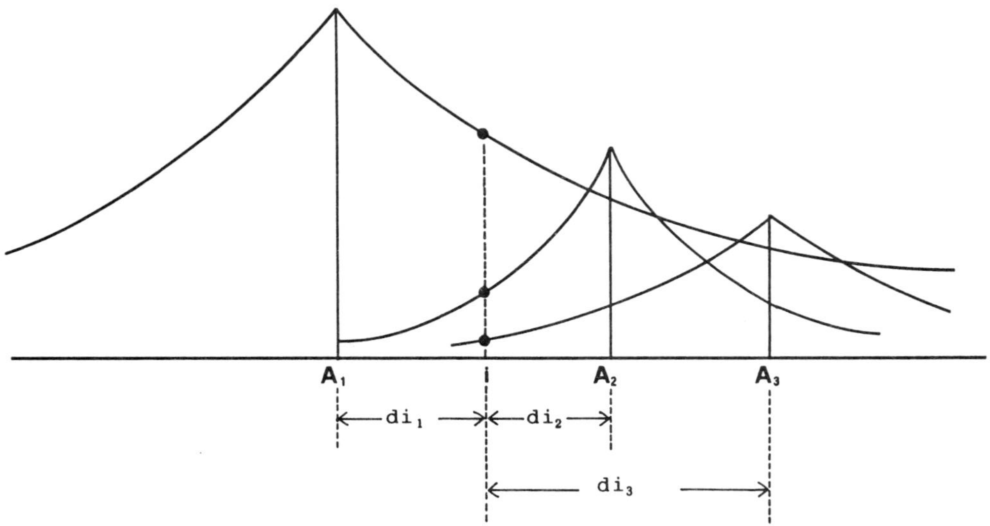

$$\text{aggregated information field w.r.t. location, i} = A_1 e^{-\alpha_1 di_1} + A_2 e^{-\alpha_2 di_2} + A_3 e^{-\alpha di_3}$$

3.3 A macroscopic model

As the next step, we develop a macroscopic model of spatial technology diffusion for the purpose of exploring current regional developments. We wish to develop a macro model applying the concept of the information field of Hägerstrand, based on the gravity model incorporating the accessibility concept.

Though it is not easy to define a proper macroscopic measure for spatial technology diffusion, Schmookler (1966) showed already in his classical research tha R & D activities are closely related to their output by means of paten numbers. These finding have also been followed by Griliches (1984) and many others. On the other hand, Andersson had developed his concept of R & D activities as a kind of public good, the information of which disperses within the economic system. (see Andersson, 1981). This idea suggests that R & D activities of a region should not only affect the region but also the other regions.

The gravity type of variables related to the accessibilities to R & D activity centres of each region will be built now in a gravity model (which may be interpreted as an application of the entropy concept employed in information theory). Also, the entropy maximising method is a statistical averaging method from a macroscopic viewpoint (Wilson, 1970). Thus based on the information field and the nature of spatial technology diffusion as a kind of public good, we intend to build a macro-model of spatial technology diffusion.

The information centers for the region are composed of various kinds of centers surrounding the location. The accessibility of a spatial point is measured by a set of accessibilities (see also Fig.3).

In our analysis we will make the following assumptions:

1) The spatial distribution of R & D activities is expressed by the regional distribution of the number of the R & D employees.

2) The distribution of the number of patents in the region is one of the most adequate approximate measures of spatial innovation distribution.

3) The spatial distribution of technical innovation (measured via the number of patents in a region) should be a function of the spatial distribution of R & D employees.

4) The contribution of a spatial distribution of R & D activities on a technical innovation in a region is expressed by a gravity model of the accessibilities of the region with respect to the R & D activity centres in other regions.

We denote the number of patents in the i-th region by N_{p_i}, the number of the researchers and employees of the research institutes in the i-th region by A_i, which expresses the R & D activity level of the region), and the distance between the i-th and j-th region as s_{ij}. The j-th region's R & D activity level influences the technological innovation of the i-th region, which is expressed by the number of patents of the i-th region, N_{p_i}.

Applying the concept of gravity to consider the impact of the centre of the j-th region's research activity on the i-th region, we formulate our model as follows:

$$N_{p_i} = N_p \frac{\sum_j A_j \exp \beta_i s_{ij}}{\sum_i \sum_j A_j \exp \beta_i s_{ij}} \qquad j=1,\ldots\ldots,n \qquad (10)$$

where $\sum_i \sum_j A_j \exp \beta_i s_{ij}$ is a balancing factor (normalisation factor), which equates the left hand side and the right hand side of equation (10), when it is aggregated in terms of i. The parameter β_i indicates how the measure of the technology innovation in a region is influenced by the R & D activity distribution, counting the distance from the region to the R & D centre in each region in our spatial system.

4. Empirical Tests

We have tested the above model using data from Japan. We have denoted administrative regions of Japan, from north to south, by number 1 to 47, and have successively calibrated the values of the respective parameters. Various calibrated values for the years 1975 and 1980 are shown in Table 1. In case of the main information centres of Tokyo and Osaka which are the nodal points of the large metropolitan areas of respectively Kanto and Kinki, the estimated values appear to be the highest ones among all regions. By comparing the values of 1975 with those of 1980, in most of the regions the estimated value of β_i appears to have risen. This means that the rates of agglomeration of technology innovation with respect to the R & D activities are most concentrated in the big metropolitan areas and have become more concentrated in terms of their information fields during the past half decade.

Table 1 Estimated values of β_1

	1975	1980
11. Saitama	-1.083	-1.111
13. Tokyo	-0.929	-0.943
26. Kyoto	-1.766	-1.793
27. Osaka	-0.690	-0.695

5. A Generalisation of the Process-Space Model

Next, we may also consider the different values of the parameters of each region. Essentially, these should be considered in a different context, since each β_i reflects the aggregate of the relationship of different attributes between the i-th and j-th region. Pederson (1971) pointed already at the relationship between city size and year of adoption of an innovation. But this means that this relationship does not only imply a time lag of adoption by the city of a given size, but also existence of a different kind of innovation adopted in the city of each size. The difference in the latter context was already previously noticed by Vernon (1979) and Davelaar and Nijkamp (1989). In the terminology of Nijkamp, there are two kinds of innovation, viz. product innovation and process innovation. The former is based on innovative ideas and the latter on innovative mass production. Product innovation is mainly metropolitan-oriented., whereas process innovation is more oriented toward medium and small size towns. Expecially in large metropolitan areas, innovation is mostly confined to product innovation, so that in our model the parameter β_i refers mainly to process innovations. Therefore, our model should be reformulated by considering the system of cities (or the rank-size rule) approach. In the case of the rank-size rule, let Pi denote the population of the i-th rank city, where P_1 is the population of the biggest size city, and Q(r) is the number of the cities of rank r, where r=1,...,m, and q(r) = 1,...,l, lm=n. Then we have the following standard expression for the rank-size rule:

$$p_i = p_1 \, r_i^{\,k} \qquad (12)$$

If we suppose that:

$$A_i = A_i(P_i), \qquad A_i = \sigma P_1 r_i^k \tag{13}$$

then:

$$N_{p_r} = N_p \frac{\sum_{q(r)} r_{q(r)}^k \exp \beta_{q(r)}^k s_{ij}}{\sum_r \sum_{q(r)} r_{q(r)}^k \exp \beta_{q(r)}^k s_{ij}} \tag{14}$$

In the case of the group of cities in the high rank category, - i.e., big metropolitan areas - technically higher order innovations should be developed usually based on an information-rich environment. On the other hand, in the case of the group of cities of the lower rank category, lower order innovation may be expected, based on a relatively poor information environment. In the case of the group of cities in the middle rank category, middle order innovations may be expected. Thus, in each category, the characteristics of innovation should be different, because of the difference of the R & D activity intensity and different innovative environments. In other words, the difference should also reflect the CBD-periphery development process of the life cycle of an industry or product-process development by Davelaar and Nijkamp (1989). Accordingly, the gradient of the information field calibrated by us should in general reflect the aggregate pattern of the various kinds of gradients.

As the accessibility to the location of knowledge related activities or R & D activities should be regarded as an important element for the spatial diffusion of technology, we have formulated here a spatial-process diffusion model along the lines of the hitherto developed models and ideas related to the diffusion and the spatial interaction of technological evolution.

Through the process-space model analyses described above, we may obtain a clearer perspective for spatial technology diffusion. In this context, there is much scope for follow-up research, e.g. by developing our model into a dynamic Volterra-Lotka differential equation system (see also Orishimo et al., 1987).

References

Adams, W. and Dirlam, J. (1966), Big Steel, Invention, and Innovation, Quarterly Journal of Economics, pp. 168-189.

Allen, B. (1982), Some Stochastic Processes of Interdependent Demand and Technological Diffusion of an Innovation Exhibiting Externalities among Adopters, International Economic Review 23, 3:595-608.

Allen, B. (1982), A Stochastic Interactive Model for the Diffusion of Information, Journal of Mathematical Sociology 8:265-281.

Andersson, A.E. (1981), Accessibility and the Economics of Public Goods, Cities in Transition: Problems and Policies, pp. 99-118, Sijthoff & Noordhoff.

Andersson, A.E. (1985), Creativity and Regional Development, Papers of Eighth Pacific Congress of the R.S.A.

Andersson, A.E. (1986), The Four Logistical Revolutions, Papers of R.S.A. 59.

Antonelli, C. (1986), The International Diffusion of New Information Technologies, Research Policy 15, 3:139-147.

Armstrong, A.H. (1947), An Introduction to Ancient Philosophy, Methuen and Co. Ltd.

Ashworth, J. (1980), Technology Diffusion Through Foreign Assitance: Making Renewable Energy Sources Available to the World's Poor, Policy Science 11, 3:241-261.

Baker, M.J., ed. (1979), Industrial Innovation, Technology, Policy Diffusion, Macmillan.

Baker, N.R. and Sweeney D.J. (1978), Toward a Conceptual Framework of the Process of Organized Innovation Technological within the Firm, Research Policy, 7:150-174.

Barker, T.C. (1982), Development and Diffusion of Technology-Okochi, A., Uchida, H. Economic History Review 35, 1:155-156.

Bartholomew, D.J. (1976), Continuous Time Diffusion Models with Random Duration of Interest, Journal of Mathematical Sociology 4:187-199.

Bass, F.M. (1980), The Relationship between Diffusion Rates, Experience Curves, and Demand Elasticities for Consumer Durable Technological Innovations, Journal of Business 53:s51-s67.

Baumgartel, H. (1983), Human-Factors, Technology Diffusion, and National Development, Journal of Applied Behavioral Science 19, 3:337-348.

Beaumont, J.R., Clarke, M. and Wilson, A.G. (1981), Changing Energy Parameters and the Evolution of Urban Spatial Structure, Regional Science and Urban Economics 11:287-315.

Beckmann, M. and Schramm, G. (1972), The Impact of Scientific and Technical Change on the Location of Economic Activities, Regional Science and Urban

Economics 2:159-174.

Blackman, A.W. Jr. (1974), The Market Dynamics of Technological Substitutions, Technological Forecasting and Social Change 6:41-63.

Blackman, A.W. Jr., Seligman, E.J. and Sogliero, G.C. (1973), An Innovation Index Based on Factor Analysis, Technological Forecasting and Social Change 4:301-316.

Blaikie, P. (1978), The Theory of the Spatial Diffusion of Innovation: A Spacious Cul-de-sac, Progress in Human Geography 2.

Brophy, D.J. (1982), Venture Capital Research, in: Kent, C.A. (ed.), Encyclopedia of Entrepreneurship, Prentice-Hall.

Brown, L.A. (1981), Innovation Diffusion: A New Perspective, Methuen, New York.

Brown, L.A. and Philliber, S.G. (1977), The Diffusion of a Population-Related Innovation: The Planned Parenthood Affiliate, Social Science Quarterly 58:215-228.

Bundgaardnielsen, M. (1976), International Diffusion of New Technology, Technological Forecasting and Social Change 8, 4:365-370.

Burns, T. and Stalker, G.M. (1961), The Management of Innovation, Tavistock Publications, London.

Butler, J. (1980), Industrial-Innovation-Technology, Policy, Diffusion-Baker, M.J. R and D Management 10, 3:131-132.

Camagni, R. (1985), Spatial Diffusion of Pervasive Process Innovation, Papers of the Twenty-fourth European Congress of R.S.A. 58.

Cameron, R. (1975), The Diffusion of Technology as a Problem in Economic History, Economic Geography 51, 3:217-230.

Carey, J. and Moss, M.L. (1985), The Diffusion of New Telecommunication Technologies, Telecommunications Policy 9, 2:145-158.

Carlberg, M. (1981), A Neoclassical Model of Interregional Economic Growth, Regional Science and Urban Economics 11:191-203.

Chen, E.K.Y. (1983), Multinational Corporations and Technology Diffusion in Hong-Kong Manufacturing, Applied Economics 15, 3:309-321.

Chow, G.C. (1967), Technical Change and the Demand for Computers, American Economic Review 57:1117-1130.

Clark, N.G. (1971/1972), Science, Technology and Regional Economic Development, Research Policy 1:296-319.

Curry, L. (1981), Division of Labor from Geographical Competition, Annals of the Association of American Geographers 71:133-165.

Dasgupta, P. and Stiglitz, J. (1980), Industrial Structure and the Nature of Innovative Activity, The Economic Journal 90, 266-293.

Davelaar, E.J., and Nijkamp, P. (1989). The Role of the Metropolitan Milieu for Innovation Incubation, Urban Studies (forthcoming).

David, P.A. (1975) Technical Choice, Innovation and Economic Growth, Cambridge University Press.

Davies, S. (1979), The Diffusion of Process Innovation, Cambridge U.P.

Deniozos, D. (1978), La Diffusion Internationale des Innovations Technologiques: La Cas de la Grece, Switzerland Management International Review 18, 1:69-81.

Denison, E.F. (1967), Why Growth Rates Differ, Brookings Institute.

Denny, M., Fuss, M., Everson, C. and Waverman, L. (1981), Estimating the Effects of Diffusion of Technological Innovations in Telecommunications: the Production Structure of Bell-Canada, Canadian Journal of Economics - Revue Canadienne d'Economique 14, 1:24-43.

Easingwood, C.J., Mahajan, V. and Muller, E. (1983), A Nonuniform Influence Innovation Diffusion Model of New Product Acceptance, Marketing Science 2, 3:273-295.

Edey, M.D. and Page, R.L. (1984), Dissemination and Diffusion of a Technology Syllabus, Programmed Learning and Educational Technology 21, 2:101-104.

Ewers, H.J. and Wettmann, R.W. (1980), Innovation-Oriented Regional Policy, Regional Studies 14:161-179.

Feder, G. and O'Mara, G.T. (1982), On Information and Innovation Diffusion: A Bayesian Approach, American Journal of Agricultural Economics, February: 145-147.

Feller, I. (1974), Technology Diffusion-Federal-Programs and Procedures-Hough, G.W. Technology and Culture 17, 3:578-580.

Feller, I. (1975), Invention, Diffusion and Industrial Location, in: Collins, L. and Walker, D.F. (eds.), Locational Dynamics of Manufacturing Activity, John Wiley.

Feller, I., Elmes, G. and Meyer, J. (1982), Spatial Aspects of the Diffusion of Technological Innovations among American Municipal Governments, Socio-Economic Planning Sciences 16, 5:225-238.

Fodella, G. (1981), Narrowing the Gap among Pure Research, Applied Research and Technological Diffusion: The Japanese Experience, Rivista Internazionale Di Scienze Economiche E Commerciali 28, 7:769.

Freeman, C. (1979), Technical Change and Employment, Frances Pinter, London.

Freeman, C. (1982), The Economics of Industrial Innovation, Frances Pinter, London.

Freeman, C., Clarke, J. and Soete, L. (1982), Unemployment and Technical Innovation, A Study of Long Waves, France Pinter, London.

Garst, R.D. (1973), Spatial Diffusion and Information Diffusion: Kenyan

Example, Proceedings of the Association of American Geographers 75-80.

Garst, R.D. (1974), Innovation Diffusion among the Gusii of Kenya, Economic Geography 50:300-312.

Globerman, S. (1974), Technological Diffusion in the Canadian Tool and Die Industry, The Review of Economics and Statistics 428-434.

Globerman, S. (1975), Technological Diffusion in the Canadian Carpet Industry, Research Policy 4, 2:190-206.

Globerman, S. (1975), Technological Diffusion in Canadian Tool and Die Industr, Review of Economics and Statistics 57, 4: 428-434.

Goddard, J.b. (1970), Technology Forecasting in a Spatial Context, Futures 12:90-105.

Gold, B. (1964), Industry Growth Patterns: Theory and Empirical Results, Journal of Industrial Economics, 8, 1:53-73.

Gold, B. (1981), Technological Diffusion in Industry: Research Needs and Shortcomings, The Journal of Industrial Economics 29, 3:247-269.

Gottinger, H.W. (1987), Economic Choice and Technology Diffusion in New Product Markets, Weltwirtschaftliches Archiv 123, 1:93-120.

Green, K. and Morphet, C. (1977), Research and Technology as Economic Activities, Butter worths.

Griliches, Z. (1957), Hybrid Corn: An Exploration in the Economics of Technological Change, Econometrica 25, 4:501-522.

Griliches, Z. (1979), Issues in Assessing the Contribution of Research and Development to Productivity Growth, The Bell Journal of Economics 10:92-116.

Griliches, Z. (1984), R & D Patents, and Productivity, University of Chicago Press.

Hägerstrand, T. (1967), Innovation Diffusion as a Spatial Process: University of Chicago Press.

Haeffner, E.A. (1973), The Innovation Process, Technology Review 75, 5:18-25.

Hayward, G. (1979), Perceived Characteristics and the Diffusion of Innovations, Industrial Innovation: Technology, Policy, Diffusion, ed. M.J. Baker, pp. 369-398, Macmillan, London.

Hayward, G., Alen, D.H. and Masterson, J. (1976), Characteristics and Diffusion of Technological Innovations, R and D Management 7, 1:15-24.

Heaton, G.R. and Hollomon, J.H. (1984), Technological Diffusion and National Policy, Research Management 27, 1:5-7.

Hill, H. (1985), Subcontracting, Technological Diffusion, and the Development of Small Enterprise in Philippine Manufacturing, Journal of Developing Areas 19, 2:245-262.

Homer, J.B. (1987), A Diffusion Model with Application to Evolving Medical Technologies, Technological Forecasting and Social Change 31, 3:197-218.

Ireland, N. and Stoneman, P. (1986), Technological Diffusion, Expectation and Welfare, Oxford Economic Papers - New Series 38, 2:283-304.

Jacobs, J. (1985), Cities and the Wealth of Nations - Principles of Economic Life, Random House.

Johansson, B. (1987), Information Technology and the Viability of Spatial Networks, Papers of the Regional Science Association 61, 51-64.

Jensen, R. (1982), Adoption and Diffusion of an Innovation of Uncertain Profitability, Journal of Economic Theory 27:182-193.

Jeuland, A.P. (1979), Epidemological Modeling of Diffusion of Innovation: Evaluation and Future Directions of Research, Proceedings of the 1979 Educators' Conference, pp. 274-278, American Marketing Association, Chicago.

Jorgensen, S. (1983), Optimal Control of a Diffusion Model of New Product Acceptance with Price-Dependent Total Market Potential, Optimal Control Applications & Methods 4:269-276.

Kalish, S. and Lilien, G. (1983), Optimal Price Subsidy Policy for Accelerating the Diffusion of Innovation, Marketing Science 2, 4:407-420.

Kamien, M.I. and Schwartz, N.L. (1982), Market Structure and Innovation, Cambridge U.P.

Karmeshu, and Pathria, R.K. (1980), Stochastic Evolution of a Nonlinear Model of Diffusion of Information, Journal of Mathematical Sociology 7:59-71.

Katz, E. (1961), The Social Itinerary of Technical Change: Two Studies on the Diffusion of Innovation, Human Organization 20:215-227.

Keynes, J.M. (1936), The General Theory of Employment, Interest and Money, Harcourt Brace.

Kenwood, A.G. and Lougheed, A.L. (1982), Technological Diffusion and Industrialization before 1914, Croom Helm.

Lakatos, I. and Musgrave, A. (eds.) (1974), Criticism and the Growth of Knowledge, Cambridge U.P.

Lande, P.S. (1978), The Interregional Comparison of Production Functions, Regional Science and Urban Economics 8:339-353.

Le Bas, C. (1985), La Diffusion De L'innovation Interne A La Firme, Revue Economique 36, 5:873-895.

Le Heron, R.B. (1973), Best Practice Technology, Technical Leadership and Regional Economic Development, Environment and Planning 735-749.

Lekvall, P. and Wahlbin, C. (1973), A Study of Some Assumptions Underlying Innovation Diffusion Functions, Swedish Journal of Economics 362-377.

Levin, R.C. and Reiss, P.C. (1984), Tests of a Schumpeterian Model of R & D and Market Structure, Z. R & D, Patents, and Productivity, ed.: Griliches.

Mahajan, V. and Muller, E. (1979), Innovation Diffusion and New Product Growth Models in Marketing, Journal of Marketing 43:55-68.

Mahajan, V. and Peterson, R.A. (1978), Innovation Diffusion in a Dynamic Potential Adopter Population, Management Science 24:1589-1597.

Mahajan, V. and Wind, Y. (1986), Innovation Diffusion Models of New Product Acceptance, Ballinger P.C.

Malecki, E.J. (1981), Science, Technology, and Regional Economic Development: Review and Prospects, Research Policy 10:313-334.

Malecki, E.J. (1982), Federal R & D Spending in the United States of America: Some Impacts on Metropolitan Economies, Regional Studies 16, 1:19-35.

Malecki, E.J. (1983), Technology and Regional Development: A Survey, International Regional Science Review 8, 89-125.

Mansfield, E. (1963), Intrafirm Rates of Diffusion of an Innovation, Review of Economics and Statistics 45:348-359.

Mansfield, E. (1961), Technical Change and the Rate of Imitation, Econometrica 29-4, 741-766.

Mansfield, E. (1963), Size of Firm, Market Structure and Innovation, Journal of Political Economy 6:556-576.

Mansfield, E. (1968), Industrial Research and Technological Innovation, W.W. Norton, New York.

Mansfield, E., Rapoport, J., Schnee, J., Wagner, S. and Hamburger (1971), Research and Innovation in the Modern Cooperation, p. 186, Norton.

Marchetti, C. (1980), Society as a Learning System: Discovery, Invention and Innovation Cycles Revisited, Technological Forecasting and Social Change 18:267-282.

Marx, K. (1867, 1885, 1894), Capital, Progress Publishers, 1960.

Mees, A. (1975), The Revival of Cities in Medieval Europe, Regional Science and Urban Economics 5.

Mensch, G. (1979), Stalemate in Technology, Ballinger, Cambridge, Mass.

Merrill, T. (1987), Will Payment Control Technology Diffusion, Hospitals 61, 13:46.

Metcalfe, J.S. (1981), Impulse and Diffusion in the Study of Technical Change, Futures October:347-369.

Minami, R., Makino, F. (1983), Conditions for Technological Diffusion: Case of Power Looms, Hitotsubashi Journal of Economics 23, 2:1-20.

Morrill, R.L. and Manninen, D. (1975), Critical Parameters of Spatial Diffusion Processes, Economic Geography 51:269-277.

Nabseth, L. (1973), The Diffusion of Innovations in Swedish Industry, in: Willams, B.R. (ed.), Science and Technology in Economic Growth, Macmillan.

Nabseth, L. and Ray, G.F. (1974), The Diffusion of New Industrial Processes: an International Study, Cambridge U.P., London.

Nelson, R. and Winter, S.G. (1977), Towards Useful Theory of Innovation, Research Policy January, 6, 1:36-76.

Nelson, R. and Winter, S.G. (1977), Dynamic Competition and Technical Progress, Economic Progress, Private Values and Public Policy.

Nelson, R. and Winter, S.G. (1982), An Evolutionary Theory of Economic Change, Essays in Honor of William Fellner, ed., pp. 57-101, Harvard U.P., Cambridge, Mass.

Nelson, R. and Winter, S.G. (1978), Forces Generating and Limiting Concentration under Schumpeterian Competition, The Bell Journal of Economics 9:524-548.

Norton, J.A. and Bass, F.M. (1987), A Diffusion Theory Model of Adoption and Substitution for Successive Generations of High-Technology Products, Management Science 33, 9:1069-1086.

Oakey, R.P. (1983), New Technology, Government Policy and Regional Manufacturing Employment, Area 15, 1:61-65.

Oakey, R., Thwaites, A. and Nash, P. (1980), The Regional Distribution of Innovative Manufacturing Establishments in Britain, Regional Studies 14:235-253.

Oishi, Y. (1985), Culture and Region: Prolegomena, Papers of Eighth Pacific Congress of the Regional Science Association 56.1-3.

Oliver, R.M. (1986), Comments on Olson Logistic Function for Technology Diffusion, Technological Forecasting and Social Change 30, 2: 189-192.

Olson, J.A. (1982), Generalized Least Squares and Maximum Likelihood Estimation of the Logistic Function for Technology Diffusion, Technological Forecasting and Social Change 21, 3:241-249.

Olsson, G. (1974), The Dialectics of Spatial Analysis, Antipode 6, 50-62.

Orishimo, I, Nijkamp, P. and Hewing, J. (ed.) (1988), Information Technology: Social and Spatial Perspective, Springer-Verlag.

Orishimo, I., Sawada, K. and Togawa, Y. (1987), Invitation to Replicator Equation, mimeo, presented at the 27th European Regional Science Conference.

Pachico, D. and Ashby, J. (1983), Stages in Technology Diffusion among Small Farmers in Nepal, Agricultural Administration 13, 1:23-37.

Paker, J.E. S. (1978), The Economics of Innovation-The National and Multinational Enterprise in Technological Change, Longman.

Pavitt, K. and Walker, W. (1976), Government Policies Towards Industrial Innovation: A Review, Research Policy 5:11-97.

Pedersen, P.O. (1970), Innovation Diffusion within and between National Urban Systems, Geographical Analysis 2:203-254.

Pred, A. (1967), The Back Ground of Hagerstrand's Work, Hagerstrand, Innovation Diffusion as A Spatial Process, University of Chicago, pp. 283-319.

Pred, A. (1975), Diffusion, Organisational Spatial Structure and the City-System Development, Economic Geography 51:252-268.

Pred. A. (1977), The Choreography of Existence: Comments on Hagerstrand's Time-Geography and its Usefulness, Economic Geography pp. 207-221.

Pred., A. and Tornqvist, G.E. (1973), Systems of Cities and Information Flows: two essays, Lund Studies in Geography B 38.

Ramo, S. (1980), America's Technology Slip, John Wiley.

Ramo, S. (1985), What's Wrong with Our Technological Society and How to Fix it, Donnelly and Sons.

Rees, J. (1979), Regional Industrial Shifts in the U.S. and the Internal Generation of Manufacturing in Growth Centers of the Southwest, in: Wheaton, W.G. (ed.) Interregional Movements and Regional Growth, pp. 51-73, Urban Institute, Washington.

Rees, J. (1979), Technological Change and Regional Shifts in American Manufacturing, Professional Geographer 31:45-54.

Reinganum, J.F. (1981), Market Structure and the Diffusion of New Technology, Bell Journal of Economics 12, 2:618-624.

Richardson, H.W. (1973), Regional Growth Theory, Wiley, New York.

Richardson, H.W. (1976), Growth Pole Spillovers: the Dynamics of Backwash and Spread, Regional Studies 10:1-9.

Richardson, H.W. (1978a), The State of Regional Economics: a Survey Article, International Regional Science Review 3, 1:1-48.

Richardson, H.W. (1978b), Growth Centers, Rural Development and National Urban Policy: a Defense, International Regional Science Review 3,2:133-152.

Richardson, H.W., (1979), Aggregate Efficiency and Interregional Equity, in: Folmer, H. and Oosterhaven, J. (eds.), Spatial Inequalities and Regional Development, pp. 161-183, Martinus Nijhoff, Boston.

Robertson, T.S. and Gatignon, H. (1986), Competitive Effects on Technology Diffusion, Journal of Marketing 50, 3:1-12.

Roessner, J.D. (1980), Technological Diffusion Research And National Policy Issues, Knowledge: Creation, Diffusion, Utilization 2, 2:179-201.

Rogers, E.M. (1962), Diffusion of Innovations, The Free Press of Glencoe, New York.

Rogers, E.M. (1976), New Product Adoption and Diffusion, Journal of Consumer

Research 2:290-301.

Rogers, E.M. and Shoemaker, F. (1971), Communication of Innovations - A Cross Cultural Approach, Free Press.

Romeo, A.A. (1975), Interindustry and Interfirm Differences in the Rate of Diffusion of an Innovation, Review of Economics and Statistics 57:311-319.

Romeo, A.A. (1977), The Rate of Imitation of a Capital-Embodied Process Innovation, Economica 44, 1:63-70.

Romeo, A.A. (1979), Diffusion and Technological Specificity: The Case of Continuous Casting, Journal of Economics 28, 1:39-53.

Rosenberg, N. (1972), Factors Affecting the Diffusion of Technology, Explorations in Economic History 10, 1:3-33.

Rosenberg, N. (1976), Perspectives on Technology, Cambridge U.P.

Rosenberg, N. (1976), On Technological Expectations, Economic Journal 86:523-535.

Rothwell, R. (1982), The Role of Technology in Industrial Change: Implications for Regional Policy, Regional Studies 16, 5:361-369.

Sahal, D. (1977), Multidimensional Diffusion of Technology, Technological Forecasting and Social Change 10, 3:277-298.

Sahal, D. (1979), The Temporal and Spatial Aspects of Diffusion of Technology, I E E E Transactions on Systems, Man and Cybernetics 9, 12:829-837.

Salter, W. (1966), Productivity and Technological Change, Cambridge University Press.

Sassu, A. (1978), Diffusion of Technological Innovations and Oligopoly, Giornale Degli Economisti E Annali de Economia 37, 5-6:363-380.

Saxonhouse, G. (1974), A Tale of Japanese Technological Diffusion in the Meiji Period, Journal of Economic History 34, 1:149-165.

Scherer, F.M. (1980), Industrial Market Structure and Economic Performance, Rand Mcnally, Chicago.

Scherer, F.M. (1986), Innovation and Growth - Schumpeterian Perspectives, MIT Press.

Schmittlein, D. and Mahajan, V. (1982), Maximum Likelihood Estimation for an Innovation Diffusion Model of New Product Acceptance, Marketing Science 1, 1:57-78.

Schmookler, J. (1966), Invention and Economic Growth, Harvard U.P.

Schumpeter, J.A. (1934), The theory of Economic Development, Harvard U.P.

Sharif, M.N. and Ramanathan, K. (1982), Polynomial Innovation Diffusion Models, Technological Forecasting and Social Change 21:301-323.

Shen, T.Y. (1973), Technological Diffusion, Substitution, and X-Efficiency, Econometrica 41, 2:263-284.

Siebert, H. (1969), Regional Economic Growth: Theory and Policy, Scranton, Penn, International Textbook.

Smith, G.C. (1976), The Spatial Information Fields of Urban Consumers, Transactions of the Institute of British Geographers, New Series 1:175-189.

Soete, L. and Turner, R. (1984), Technology Diffusion, and the Rate of Technical Change, Economic Journal 94:612-623.

Sonis, M. (1982), Diffusion of Competitive Innovations, Instrumental Society of America, Vogt. and Nickle ed. Modeling and Simulation, 12. part 3.

Sonis, M. (1983), Dynamic Choice of Alternatives, Innovation Diffusion and Ecological Dynamics of the Volterra-Lotka Model, Innovation Diffusion and Ecological Dynamics, pp. 29-43.

Sonis, M. (1983), Spatio-Temporal Spread of Competitive Innovation: An Ecological Approach, Papers of the Regional Science Association, 57.

Solow, R.M. (1957), Technical Change and the Aggregate Production Function, Review of Economics and Statistics 39:312-320.

Souder, W.E. and Quaddus, A. (1982), A Decision-Modeling Approach to Forecasting the Diffusion of Longwall Mining Technologies, Technological Forecasting and Social Change 21, 1:1-14.

Stead, H. (1976), The Cost of Technological Innovation, Research Policy 5:2-9.

Stewerd, J.H. (1972), Theory of Culture Change - the Methodology of Multilinear Evolution, University of Illinois Press.

Stohr, W. and Todlling, F. (1977), Spatial Equity - Some Anti-theses to Current Regional Development Doctrine, Papers of the Regional Science Association 38:33-53.

Stoneman, P. and Ireland, N.J. (1983), The Role of Supply Factors in the Diffusion of New Process Technology, Economic Journal S, 66-78.

Suarez-Villa, L. (1987), Entrepreneurship and the International Diffusion of Innovations in Manufacturing, A General Approach, Rivista Internationale di Scienze Economiche e Commericiali 34.

Sweeney, G.P. (1987), Innovation Entrepreneurs and Regional Development, Frances Pinter.

Takeuchi, H. and Nonaka, I. The New Product Development Game. Jan.-Feb. Harvard Business Review, pp. 137-145.

Talaysum, A.T. (1985), Understanding the Diffusion Process for Technology-Intensive Products, Research Management 28, 4:22-26.

Tann, J. (1973), Fuel Saving in the Process Industries During the Industrial Revolution: A Study in Technological Diffusion, Business History 15, 2:149-159.

Tapiero, C.S. (1983), Stochastic Diffusion Models with Advertising and Word-of-Mouth Effects, European Journal of Operational Research 12:348-356.

Teotia, A.P.S. and Raju, P.S. (1986), Forecasting the Market Penetration of New Technologies Using A Combination of Economic Cost and Diffusion Models, Journal of Product Innovation Management 3, 4:225-237.

Thomas, M.D. (1975), Economic Development and Selected Organizational and Spatial Perspectives of Technological Change, Economie Appliquee 29:379-400.

Thomas, M.D. (1975), Growth Pole Theory, Technological Change, and Regional Economic Growth, Papers of Fourteenth European Congress of the Regional Science Association 34:3-25.

Thomas, M.D. and LeHeron, R.B. (1975), Perspectives on Technological Change and the Process of Diffusion in the Manufacturing Sector, Economic Geography 51:231-251.

Thwaites, A.T. and Oakey, R.P. (ed.)(1985), The Regional Economic Impact of Technological Change, Frances Pinter, London.

Utterback, J.M. (1974), Innovation in Industry and Diffusion of Technology Science 183:620-626.

Utterback, J.M. (1979), The Dynamics of Product and Process Innovation in Industry, in: Hill, C.T. and Utterback, J.M. (eds.), Technological Innovation for a Dynamic Economy, Pergamon.

Utterback, J.M. and Abenathy, W.J. (1975), A Dynamic Model of Process and Product Innovation, Omega 3:639-656.

Vandermeersch, L. (1986), Le Nouveau Monde Sinse, Presses Universialirs de France.

Vernon, R. (1979), The Product Cycle in a New-International Environment, Oxford Bulletin of Economics and Statistics 41:225-267.

Wilkins, M. (1974), Role of Private Business in International Diffusion of Technology, Journal of Economic History 34, 1:166-188.

Wilson, A.G. (1970), Inter-Regional Commodity Flows: Entropy Maximizing Approaches, Geographical Analysis 255-282.

Wilson, A.G. (1971), A Family of Spatial Interaction Models, and Associated Developments, Environment and Planning 3:1-32.

Yapa, L.S. (1975), The Green Revolution: A Diffusion Model. Sept. Annals of the Association of American Geographers.

Yapa, L.S. (1975), Analytical Alternatives to the Monte Carlo Simulation of Spatial Diffusion, Annals of the Association of American Geographers 65:163-176.

Yapa, L.S. (1976), On the Statistical Significance of the Obsered Map in Spatial Diffusion, Geographical Analysis 8:255-268.

15 Technology, trade and differential growth in the European Community

W. MOLLE

1. Introduction

1.1 Objectives

Many people in Europe believe that the wide differences in wealth among EC member countries are becoming progressively wider. For one thing the more developed countries presumably keep abreast of technological advance by specialising in technology-intensive products for which they have a comparative advantage; for another, they are supposed to achieve economies of scale through process innovations. The poorer countries, on the contrary, unable to realise the industrial structures needed for similar product and process innovations, are, in that view, condemned to stay behind.

The possibility that trade liberalisation would cause adverse spatial welfare effects was already voiced in the beginning of the European integration process. In the 1970s, concern about the effects of the enlargement gave rise to the creation of the European Regional Development Fund, intended to help lagging regions. The Fund has tried to do so also by the more classical instruments of regional policy, such as aid to industrial investment and infrastructure. Since 1980 it has also for that purpose strengthened the regions' technological base. Recently, concern about the spatial effects of EC policy has been growing again as negative regional consequences are expected from the prosposed measures to complete the internal market by 1992.

The popularity of this dominant philosophy is in stark contrast with some real developments. As a matter of fact, from a very long-term analysis (stretching over two centuries), the differences between the wealth levels of the countries in Europe appear to have diminished rather than increased. Indeed, many countries have been able to overcome an initial handicap in their trade specialisation and for some time were growing faster than the then leading countries (Molle and van Haselen, 1987).

How could this happen in a period of increased international competition, fast technological change, and sometimes profound societal change? Since international policies for the transfer of money or technology were practically absent at the time, we cannot give the credit to farsighted governments. There must be something else then. Were the assumptions of the philosophy cited above incorrect, and does the economic system work differently?

The objective of the present analysis is to establish the empirical basis for each link of the causal chain technology change - economic growth. We will do that by a critical examination of the relevant studies. Our perspective is international rather than interregional.

1.2 A theoretical framework

Looking closely at the claims referred to in the previous section, we observe that the following links qualify for further critical examination:

1. the differences in wealth;
2. the role of international trade and specialisation of production;
3. the possibility of fast growth from a low starting point;
4. the role of technology, and finally
5. the role of public policy in correcting the situation.

The theoretical framework best adapted to tackle this set of elements seems to be the technology-gap approach. It breaks the problem up into the following sequence of relations:

1. there is a relation between the levels of a country's or region's economic and technological development;
2. technological change permits the manufacture of new products or the cheaper manufacture of existing ones, thus giving a competitive advantage to the firm marketing them and improving the export position of the country where this firm is located;
3. hence, innovation by firms working in a certain country or region stimulates economic growth in them;
4. a country or region facing a technology gap can catch up with leading countries by following an imitation strategy;
5. a country's potential to mobilise resources to transform its structures determines how far it can benefit from such a strategy;
6. to stimulate innovation by public policy measures means to stimulate economic growth.

Much work has gone into analysing the dynamics of each stage in the above sequence. New ground had to be explored because meanstream economics was suspiciously silent on some of the questions crucial to this approach. These questions relate to the nature and the dynamics of the innovative process (how and where is innovation brought about, how can it be measured), to the transfer of technology (how is it transferred to different parts of the same firm, to different firms, between research institutes and firms), and finally to the factors that can be seen as facilitators of, or constraints on, innovation and transfer (Madeuf, 1981).

1.3 Organisation of the chapter

In the following sections we will go into the points raised above. First we will give an idea fo the relation between economic development and innovation potential. Next we will explore the relation between innovation activity and trade performance. Then we will look at the influence of technological change on growth rates. Subsequently, we will discuss how far the mechanisms for the transfer of technology among European countries work. In the final section we will draw some general conclusions.

In earlier analyses of the dynamics of spatial growth in Europe we have found that the development of regions had been largely determined by the development of the countries to which they belong (Molle et al., 1980, van Haselen and Molle, 1980). The dominance of the national factor over the regional one is apparent not only from an aggregate concept like GDP/P, but also from a wide variety of other indicators, such as employment structure. If we assume that the different roles of technology in regional development in Europe are likewise largely determined by national differences, to analyse the matter first on the national level and only after that on the regional level seems the proper way to proceed. Anayway, for many purposes the countries of Europe can be looked upon as a kind of mega-regions.

2. Technology and Economic Levels of Development

2.1 The national level

The hypothesis of a relation between the level of economic development and innovative activity has been put forward in particular by international economists, who had the differences between developed and third world countries in mind (see among others Krugman, 1979, and for more details, Malecki's contribution to this volume). To find out how far this hypothesis applies to the EC, we have combined indicators of economic development and technology in the following table.

Table 1 Levels of economic and technological development, 1980-1985

	GDP/P index 1985	Annual % GDP growth 1950-73	1973-83	Permillage of labour force engaged in R&D	R&D outlay in % of GDP	Patents ??
FRG	135	5.6	1.9	13	1.7	2.2
France	122	3.6	1.6	11	1.2	1.3
UK	105	1.8	1.3	12	1.4	1.1
Denmark	149	3.4	1.7	7	0.8	0.8
Netherlands	114	4.4	1.1	10	1.4	0.9
Belgium	106	2.5	1.6	7	1.1	0.3
Italy	82	4.2	1.6	5	0.7	0.6
Ireland	68	2.9	2.0	4	0.6	0.1
Spain	56	5.3	1.1	2	0.3	0.2
Greece	43	6.7	1.5	2	0.2	*
Portugal	28	4.3	1.3	2	0.2	*
EC	100	3.9	1.5	9	1.1**	1.0**
USA	151*	4.6**	1.8	11	1.4	3.4
Japan	103*	5.0**	3.6	12	1.7	2.1

Sources: Eurostat, OECD, UN,
* 1982; ** estimates.

The association between the level of economic development measured by relative GDP per capita, and the three indicators of innovative activity, is very clearly apparent in the table, although the situation of some of the smaller EC countries somewhat blurs the picture. Other data on the development of R&D activities teach us that three countries of the EC, namely, Germany, the UK and France, account for more than three quarters of industry-financed R&D in the entire EC. That concentration in a few countries has been a very stable factor; within the group, the shares for both the FRG and France tended to increase while that of the UK decreased.

Another interesting fact is that by some indicators the EC is performing less well than its main competitors, the USA and Japan. Moreover, since the mid 1970s innovating activities have grown slower in Western Europe than in the USA and Japan. Does that mean that the EC is losing the technological race? Not necessarily, for the effect matters more than the input. Patel and Pavitt (1987), who analysed the commercialisation of specific innovations and evaluated the fundamental character of some innovations by industrial peers found that there is no room for such generalisations. Countries compete along a long and differentiated technological frontier, and their relative position on this frontier differs from sector to sector.

2.2 The regional level

Recently a group of experts finished a study for the EC Commission on the location of R&D activities in the regions of the EC (NBST 1987). This study was the first to give a systematic survey of the distribution of both BERD (business expenditure on R&D) and GERD (gross expenditure including government). The figures show that BERD is concentrated not only on the national level but also on the regional level. That applies in particular to Italy, Germany and Denmark, and to a lesser extent to France, where a Mediterranean front is increasingly providing a counterweight to the Paris region. The regions of concentration tend to be the richer regions of their countries, which means that hypothesis 1 is still valid for the regions of the EC. (Similar evidence is reported in the contribution of Thwaites and Alderman to this volume).

Analysing the location tendencies of R&D establishments in the EC (Molle et al., 1989) we found a clear tendency to conserve the present pattern, with the South East of England ranking high as a potential location for R&D establishments of European scope.

3. Technology, Competitiveness and Trade Performance

3.1 EC trade, theory and empirics

The neo-classical theory of international trade based on factor proportions, although elegantly elaborated by Hecksher, Ohlin and Samuelson, proved hardly relevant to the actual patterns of trade, whereas far cruder concepts based on technology and related to the hypotheses of the product life cycle and filtering-down processes seemed to come much closer to explaining the patterns of trade observed in reality (Hufbauer, 1970). In a review article, Cheng (1984) concluded that to try and develop a model integrating technological innovation and diffusion in a factor-endowment model of international trade and investment would be a long-term proposition, and that for the short term the goal should rather be to extend our knowledge through informal models and empirical work in the field of technology and trade.

One basic assumption of the technology-growth relation is the one formulated as point 2 of the sequence in section 1.2, stating that R&D leads to innovations, which in turn lead to a better competitive position and therefore stimulate exports. The influence of the R&D effort on the rate of technological progress is now fairly well documented (Freeman, 1982). On the other hand, the relation between technological progress and trade performance is still very poorly understood. The first part of this relation (innovation - cost cutting) has been studied by Mansfield (1986), who showed that the growth of productivity was related to the rate of growth of R&D outlays made by firms or industries (see also Hughes, 1986). More specifically, Rothwell (1979-1980) showed that R&D is associated with export performance. So, although the indications of a relation between R&D or technology and export success are clear, there is much uncertainty about the way this actually comes about. Indeed, in their recent review of the situation Dosi and Soete (1988) identified a whole list of theoretical, empirical and normative gaps.

These gaps are also highly relevant to the European Community. As a matter of fact, the basic idea of the EC was to establish a zone of free trade that was to promote specialisation and through the use of the potential for economies of scale bring more trade and welfare to the member countries. The effects of the creation of the EC on the trade patterns has been studied in some depth. There has indeed been a tremendous increase in the total volume of trade. However, contrary to what many believed at the outset, a considerable intra-industry specialisation rather than a inter-industry specialisation has developed (see, among others, Drabek and Greenaway, 1984; Bergstrand, 1983).

Several factors have been proposed as explanations of the upsurge of ITT in intra-EC trade. One of them is technology. If R&D produces a rapid turnover of products protected by patents, countries tend to specialise in different segments of the market. However, as far as I know there is as yet no good empirical study available that measures the influence of the factors we think relevant. A recent study by Balassa and Bouwens (1988) of the extent of ITT between pairs of countries in Europe identifies three groups of relevant factors. ITT between two countries was found to be the more important as the countries involved were more developed (specialisation effects), as there were fewer trade impediments between them (integration effect), and as the distance and communication costs were lower. An additional variable that bears similarity to the technology-gap idea, about the influence of differences in level of economic development between pairs of European countries, was found to have a negative influence (the larger the difference, the smaller ITT).

3.2 Classification by technological intensity

The most common way to analyse the relation between technology level and trade performance is to classify products by category of R&D intensity- usually on the basis of the different inputs of R&D effort by industry - and go on to calculate the market shares of countries for these products. Analyses of that type are regularly made by the OECD. Table 2 gives an idea of the results of such an analysis.

Table 2 Coverage ratio for the products of high, medium, and low R&D-intensive industries by trade block, 1985

	high	medium	low	total
EC	1.01	2.11	1.09	1.36
USA	0.91	0.58	0.32	0.56
Japan	5.38	4.95	1.44	3.32

Source: OECD: STI Indicators Newsletter, no. 10, Paris, 1987.

These figures show the EC to be specialising in medium-R&D goods, while in the high-R&D category EC exports barely cover imports. The latter figure had significantly dropped in the previous years.

Another OECD study (1988) gives the details for individual countries and sectors of activity (see Table 3). The table reproduces the figures of trade performance (coverage ratio of imports by exports) in specific capital goods with a very high technology content for the EC countries.

Table 3 Export-import ratio of R&D-intensive industries by category and country 1985

Country	Office machines	Communication equipment	Electrical transmission	Scientific instruments
Germany	0.9	1.1	2.0	1.4
UK	0.7	0.9	1.3	0.9
France	0.7	1.2	1.4	0.8
Netherlands	0.7	1.3	0.9	1.0
Belgium	0.5	0.9	0.8	1.5
Denmark	0.3	0.8	1.2	0.8
Ireland	1.7	1.2	0.6	1.8
Italy	0.8	0.8	1.0	0.6
Spain	0.4	0.3	0.5	0.2
Portugal	0.7	0,6	0.4	0.3
Greece	0.0	0.1	0.2	0.0
USA	2.1	0.7	1.1	0.9
Japan	4.3	7.1	4.0	6.2

Source: OECD 1988: Main Science and Technology Indicators, 1981-1987, Paris.

Compared to the USA and particularly Japan, the EC countries show rather low figures. Only Germany stands out as a major net exporter of capital goods. It is followed at some distance by the majority of EC countries, where the situation is more or less balanced. A third group can be distinguished, consisting of the three new member countries, which are very large net importers of advanced equipment. Several studies have refined this type of analysis (some using variants of the Revealed Comparative Advantage indexes), but they are all more descriptive than explanatory.

3.3 Attempts at explanation

An explanation of the shares of OECD countries in total OECD trade in 40 products was attempted by Soete (1987). His static cross-section model contained such variables as the proportion of the countries' patents in USA as an indicator of their innovative capacity, gross fixed capital formation divided by employment as an indicator of factor proportions, population as a size variable and finally distance as a resistance variable. The results indicated forcefully how crucial the technology variable is for explaining the inter-country variation in export performance in most industries; patents were indeed more important than the factor-proportions variable, and its elasticity with increasing technology intensity.

A static analysis on the sector level cannot give definitive answers to the question posed earlier. Following Rothwell (1979), the relation between innovation and trade performance is probably better tested on the micro level than on the meso level. A study of the Dutch machine industry specialising in the food sector was recently carried out by Van Dijken (1988), who analysed the growth of exports of a sample of firms and tried to explain it from corporate characteristics. He found a very clear positive relation between export growth and the innovative profile of the firm, indicated by two variables: the proportion of total turnover the firm devotes to R&D, and the in-house initiating of technological change of products. Apart from these variables, some other corporate characteristics not related to technology proved to be significant. The evidence of the study thus clearly supports the third hypothesis of the introduction, namely, that innovation leads to a better competitive position and a better export performance.

3.4 Regional aspects

Trade was introduced into regional economics with the economic-base theory, which relates the development of a region to the demand for its export products. In that theory, the role of technology was to determine, jointly with the region's characteristics, the location of the basic activity. This crude concept of the economic base was soon generalised to that of the interregional trade multiplier (Isard, 1960), a concept that essentially was still of a demand-oriented or Keynesian nature.

Notwithstanding some theoretical and empirical probes, the determinants of regional trade patterns remain largely unexplored. One important reason for that state of affairs is that the absence of relevant data precludes sophisticated methods (such as the general equilibrium models now utilised for international trade) and barely allows the testing of relatively simple static integration effects on the regional level (Broecker and Peschel, 1988). That no models integration technology and trade patterns on the regional level have yet become operational is not surprising in the circumstances.

Unfortunately, we cannot fall back on the locational profile of a certain activity to develop the approach. As a matter of fact, more and more activities are now of the light manufacturing type, for which hardly any specific location factors can be detected. Indeed, technological development has made economic activity more footloose.

4. Technology and Growth

4.1 Different approaches

In the previous section we looked into the sequence technology lead -> competition advantage -> export growth, and found that there is indeed a positive relation. The logical conclusion of this reasoning is growth of GDP through the export multiplier. As far as we know, no studies have been made of the final link in this chain. The studies we have found relate the differences in growth among countries directly to differences in technology, bypassing the elements from the previous section.

Now, the question why growth rates differ is a classical one in the economic literature (Denison, 1967). Demand factors and the availability of natural resources and production factors have long been considered to determine economic growth. Solow's (1987) empirical implementation of the aggregated production-function model showed, however, that technical change acounted for a very large portion of US growth. Denison (1967), comparing the growth of several countries, found that 'advances in knowledge' made a substantial contribution to growth.

Although the technology variable had thus been identified, the way it influences growth proved very difficult to grasp. The standard macroeconomic terminology contained such distinctions as embodied and disembodied technical change, clay or putty-clay capital goods, which have all proved of very little practical relevance (see, among others, Gold, 1987). Indeed, the deceptive results of explanatory models based on the classical factors (production factors and demand factors) have forced researchers to look for other analytical approaches.

An approach that has drawn a fair amount of attention is the one based on the views of technology as presented by Schumpeter (1934) put in the framework of the Kondratief cycle (see, among others, Kleinknecht, 1987; Freeman, 1985). In that approach, technical change is the key factor in the dynamics of the subsequent waves of growth of developed countries. Although much of their work was oriented to the increase in growth on the world level that would ensue from a new wave of innovations, yet some strands of thought were also devoted to the aspect of differential growth among countries. One is the technological-gap approach, briefly referred to in the introduction.

4.2 The technology-gap approach

Central to the technology-gap approach is the so-called catching-up hypothesis, representing the view that economic growth is the result of two conflicting forces. On the one hand, innovating countries try to keep their lead, while on the other imitating countries try to catch up. Although the catching-up hypothesis is generally applied in political and academic discussions, very little empirical reserach has so far been carried out into its relevance to developed countries (see, among others, Baumol, 1986, who found evidence for catching up only between countries with similar systems and not between different groups of countries. We will present the results of the few studies that have tried to test the hypothesis in a context that is relevant to Europe.

Pavitt and Soete (1982), following Soete (1981), estimated a model which used cross-sectional data referring to the development of fourteen OECD countries through a long period, namely, from 1890 to 1977. Their results are very inconclusive. The gap variable (relative GDP per capita) does not influence growth with GDP as dependent variable, and only slightly when productivity growth was taken as an alternative dependent variable. The independent variable 'technology' measured in terms of patents (either the level or the growth of patents) or in terms of R&D showed opposite signs in different periods. Such results cannot be taken as a support for the technology-gap theory.

Fagerberg (1987) constructed a different model, using pooled time series (1963-1983) and cross-sectional data (20 OECD countries plus six NICs) to estimate it. GDP growth was found to be dependent on the gap (GNP/P) with a negative sign, on the growth of patent applications by i residents in all other countries (positive relation), and no the level of investment (positive relation). So we may conclude that the scope for imitation (GDP/P), the growth of innovative activity (patents or R&D), and the efforts to close the gap (inv) all help to explain differences in economic growth.

In a second paper, Fagerberg (1988) re-estimated the model for the 1973-1983 period, with much the same results. Additional hypotheses on the autonomy of changes in the industrial structure and the degree of openness to the outside world could not be substantiated.

In his first contribution, Fagerberg had applied his model as well to a sample of European OECD countries only. That version produced the same type of results, albeit much less significant. Apparently these fairly simple (cross-section) models are not well suited to explain the relatively small differences in GDP/P levels and in GDP growth rates between European countries (see Table 1). The fact remains that some of the more backward European countries have been able to realise high levels of GDP/P with moderate levels of innovative activity. Fagerberg concludes that "to explain differences of growth between these countries in the post-war period, a much more detailed analysis of economic, social and institutional structures need be carried out."

Thus, we seem to rejoin an approach by Olson (1982) and Heitger (1987), who relate stagnation to institutional sclerosis preventing adaptation to new circumstances. This approach indeed produces a fair result for differences among countries in Europe. (By the way, the idea behind the EC programme for breaking down the remaining trade barriers within the EC by 1992 is to shake up protected situations, thus forcing sectors of the economy into more competitive situations.)

4.4 Regions

In regional economics, too, the attention has long been on (inter)regional growth models inspired by macro-models (see, among others, Albegov et al., 1980; Issaev et al., 1982), which tended to include technology only as an afterthought or as a residual (see for a description of the technology factor in these models, Malecki, 1983). To my knowledge, a technology-gap approach to regional development, explicitly introducing technology as a factor in its own right has not yet been applied to Europe. As other contributors to this volume have dealt with the general dynamics of technology change and regional development in some detail, we will not develop the subject further here.

5. Imitation and the Transfer of Technology

5.1 Different channels of transfer and stages of integration

The study by Fagerberg mentioned in the previous section gave evidence that backward countries are catching up, but did not elaborate on how they do it or on the part technology plays. Transfer and imitation are the keywords there. The dynamics of imitation and the diffusion of technology have been the objects of many studies (Bradbury, 1978, for example). Many of these studies distinguish three channels through which technology is diffused: it can be embodied in capital or humans, or transferred as disembodied technology. Capital-embodied technology can be dispersed through trade in investement goods, or by direct investment, to be understood as a combined goods and capital transaction. Human transfer is related to the free movement of labour. Disembodied technology, finally, is transferred mostly in the form of a service. The three forms of transfer are often combined to make up technology packages: for example a turnkey project involving the building of a plant, the use of patents, the training of local staff, etc.

For these concepts to be applied to the European Community, the particular dynamics of integration have to be taken into account. Traditionally, three stages are distinguished. At the first stage, goods and service markets are integrated, that is to say, all barriers to trade are broken down. Next comes the stage of integrating the markets for production factors, labour and capital, and entrepreneurial skills. The integration of economic policies, such as macroeconomic and monetary policies, marks a higher level of integration. The three channels through which technology is diffused can be related to the stages of integration (e.g., goods trade is associated to capital embodied, labour to human embodied, etc.).

Integration is supposed to stimulate innovative behaviour of firms, but we know little about how it affects the relocation of plants to the less affluent areas.

5.2 Investment goods; capital-embodied technology

Capital-embodied technology is transferred through trade in equipment, tools and intermediate goods; here we tread on ground that has already been well explored in numerous diffusion studies both on the international and the interregional level (see, among others, the review in Ewers and Wettmann, 1980). A review specifically based on the situation in the EC would be in order, however.

5.3 Human-embodied technology

Technology circulates through technical assistance, training programmes, professional mobility, and personal contacts of experts. For none of these categories is there any systematic and quantified information available. Presumably, however, for most such transfers the origin is in countries and regions with a high density of R&D activity and the destination in areas with a low R&D density. For two categories some indications are available from empirical studies.

On the European level, professional mobility has been studied extensively for the category of **foreign labour**. One hope cherished by the, mostly poor, sending countries in the Mediterranean basin was for those working abroad to acquire skills that they could use on return. The evaluation of the practical situation in many countries (Macmillan, 1982) showed that this hope was by no means fulfilled. Most returnees had taken up activities that were unrelated to the skills they had acquired, for lack of job opportunities on the one hand, and on considerations of social status on the other. The European integration process is giving a specific dimension to the problem of migrant workers. Straubhaar (1984) believed that the free movement of workers in the enlarged Community would bring about huge flows of migrant workers from the new members to the developed regions of the EC. Contrary to that view, Molle and Van Mourik (1988) argue on theoretical and empirical considerations that the further integration of Europe is likely to lead to less rather than more migration. As migration anyway did nothing to contribute to the transfer of technology to backward regions, the deception need not be too great.

Personal contacts of experts have received increasing attention lately. Experts appear to be building networks for the exchange of information. As Von Hippel (1987) reports, professional colleagues are even prepared to reveal their proprietary know-how to rival firms on an informal and highly selective basis. That leakage phenomenon might in fact explain the rapid adoption of innovations by competitors. It seems to work best where there is close geographical proximity. Could that hold a plea for the renewed promotion of functional complexes on the model of Sillicon Valley?

5.4 Services or disembodied transfers

The dissemination of disembodied technology operates via patents and licences, feasibility studies, blueprints, operating instructions, manuals, etc. Until recently there was very little information available on the subject, but now a large portion of the disembodied transfers are measured by the technological balance of payments (TBP) (see for a description of TBP, Madeuf, 1984). The TBP registers the flows of proprietary knowledge into or from a country. The flows consits of money paid or received for the use of trade marks, designs, inventions, know-how, and closely associated technical services. Although the coverage may vary from country to country, the data are likely to be representative of the structural situation of each country. Table 4 gives some key data for the European countries.

Table 4 Disembodied technology transfer indicators 1980

	Payments % R&D	Receipts % R&D	TBP coverage
Germany	10	5	0.5
UK	11	13	1.2
France	15	13	0.9
Netherlands	42	27	0.6
Belgium	*	*	0.5
Denmark	23	34	1.1
Italy	28	10	0.3
Spain	158	50	0.2
Portugal	154	25	0.1
EC			
USA	1	13	25.7
Japan	7	5	0.8

Source; OECD STI 2 (1986) ; STI Newsletter 10 1987

In the last column we have given the coverage ratio, that is, the receipts divided by payments. An instant observation is that the EC as a whole covers only two thirds of its outlays. Japan also displays a deficit, albeit a smaller one. The opposite is true of the USA. Within the EC only the UK has a significant surplus of its TBP; all other countries are more (Mediterranean countries) or less (northern countries) dependent on the outside for their technology.

Combined with the figures of Table 1, the TBP data give a clear idea of one of the mechanisms for growth in Europe: the southern countries buy technology from the leading countries of the northern part of the EC and the rest of the world in order to start up production processes for products they have not fully developed themselves.

5.5 Packages of transfers, strategies of multinational firms

The rapid development of some European regions in the 1960s was brought about by the influx of American multinationals desirous to be present in the EC (in Flanders, for instance). They brought their superior products and production techniques, coupled with a worldwide marketing organisation and superior management techniques (Klaassen and Molle, 1983). Nowadays much stress is laid on so-called endogeneous growth because incoming investment is growing at a rapid pace.

There is only an eclectic theoretical basis for the behaviour of multinationals, allowing for many different suppositions. However, in the EC the data on direct investments by origin and destination are so poor that these suppositions are unlikely to be rigorously tested in the near future. There is nothing for it but to turn towards more case studies and try to draw generalisations from them.

6. Conclusions

The **results** of the previous sections are clear. There are differences in wealth level among the countries of the EC which largely coincide with the differences in technological level. Innovation is important to obtain a competitive advantage in international trade. There is a relation between low wealth levels and fast growth in Europe. No relation between technology and growth could be established. There was evidence, however, that the rate of investment,which can be seen as embodied technical progress, has had a positive influence on growth. The conclusion must be, then, that it is in particular the transfer of technology by several means that permits countries to catch up. Unfortunately, our knowledge of one presumed major transfer channel, namely direct investment by multinational firms, is still very much obscured.

In the light of the progress of European integration, these results hold a special significance for our view of future developments.

For one thing, the goods trade is likely to grow still more in relative terms, while changing more and more to the intra-industry type, which cannot well be explained by elements of technology-gap theory. We may in that respect cite Vernon (1979) who thinks that there is no reason any more to differentiate among developed countries because the transfer of standardised production to low-wage countries in the third world now takes place as soon as the initial stage is over, bypassing intermediate regions.

Direct investment is increasing on a considerable scale now, and we expect this to continue in the future, with multinationalisation on the wax, and firms seeking the most advantageous location for each of their plants on the European scale. That also makes complex packages of technological transfer more likely, both within one company and between companies.

We also foresee more human-embodied and disembodied transfers as transaction costs drop through Europe-wide patenting and the development of European networks of researchers, sponsored by European programmes. And we expect the international mobility of workers to extend as well, though probably on a limited scale only.

There are still many **open questions.**

The first concerns the specific dynamics of the catching-up process: which factors promote success and which factors failure? Is the technological factor dominant or are there societal factors that carry more weight? Within technology, what is the role of the various forms of transfer, and how great is the need for autonomous innovative activity?

The second concerns the efficiency of the markets for technology in Europe. How can public policy help to increase the allocational efficiency on these markets. Can systems be developed that have not only a beneficial effect on the national level, but send direct impulses to the regions?

The third question is about services. Most studies focus on the manufacturing sector and the trade in manufactures. How can a competitive advantage in services be capitalised on through innovation? How can it be realised in the market? What role do regions play?

Finally we have to stress the **lack of empirical basis** for a number of propositions that are now beginning to become a standard part of technology and regional policies.

References

Albegov, M., Andersen, A.E., and Snickars, F. (eds.), Regional Development Modelling, Theory and Practice, North Holland, Amsterdam.

Balassa, B., and Bauwens, L. (1986), The Determinants of Intra-European Trade among Exporters of Manufactured Goods, European Economic Review 32.7: 1421-1439.

Baumol, W.J. (1986), Productivity Growth, Convergence and Welfare; What the Long-Run data Show, American Economic Review 76.5:1071-1085.

Bergstrand, J.H. (1983), Measurement and Determinants of Intra-Industry International Trade, in: Tharakan, P.K.M. (ed.), Intra Industry Trade, Empirical and Methodological Aspects, pp. 201-255, North Holland, Amsterdam.

Bradbury, F. R. (ed.) (1978), Technology Transfer Practice of International Firms, Sijthoff and Noordhoff, Alphen a/d Rijn.

Broecker, J., and Peschel, K. (1988), Trade, in: Molle, W. and Cappellin, R. (eds.), op.cit., pp. 127-152.

Cheng, L. (1983), International Trade and Technology; a Brief Survey of the Recent Literature, Weltwirtschaftliches Archiv 120:165-189.

Denison, E.F. (1976) (assistend by J.-P. Pouillier), Why Growth Rates Differ, Brookings Institution, Washington.

Drabeck, Z., and Greenaway, D. (1984), Economic Integration and Interindustry Trade, the CMEA and EEC compared, Kyklos 37:444-469.

Dijken, J. van, et al. (1988), De Nederlandse machine-industrie voor de voedings- en genotmiddelenbranche en haar exportinspanningen, mimeo, NEI, Rotterdam.

Dosi, G., and Soete, L. (1988), Technical Change and International Trade, in: Dosi, G. et al. (eds.), Technical Change and Economic Theory, pp. 401-432, Pinter, London.

Ewers, H.J., and Wettmann, R.W. (1983), Innovation Oriented Regional Policy, Regional Studies 14:161-179.

Fagerberg, J. (1987), A Technology Gap Approach to Why Growth Rates Differ, Research Policy 16:87-99.

Fagerberg, J. (1988), Why Growth Rates Differ, in: Dosi, G. et al. (eds.), Technical Change and Economic Theory, pp. 432-458, Pinter, London.

Freeman, C. (1982), The Economics of Industrial Innovation, 2nd ed., Pinter, London.

Freeman, C. (1985), The Role of Technical Change in National Economic Development, in: Amin, A. and Goddard, J.B. (eds.), Technological Change, Industrial Restructuring and Regional Development, pp. 100-114, Allen and Unwin, London.

Gold, B. (1987), Technological Innovation and Economic Performance, OMEGA 5:361-371.

Haselen, H. van, and Molle, W. (1980), Regional Disparity and Assisted Areas in the EC of Twelve, Revista de Estudios Regionales 6:153-198.

Heitger, B. (1987), Corporation, Technological Gaps and Growth in OECD Countries, Weltwirtschaftliches Archiv 123:463-473.

Hippel, E. von (1987), Cooperation Between Rivals; Informal Know how Trade, Research Policy 16:291-302.

Hufbauer, G. (1970), The Impact of National Characteristics and Technology on the Commodity Composition of Trade in Manufactured Goods, in: Vernon, R. (ed.), The Technology Factor in International Trade, Columbia University Press, New York.

Hughes, K. (1986), Exports and Technology, Cambridge University Press, Cambridge.

Isard, W. (1960), Methods of Regional Analysis, MIT, Cambridge.

Issaev, B., Nijkamp, P., Rietveld, P. and Snickars, F. (1982), Practice and Prospect of Multiregional Economic Modelling, North Holland, Amsterdam.

Klaassen, L.H., and Molle, W.T.M. (eds.) (1983), Industrial Mobility and Migration in the European Community, Gower, Aldershot.

Kleinknecht, A. (1987), Innovation Patterns, in Crisis and Prosperity, MacMillan, London.

Krugman, P. (1979), A Model of Innovation, Technology Transfer and the World Distribution of Income, Journal of Political Economy 87:253-266.

Macmillan, M. (1982), The Economic Effects of International Migration: a Survey, Journal of Common Market Studies XX:245-267.

Madeuf, B. (1981), L'ordre technologique international; production et transferts, Documentation francaise no. 4641-4642, Paris.

Madeuf, B. (1984), International Technology Transfers and International Technology Payments: Definitions, Measurements, and Firms' Behaviour, Research Policy 13:25-140.

Malecki, E.J. (1983), Technology and Regional Development; a Survey, International Regional Science Review 8,2:89-125.

Mansfield, E. (1968), The Econometrics of Technical Change, London.

Molle, W., et al. (1980), Regional Disparity and Economic Development in the European Community, Gower, Aldershot.

Molle, W. (1983), Technological Change and Regional Development in Europe (Theory, Empirics, Policy), Papers and Proceedings of the Regional Science Association (European Congress), pp. 23-38.

Molle, W., and Van Haselen, H. (1987), Divergence or Convergence between the Centre and the Periphery of Europe; an Analysis of Very-Long-Term Trends, NEI: FEER 1987/4, Rotterdam.

Molle, W., and Van Mourik, A. (1988), International Movements of Labour under Conditions of Economic Integration: the Case of Western Europe, Journal of Common Market Studies XXVI,3:317-342.

Molle, W., and Cappellin, R. (eds.) (1988), Regional Impact of Community Policies in Europe, Gower, Aldershot.

Molle, W., Beumer, L. and Boeckhout, I. (1989), The Location of Information Intensive Industries in the European Community, to be published.

NBST (1987), STRIDE, Science and Technology for Regional Innovation and Development in Europe, Final Report, mimeo, Brussels.

Olson, M. (1982), The Rise and Decline of Nations, Economic Growth, Stagflation and Social Rigidities, Yale, New Haven.

Patel, P., and Pavitt, K. (1987), Is Western Europe Losing the Technological Race? Research Policy 16:59-85.

Pavitt, K. and Soete, L. (1982), International Differences in Economic Growth and the International Location of Innovation, in: Giersch (ed.) Emerging Technologies, Consequences for Economic Growth, Structural Change, and Employment, Mohr, Tuebingen.

Rothwell, R. (1979), The Relationship between Technical Change and Economic Performance in Mechanical Engineering; Some Evidence, in: Baker, M.J. (ed.) Industrial Innovation; Technology Policy, Diffusion, MacMillan, pp. 36-59.

Rothwell, R. (1980), Innovation in Textile Machinery, in: Pavitt, K. (ed.) Technical Innovation and British Performance, pp. 126-141, Methuen, London.

Schumpeter, J.A. (1934), The Theory of Economic Development (transl. OPIE), Oxford University Press, Oxford.

Soete, L. (1981), A General Test of Technological Gap Trade Theory, Weltwirtschaftliches Archiv, pp. 638-659.

Soete, L. (1987), The Impact of Technological Innovation on International Trade Patterns: the Evidence Reconsidered, Research Policy, pp. 101-130.

Solow, R. (1957), Technical Change and the Aggregated Production Function, Review of Economics and Statistics 39:312-320.

Straubhaar, T. (1984), The Accession of Spain and Portugal to the EC from the Aspect of the Free Movement of Labour in an Enlarged Common Labour Market, International Migration Quarterly 22:228-238.

Vernon, R. (1979), The Product Cycle Hypothesis in a New International Environment, Oxford Bulletin of Economics and Statistics 41:255-267.

PART D
POLICIES REVISITED

16 Local development: patterns and policy implications

G. GAROFOLI

1. The Patterns of Regional Development in Europe

There has been a substantial reversal of the core-periphery relations in Europe in the last 10 to 15 years. This has been the consequence of a trend inversion in location patterns of industries that has involved a shift in the pattern of regional development; in fact there has been a shift from a pattern involving increasing spatial concentration of production (typical of the development during the 1950s and 1960s) to a new pattern characterised by a territorial diffusion of the industrialization process. This phenomenon has neither produced a reduction in regional disequilibria in the European Community as a whole nor in the different countries, because depressed regions have met new problems in their process of transformation and development. Intermediate regions (almost everywhere in Europe) have experienced an improvement in industrial employment, compared with a reduction of employment in most developed regions. Almost everywhere this phenomenon has been combined with the crisis of large firms and with a new role of small firms.

This phenomenon has been studied in various countries and different interpretations have been proposed. It is very noticeable in Italy where some scholars have advanced a new interpretation of the change in the core-periphery relation, stressing the degree of autonomy in the development in intermediate regions and the new organization of production mainly based on new inter-firms relations. At least in the beginning, interpretations put forward in other countries, such as France and Great Britain, were different. In these countries this phenomenon was considered as the consequence of the decentralization of production and 'branch circuit' organization with a new form of 'dependency' of the periphery in relation to central areas (cf., Lipietz, 1980 and Damette, 1980).

Nevertheless, it is now possible to understand the real crisis of 'central regions', of more developed regions, whose development was organized on the crucial role of large firms and with a process of increasing urbanization.

At the same time it is possible to discover novel 'forms' of development in intermediate regions; in these regions new patterns of development are arising, which are different from those followed in advanced countries and regions. In these cases, there is a crucial role played by small firms and new entrepreneurship, a process of diffuse development without a process of strong urbanization, with a co-presence of small firms and small towns, with a very articulated social structure, not dichotomized as in the regions with large firms and an ancient process of development.

The backward regions in various countries, especially in South Europe, have not changed their position; they met the problems of the economic crisis without having, at least until now, opportunities of new forms of development.

These new patterns of regional development have arisen, on the one hand, by careful attention to the endogenous factors of development and to the exogenous ones and also to the relationships between the local and the international economic system; on the other hand, these phenomena have determined tremendous changes in the interpretations of the spatial articulation of production. The new patterns of regional development have caused, for example, the 'breakdown' of the dualistic paradigm, with the opposition of the centre to the periphery; but at the same time they have shown the explanatory inadequacy of the 'filtering down theory' (Thompson, 1968; Berry, 1973). According to the latter theory, industrial relocation from the centre to the periphery should occur following the predictions of the theory of the new product life cycle, with the progressive movement of standardized products towards the periphery; in this case the process of development should be a physiological process of spillover from the centre to the periphery. Both the dualistic paradigm and the filtering down theory consider the territory as a passive element in the process of development: the crucial factors of change are exogenous to the local and regional economic and social system, there is no opportunity for an autonomous process of development. There is not a substantial difference, furthermore, with the hypothesis of the decentralization of production and with the 'branch circuit' hypothesis; in both cases the process of development is an extraverted process.

The new regional patterns of development have been very often considered strictly linked to a substantial change in the mode of organization of production which was realized during the 1970s, with a progressive crisis of the Fordism pattern and the increasing role of the flexible specialization (Piore and Sabel, 1984).

The Fordism pattern of organization of production, following the French school of regulation, is determined by the combination of Taylorism and Keynesian state. It is based on mass production, in which economies are obtained through fixed capital and labour productivity increases within the production process. Standardized products (with long run production) are obtained, using monopurpose machinery and semiskilled workers, with fragmented and standardized tasks. The prevailing form of the market is the oligopolistic one and the management of the economy is organized at a national level, especially with the goal to offset the immanent tendency to over-production. Finally, competition among firms is carried out through full capacity utilization and cost cutting.

The flexible specialization model is based instead on flexible automation: economies are obtained through working capital productivity increases. Differentiated products, with small batches of production, are obtained using mainly general or multipurpose machinery and skilled workers, with a close integration of mental and manual tasks. The products are very specialized for niche markets. Competition is combined with co-operative links among firms: the horizontal integration of production is based on thick networks among firms and sub-contracting relations; all this leads to a 'just in time' organization. The forms of social regulation are mainly established at a local level, with an important role of specific local institutions. Competition among firms is mainly carried out through innovation and the introduction of new products. The response of the organization system to falling markets is very flexible and complex; it is carried out through diversification of production, innovation, sub-contracting and lay-offs.

Clearly with the second pattern of organization of production the role of local social forces is quite important and the opportunity arises for autonomous processes of development at local and regional levels.

2. Territory and Local Patterns of Development

The new patterns of regional development in Italy and in other European countries have obliged economists and other social scientists to reflect with more attention on the relationship between economic development and territory, looking at the specific characteristics of local patterns of development.

The attention given to local models of development is not only the result of the differentiation of local economic situations and the emergence of new forms of industrialization and development, but above all it is a consequence of the crisis of the 'functionalist paradigm' of development which considered space as simply the place where the effects of the process of general development occurred. Or in other words the place where 'forces and economic functions occur, which as a consequence of their internal logic define the activities' location and hence the structure and hierarchy of economic regions' (Pecqueur, 1987, p. 85).

Polarized development, concentrated territorially, and the diffusion of development 'from above' have for some time been the dominant theoretical paradigm on which development strategies and regional policies have been based. Development was made possible by the continuous expansion of large scale industry, prevalently localized in large urban centres since the typology of technological innovation, economic policies and the development of infrastructures allowed the formation of large scale economies and growing external economies (Garofoli, 1983d, 1984a and 1987).

The development policy based on the 'transferal' of the model founded on big business (by implicitly accepting the theory of automatic mechanisms of income multiplier and new economic linkages) has characterized the greater part of government interventions for a number of decades both in developed and developing countries.

In the last 10 to 15 years the growing inability of the functionalist model to explain the ever more complex redistribution of productive activities over territory and the emergence of autonomous patterns of development in many relatively peripheral regions have progressively attracted scholar's attention to the territorial dimension of development and to the categories of **environment** (or 'milieu') and **territory** seen as the 'sedimentation' of specific and interrelated historical, social and cultural factors in local areas which generate significantly different processes of development directly due to local specifications.

These considerations lead us to a new paradigm for the research of economic development mainly based on territory, while at the same time consideration regarding the local characteristics of models of endogenous development, including its internal control, gives rise to the paradigm of development 'from below' (Stöhr, 1978, 1981 and 1984). During the late 1970s and the early 1980s the literature, at an international level, on endogenous and 'self-centred' development[1], on territorial and 'agropolitan' development (Friedmann and Douglass, 1975; Friedmann and Weaver, 1979), on the mobilization of 'indigenous potential' (CCE, 1981), etc., began to increase noteably.

The end result of these varied reformulations of the problem of development has been above all a different concept of space held by economists: space is not only the distance between different places, something which conditions the exchange of goods and a source of cost for economic agents, as in the traditional theories of industrial location; in these new interpretations it is the distinguishing feature of territory, of a strategic factor of development opportunities and of characteristics that it could assume. Territory represents a clustering of social relations, it is also the place where local culture and other non-transferable local features are superimposed. It is the place where men and business establish relationships, where public and local institutions intervene to regulate society. Territory represents the meeting place of market relationships and social regulation forms, which determine different forms of production organization and different innovative capacities (regarding both products and processes) leading to a diversification of products presented to the market not simply on the basis of relative cost of factors.

The relationship between business systems and environment is then highlighted and constitutes the basis for the frequent existence of **external economies to the firms, but internal to the area.** These external economies make the adoption of different productive techniques possible together with organization processes and strategies **different** even in areas where the same goods are produced independently of the simple relationships between the prices of 'factors' or those of the usable inputs (Garofoli, 1984c).

Local specificities are mainly based on **localized networks among firms** and also on the specific relationships between economic structure and the environment and the local 'milieu'. All this leads to the following points:

- the existence of different patterns of local development;
- the crucial role of local policies of development (cf., especially local solidarities and the forms of social regulation at local level).

An analysis of the typology of local development immediately throws into relief the differences between patterns of 'extraverted' or 'exogenous' development (dependent on decisions mainly taken by economic agents outside the local area) and patterns of 'self-centred' or 'endogenous' development (based on variables mainly controlled by agents inside the local area) (Garofoli, 1983d and 1984a). The key variables which are the basis of the two groups of development models are obviously very different; the specific nature of the model, the interaction between the economic structure and the environment, the ability to 'lead' the development process are clearly much higher in (or even exclusive to) the second group of patterns.

Local development is characterized by endogenous factors as well as by exogenous variables; processes of development mainly based on exogenous factors more than on endogenous ones do exist. But also among the endogenous patterns of development there is a wide differentiation according to the time scale of the industrialization process and to the organization of production.

The definition of local development used here is subsequently less restrictive than that proposed by Coffey and Polese (1984) who have identified local development with endogenous development, with a determinant and exclusive role of local agents. According to these authors local development consists of a unique pattern based on a sequential set of the following four stages of development (Coffey and Polese, 1984):

a) the emergence of local entrepreneurship;
b) the expansion of local firms beyond the region (both with exports and with direct investments in other regions);
c) the maintenance of local enterprises under local control;
d) the achievement of an economic structure with control functions and business services under local control.

Therefore, in their view, there is a predeterminate and obliged path of development that assumes the nature of a normative model.

The continuous interaction, in the process of development of a region, between endogenous factors and exogenous ones, combined with the effects on the territory of economic policies, entails such wide changes in the socio-economic structure at a local level that it could not be interpreted only as different stages of a unique process of development.

Therefore, there is a plurality of paths of development and local models and prefixed stages of development do not exist. It is possible, moreover, to identify at least a number of groups of local models of development with some variants (both according to the time scale of the development process and to structural characteristics), especially concerning the position of the area in the territorial division of labour[2].

It seems quite important to reflect in more detail on endogenous models of development also because, as has been stated before, in these cases there is a higher possibility to control the process of development locally and improve opportunities for local economic policies.

Moreover, the literature on endogenous development has been so extensive in the last few years, especially in Southern European countries, not only for the necessity to understand the new patterns of regional development, the presence of significant local growing clusters of small firms and the phenomenon of recent industrialization in rural areas[3], but also for the importance to reflect on development policies carried out at a local level in favour of transformation and continuing innovation in contemporary cases of endogenous development.

A model of endogenous development capable of guaranteeing the autonomy of the process of transformation of the local economic system and which is relatively self-sustaining must be based on local characteristics and on the ability to control certain fundamental variables. In particular, it would seem to be based on:

1) use of local resources (e.g., work, local historically accumulated capital, entrepreneurship, specific knowledge of production processes, specific professional tasks and material resources;
2) the ability to check the accumulation process locally;
3) the capacity to innovate;
4) the existence of (and the ability to develop) productive inter-dependences, both intra-sectoral and intersectoral, at a local level.

Endogenous development does not mean 'closed economy', even if many statements found in some contributions to this topic underline the importance of concentrating on the expansion of the internal market (regional and interregional) which should be covered by local production, solving the specific problems of the area and satisfying the basic needs of local communities (cf., e.g., Friedmann and Weaver, 1979; Stöhr, 1984) by choosing strategies of selective economic protection and 'territorial closure', which would allow the continuous expansion of the use of local resources (especially work) as opposed to the fetishism of economic growth and efficiency (Friedmann and Weaver, 1979, p. 195).

Endogenous development means in effect: a) the capacity to transform the social-economic system; b) the ability to react to external challenges; c) the promotion of social learning; d) the ability to introduce specific forms of social regulation at a local level which favour the aforementioned points. Endogenous development is, in other words, **the ability to innovate** at a local level.

3. The Structural Characteristics of Small Firms' Systems

Among the endogenous local patterns of development we shall deal subsequently with the local systems of small firms which represent the most interesting cases, especially since they represent the 'sublimation' of the model of flexible specialization.

As has been stated before, the relationship between the economy and the economic milieu and, in the context of strictly economic relations, the relationship between firms are determinant elements of the endogenous patterns of local development.

If we limit the analysis to economic relations[4] it seems possible to identify the following structural characteristics of industrial districts or local productive systems (Garofoli, 1981 and 1983a):

1) A high level of division of labour between the firms in the local productive system which gives rise to very close input-output relations, both infrasectoral and intersectoral.
2) A remarkable specialization of production at the level of firm and plant which limits the field of activity, stimulates the accumulation of specialized knowledge, facilitates the introduction of new technologies and increases -through the rise of labour productivity- the economic autonomy of firms and sub-systems in the area.
3) The large number of local agents (the 'plurality of the protagonists') which leads to diffuse adoption of 'trial and error' behaviour; this implies a greater likelihood of finding satisfactory solutions to the problems of the economic system, at least by some operators, followed by immediate imitation by other economic operators.
4) The existence of an efficient system of transmitting information at a local level which guarantees the rapid and efficient circulation of information about:
    a. outlet markets;
    b. alternative technologies;
    c. new raw materials;
    d. components and intermediate products which can be used in the production cycle;
    e. new commercial and financial techniques which can be used.

   This helps to transform the knowledge of each individual economic operator into the 'common economic heritage' of the area.
5) The high level of skilled workers in the area, as a result of a historical sedimentation of knowledge of the production cycle and of the technologies used.
6) The increase in 'face-to-face' relationships between economic operators (especially between suppliers and users of intermediate products and of services to businesses). This favours the diffusion (through the different divisions of the production cycle) of technological and organizational improvements which increase the total efficiency of the local system. The effectiveness of this mutual support and co-operative links among local firms reaches its highest level in the interface contacts between innovators and users of innovation, in local systems

where there also exists the production of machine tools for the typical local product.

When the local production system reaches a high level of development, of division of labour between firms and integration of production, the definition of the local system on the basis of the sector of specialization becomes more and more restrictive; the system gradually expands to different industries and sectors and creates the phenomenon which apparently or superficially could be considered as a consequence of a process of despecialization and diversification. In reality, however, it must be interpreted as a strengthening and deepening of the original productive system. Local input-output relations between firms belonging to different product sectors are more frequent; this gives rise to growing intersectoral integration within the local system (Garofoli, 1983c).

All this is the consequence of the gradual growth and development of industries closely linked to the original sector of specialization and directly stimulated by it. Often these new industries are technologically and economically more evolved (there is a higher value added content in production, higher labour productivity and more intensive use of capital in relation to labour). In addition, they have a greater capacity for technical innovation and a greater control of the market. The combined operation of these processes improves the position of the local productive system in the regional and international division of labour and progressively creates a distance between it and firms and areas which are purely imitative. Thus it is assured a greater ability to defend the local system against outside competition (firms in other areas, both at home and abroad).

3.1 A typology of local productive systems

A comparative analysis of the productive structure, of the characteristics of local social formations, of the productive trends and qualitative transformations of the small firms' systems shows the opportunity to utilize a typologic classification of the specialized production areas based mainly on small firms. It seems possible to propone this kind of typology (Garofoli, 1983a): a) areas of specialized production; b) local productive systems; c) system areas.

a) Areas of specialized production

This kind of area is characterized by a horizontal structure of the local system: there exists a large number of small and medium sized firms producing the same product or productive component. Then there are scarce relationships among local firms, with the exception of some capacity decentralization for the less innovative and more labour intensive phases of production, especially with home work and putting out system. This kind of agglomeration of firms seems to depend upon the following conditions:

- flows of information (about technologies and outlet markets);
- formation of specific tasks in the local labour market;
- sprouting of new firms with imitative strategies; new entrepreneurs come from previous experiences as dependent workers in the already existing local firms.

Very often this kind of local pattern of development has been formed through the decentralization of production and/or relocation processes of external firms. The following are the main conclusive remarks about this kind of development pattern:

- the prevalence of imitative strategies, at the level of firms, mainly based on production cost cutting;
- co-operation among local firms does not exist, because of the excessive burden of competition;
- usually, the perception of local identity does not exist;
- it is quite difficult to organize local economic policies.

b) Local productive systems

This kind of area is characterized by the presence of infrasectoral interrelations, which means the formation of a network among local firms with exchange of complementary items at local level.

Other characteristics of this kind of development pattern are:

- a strong territorial integration;
- a continuous formation of new subsectors, very specialized by product components and specific working phases;
- a high rate of new firms formation;
- transfer of technologies among sub-sectors (this implies an innovative fall out) and co-operative links among firms;
- continuous improvement of labour productivity, both at a firm level and at local system's level;
- formation of specific tasks for machinery maintenance and repair and for small adjustments to existing machinery.

The main conclusive remarks with reference to the strategies and policies that could be carried out are:

- the perception of local identity exists;
- there is a diffuse demand for local economic policies, both from local

firms and from trade unions;

- the pattern of development is an extensive type, which means that there is a continuous increase both in the number of firms and in industrial employment;
- there is great opportunity to organize local economic policies, because social bodies and local institutions have realized the existence of specific characteristics and problems to be protected and faced.

c) System areas

This is the most advanced kind of area among small firms' systems; the main characteristic is the presence of a very strict division of labour among local firms which progressively diversifies the local systems, inducing an increasing complexity: the interrelations among firms are very numerous and they operate both at intersectoral level and intrasectoral level.

The complexity of the area is often so high as to inglobate local production of machine tools for typical local products; this means control of the productive techniques and then a technological autonomy.

Other characteristics of this kind of area are:

- a continuous formation of new industries and sub-sectors, which are all integrated in a local productive system;
- the existence of innovative fall out due to local firms belonging to different industries and sub-sectors;
- the existence of local solidarities and local institutions, historically sedimented in the area;
- local economic policies have been and are usually introduced and implemented;
- business services have been usually introduced at local level (however, often to an insufficient degree).

Furthermore in this kind of area we find:

- a capacity to control the 'technical frontier';
- an intensive pattern of development, which means industrial employment does not increase anymore;
- a 'self-centred' pattern of development: the opportunity exists to control (from inside the area and from local agents) the transformation of the local production system and its position in the territorial division of labour, choosing the path of innovation and of introduction of high quality and differentiated new products.

4. The Strategic Variables for the Strengthening of Local Productive Systems

As the productive system becomes more complex and makes external economies increasingly important to firms[5], when the productive system has made the most of specific local factors, the process of development has acquired all the characteristics of a 'self-centred' pattern of development. The process of development becomes completely endogenous when the best part of the resources used are local (local entrepreneurs, highly-skilled workers with local training, financial resources accumulated locally, technological innovation introduced locally)[6]; at this point, the local system[7] may be capable of 'controlling' its own process of development and transformation even if this obviously does not mean that the survival of the local system is assured[8].

The conditions for survival change constantly: the industrial district is a pattern of continuous change and not of mere adaptation, both in the interrelationships within the area (relationships among firms, inter-relationships with the economic milieu and with the institutions) and those with the outside forces (the market, the other competing areas, the other territorial systems). All this has obvious consequences on the placing of the local system in the territorial division of labour. Change and innovation are therefore the conditions for the survival of the local system. This being the case, it is necessary to transform the local system, which entails an ability to understand the relative position of the local system and to predict the scenario in the medium-to-long term. It is difficult for individual small firms to have this ability, thus it becomes evident that decisions need to be taken at the consortia level[9], and that suitable local economic policies are called for, in order to take more advantage of the relations between the local institutions and the local economy.

The fundamental condition for the strengthening of local productive systems is the achievement of an increasing 'systemic' structure which reinforces the economic links between firms and the relationship with the economic milieu, so that local specific characteristics become the most important factor in location and development.

The crucial variables for the strengthening of the local system are both endogenous (completely controlled within the area) and exogenous.

Among the variables which could be controlled within the local system, we can number:

a. Technical-organizational innovation which increasingly takes on (in small firm systems) the characteristics of a continuous process (a large number of technological changes, all of them small, cumulative and interdependent); and then an incremental process of innovation (à la Rosenberg) rather than one of giant leaps (à la Schumpeter); technological innovation in local systems is not only the result of a process of territorial diffusion but very often it is itself the product of the local system with original introduction of techniques, especially when there is the production of machine tools for the typical local product. The shifting of the technological frontier is therefore crucial for the consolidation and survival of the local system.

b. The information system: a rapid and efficient circulation/distribution of information, an opportune knowledge of the markets, the diffusion of information on technology and on inputs, are in fact the basis of a correct assessment of the prospects of development for local enterprises.

c. The ability to control the market; the strengthening of the commercial capacity of the system of local firms is one of the crucial factors for the autonomy of industrial districts and at the same time a continuous opportunity for the introduction of new products.

d. The forms of social regulation, which operate outside the market and depend upon a useful integration between the institutions and the local economy (cf., Sabel, Zeitlin, 1982; Zeitlin, 1985). It is enough to think of the crucial role, for the development of firms and the local system, which can be performed by quality control systems, technological centres, business service centres established on the basis of agreements between local public bodies and consortia of firms; centres for professional training and institutions which coordinate the worlds of school and work; institutions which regulate competition between enterprises, favouring competition in new products and processes rather than competition based on costs cutting and, indirectly, labour costs cutting, that means without brutal price wars.

But when we think of the crucial variables in the success of local productive systems or industrial districts we must remember the exogenous ones; let us think, for example, of the crisis in the markets for standardized, mass-produced products (Sabel, 1983), which has favoured the success of flexible specialization, and of the effects of new technologies better suited to small sized firms (micro-electronics, informatics and numerical control machines) (Friedrichs, 1982; Sabel, 1983)[10].

5. The Policy Implications

Weak points also exist within small firms' systems -and so moreover in other local patterns of development- and the historical importance of the forms of social regulations at the local level claim for local policies supporting of both innovation and a continuous process of transformation.

According to the wide articulation of local patterns of development it is necessary to take into account local specificities, problems and the perspectives of different areas so as to have territorially differentiated local economic policies.

This means taking into account the relationships between economy, society and local institutions that are at the basis of the forms of social regulation and represent the 'trait d'union' between the state and the market, between the rules of competition and economic efficiency and the necessity of public intervention in the economy and in society.

Local development must be fullfilled especially by making the most of local resources -both material and human- through the improvement of their productivity and through a process of consolidation and strengthening of the existing productive structure, especially facilitating the access to innovation, information and strategic business services.

The main objectives for policy intervention are:

- to improve **the external economies** of the local system, strengthening the **networks among local firms** (cf., co-operative links among local firms vs. competitive relationships);
- to improve the productivity of local resources;
- to induce technological and organizational innovation and facilitate the transfer of techno-scientific knowledge into concrete adaptations of local production;
- to improve the relationship with the outlet market;
- to induce the diversification of the local economy, mainly the formation and development of business services (with the objective of improving the ratio between value added and gross production at a local level);
- to improve the economic, technical knowledge and social learning;
- to promote local solidarities and interfirm co-operation;
- to improve local capacity to solve local problems (i.e., reinforce the capability to introduce local planning projects) (cf., Stöhr, 1984).

Local policies must be based, firstly, on a good knowledge of local economic and social structure and their perspectives (diagnostic capability), then on a co-ordination capacity between local agents (both public and private, including both social bodies and local institutions); and then it is a challenge to the planning capability at the local level.

I shall deal here with three main instruments of economic intervention at a local level, which seem to me particularly strategic for sustaining local development, especially in the cases of endogenous development:

- local development agencies;
- technological centres and technological diffusion agencies;
- business services centres.

5.1 The local development agency

A typical characteristic of a large number of local patterns of development -and especially of the small firms' systems- is represented by the wide 'hiatus' between the production capacity of the local system and the strategic capability to reconvert the local economic structure: the typical advantages of local small firms (i.e., the productive flexibility) enter into collision with the outlooks capacity in the middle-long run.

For this the role of corporate initiatives, dealing with common problems of the local firms, is important. In this sense some economic and social operators seem to be very important; they could be defined as 'pivots' of the local economy, being able to catalyze the community's interests and to identify innovative strategies for the entire local productive system.

This kind of experience is very well-known in some European countries; it is sufficient to think of the Expansion Committees and Development Agencies in France (where the decentralization laws of 1982 and 1983 granted more autonomy to the local councils, legitimizing some of their economic activities and with the formation of the so-called 'Contrats de Plan' and 'Contrats de Pays') (cf., Mifsud, 1984; Pecqueur, 1986 and 1987; Courlet-Pecqueur-Rousier, 1987); the Industrial Development Officers and the Regional Development Agencies in Great Britain; the Sagazde in the Basque country in Spain (cf., also, at local level the initiatives of the Industrial Promotion Department in Lebrija (Andalusia), and of the 'industrialde', a sort of industrial park, in Oñate (Basque country), both studied by V.A., 1987b, and A. Vasquez Barquero, 1988).

Local development Agencies are usually very flexible structures that function as promotion and implementation of economic activities at a local level, with a **role of interface** between public and private actors and between the demand and the supply of business services, especially promoting the circulation of information. The principal aim is to stimulate the community's participation in local development projects, through the creation of new associations, to steer social and economic actors towards a global solution of the community's problems, to generate understanding and agreement capability among local actors and to develop new forms of solidarity (cf., for example, the development of contacts between science laboratories and local entrepreneurs as in the Savoy Technolac operation and the creation of discussion groups with the main local economic actors, both public and private, in various Economic Expansion Committees in France, studied by Courlet-Pecqueur-Rousier, 1987).

The main objective of the local development Agencies, during the 1980s, concerns the turning into account the local resources, through the management and the improvement of the existing economic system rather than attracting external firms, with the supply of infrastructures and industrial estates as happened in the previous decades. This means that the principal results of the initiatives of local development Agencies, in the various European countries, have concerned the following spheres:

- the capability to diagnose the local economic situation and to evaluate its potentialities;
- promoting the local industrial environment;
- reinforcing the links among local firms, developing relationships between them on the basis of mutual interests (for example facilitating the sub-contracting linkages);
- promotion of the scientific and technological environment;
- establishing links with the outside world, balancing internal and external dynamics, with outward-oriented activities (seeking new outlet markets, developing exchange programmes, co-operation and partnerships with external firms, in other regions and countries);
- mobilizing internal savings towards local industrial firms, which could then make use of extra local funds.

5.2 The technological centres

The local policies aimed at promoting innovation must be very articulated and territorially differentiated according to the differences both in local economic structure and in objectives. It is possible to think of at least four kinds of local innovation policies:

a) Technological parks in urban areas with a developed industrial fabric and with the presence of techno-scientific infrastructures (research laboratories, university, ...), with the aim of facilitating the adaptations of the research results into industrial production, allowing the introduction of new products and new processes, stimulating the creation of new firms in innovative sectors (cf., for example, the case of Cambridge Science Park)[11].

b) Technological centres in industrial districts and system areas with the aim of horizontal transfer of new technologies to sectors that have not yet used them and, mainly, to facilitate the shift of the **technological frontier** in the areas in which machine tools for local typical products are produced. The settlement of co-operative initiatives and technical centres and consortia among firms is necessary to use a sufficient amount of technical and financial resources to obtain a **technological jump** which could not be pursued by an individual small firm producing machine tools; all this will increase the productivity of the final products' sector and will improve the capability of exports of innovative technology[12].

c) In areas of diffused industrialization with scarce networks among firms the use of horizontal means of intervention would be convenient, promoting the diffusion of technological innovations and new processes already used in different industries or regions.

d) In the less industrialized areas it is important mainly to promote both the formation of new firms, through policies of professional schemes and managerial capacity formation, and the strenghtening of existing firms through policies which could promote the demand of innovation and business services; furthermore it is important to promote the circulation of information and channels of communications with the experiences of other areas and regions.

5.3 Business services centres

There are a number of common objectives to different areas and typologies of local productive systems in relation to a good diffusion of business services, namely:

- explicitating latent demand of business services by local small firms;
- taking information about the effective demand of business services to understand the inadequacy and shortage of local supply and facilitate the creation of new local firms in business services; this is very important when we think about the significance of 'cultural coherence' between seller and user of business services;
- promoting the capability of projecting business services agencies at a local level.

The experiences of the business services centres are very varied: there are cases in which a direct production of business services is offered and others which sustain the services demand (as in the case of Rkw -Rationalisierungs Kuratorium der deutschen Wirtschaft- in West Germany). The management pattern of business services centres organized on the role of **interface** between their demand and the supply seems to be better. The main condition for efficiency in the management of these centres is the participation of local firms, both in the financing and in the management and control. This is an effective condition to realize the production of useful services utilized by local firms and to fulfill the goal of self-financing of the centre (cf., in this respect, the case of Citer in the knitting area of Carpi in Italy).

The public support of this kind of initiatives and the support by the category associations are necessary, nevertheless, in the first stages of the organization of the centre owing to the **lack** of the business services market at a local level (cf., Brusco, 1984).

Notes

1. The concepts of 'self-centred' and 'extraverted' development were first introduced by Samir Amin (1973) and then adopted in regional analysis by Alain Lipietz (1980) and Garofoli (1983d). Self-centred development represents a process of endogenous development in which the process of accumulation and the main decisions on the use of surplus are internally controlled in national or regional economy in opposition to a process of extraverted development that is exogenously controlled.

2. Cf., for a discussion and an identification of different patterns of local development, Garofoli (1988).

3. The literature on the Italian situation is very extensive: Bagnasco (1977), Bagnasco-Pini (1981), Becattini-Bianchi (1982), Brusco (1982 and 1986), Fuà (1983), Garofoli (1981, 1983a, 1983d, 1984a and 1987). With regard to the Spanish situation there have been numerous researches during this decade as well; cf., for example, Vasquez-Barquero (1983, 1987 and 1988), Granados (1984), Guinjoan-Santacana (1985), V.A. (1987a); in this last work, moreover, 83 areas of endogenous development in Spain have been identified.
Concerning the French situation cf. Houssel (1984), Courlet (1988), Courlet-Judet (1986). With regard to the Portuguese situation cf., Lewis-Williams (1984), Silva (1987), Ferrao-Mendes Baptista (1988). Finally, regarding the Greek situation cf., Kafkalas (1984) and Hadjimichalis-Vaiou (1987).

4. In this paper we will not deal with the typical features of peripheral social formation which are the basis of the flexibility of production of small firm systems. Cf., on this subject, Bagnasco (1977) and Bagnasco, Pini (1981).

5. In these cases the economies of agglomeration often prevail over the diseconomies of urbanization in such a way as to continue to favour localisation in an area of growing clusters of firms (even if there are rising location costs and inadequate levels of territorial infrastructure and transport) (Cf., Garofoli, 1983a, Chapter 4).

6. This obviously does not mean that the local system is closed to the outside world; one of the specific features of these systems is the

great and growing openness to the international market; the efficiency of the local system is assured precisely by its presence on the international market. Cf., for a different definition of endogenous local development, Coffey and Polese (1984).

7. The local system, here, is for the local social bodies and local institutions.

8. Note, in this situation, the often irreversible crisis experienced by many industrial districts abroad (cf., Garofoli, 1983b).

9. This means with extensive co-operation among competitors either directly or indirectly, through joint participation in the institutions regulating the local economy (cf., Sabel, 1987).

10. We should bear in mind the growing importance of the horizontal diffusion of innovations between sectors which also paves the way for the introduction of innovations outside the enterprises and sectors which invest in R & D (Momigliano, 1982 and 1984); all this has largely improved the capacity for innovation of small firms.

11. Cf., for an analysis of Cambridge Science Park and other UK Science Parks, Monck et al., 1988.

12. Cf., for an analysis of the technological centres in the small firms' systems in Italy, Garofoli (1984b), Bianchi et al., (1986), Colnaghi (1987).

References

Amin, S. (1973), Le développement inégal, Les Editions de Minuit, Paris.

Bagnasco, A. (1977), Tre Italie. La problematica territoriale dello sviluppo economico italiano, Il Mulino, Bologna.

Bagnasco, A., Pini, R. (1981), Sviluppo economico e trasformazioni socio-politiche dei sistemi territoriali a economica diffusa. Economia e struttura sociale, Quaderni della Fondazione G. Feltrinelli, no. 14, Milan.

Becattini, G. (1979), Dal 'settore' industriale al 'distretto' industriale. Alcune considerazioni sull'unità di indagine dell'economia industriale, Rivista di Economia e Politica Industriale, V., no. 1, Jan-Apr.

Becattini, G. (1981), Introduction to Marshall, Antologia di scritti economici, Il Mulino, Bologna.

Becattini, G. (1984a), Ispessimenti localizzati di esternalità, Banca Toscana - Studi e informazioni, no. 1.

Becattini, G. (1984b), L'economista e l'ambiente, Giornale degli Economisti e Annali di Economia, XLIII, no. 3-4, March-Apr.

Becattini, G. (1987), Il distretto industriale marshalliano: cronaca di un ritrovamento, Introduction to Becattini, G. (ed.) Mercato e forze locali: il distretto industriale, Il Mulino, Bologna.

Becattini, G., Bianchi, G. (1982), Sulla multiregionalità dello sviluppo economico italiano, Note Economiche, no. 5-6.

Berry, B.J.L. (1972), Hierarchical Diffusion: the Basis of Development Filtering and Spread in a System of Growth Centres, in: Hansen, N.M. (ed.), Growth Centers in Regional Economic Development, The Free Press, New York.

Berry, B.J.L. (1973), Growth Centers in the American Urban System, Ballinger, Cambridge (Mass.).

Bianchi, P., Bellini, N., Giordani, M., and Pasquini, F. (1986), Servizi reali e politica industriale a livello locale, Stato e Mercato, no. 16, April.

Brusco, S. (1982), The Emilian model: productive decentralisation and social integration, Cambridge Journal of Economics, vol. 6, no. 2, June.

Brusco, S. (1984), Quale politica industriale per i distretti industriali?, Politica ed Economia, XV, no. 6, June.

Brusco, S. (1986), Small firms and industrial districts: the experience of Italy, in: Keeble, D. and Wever, E. (eds.), New Firms and Regional Development in Europe, Croom Helm, London.

Coffey, W.J. and Polese, M. (1984), The concept of local development: a stages model of endogenous regional growth, Papers of the Regional Science Association, no. 54.

Coffey, W.J. and Polese, M. (1985), Local development: conceptual bases and policy implications, Regional Studies, vol. 19, no. 2.

CCE (1981), The mobilization of indigenous potential, internal documentation of regional policy in the Community, Bruxelles, no. 10, Sept.

Colnaghi, A. (1987), Innovazione tecnologica e sistemi produttivi locali, Economia Marche, VI, no. 2, Dec.

Courlet, C. (1988), Nouveaux dynamismes spatiaux et politiques industrielles locales: l'example français, paper presented at the workshop on 'Depressed Regions in the Mediterranean European Countries and Endogenous Development, Scilla (Reggio Calabria), June.

Courlet, C. and Judet, P. (1986), Nouveaux espaces de production en France et en Italie, Les Annales de la recherche urbaine, no. 29, janvier.

Courlet, C., Pecqueur, B. and Rousier, N. (1987), Local Industrial Policies in favour of Small and Medium-Sized Firms: the French case, paper presented at the International Symposium Nouveaux Regards sur l'Industrialisation, Tunis, 14-17 déc.

Damette, F. (1980), The regional framework of monopoly exploitation: new problems and trends, in: Carney, J., Hudson, R. and Lewis, J. (eds.) Regions in crisis, Croom Helm, London.

Ferrao, J. and Mendes Baptista, A. (1988), Endogenous Development in Portugal: Problems and Perspectives, paper presented at the workshop on Depressed Regions in the Mediterranean European Countries and Endogenous Development, Scilla (Reggio Calabria), June.

Friedmann, J. and Douglass, M. (1975), Agropolitan development: towards a new strategy for regional development in Asia, in United Nations Centre for Regional Development, pp. 333-387.

Friedmann, J. and Weaver, C. (1979), Territory and Function: The Evolution of Regional Planning, Edward Arnold, London.

Fuà, G. (1977), Sviluppo ritardato e dualismo, Moneta e Credito, vol. XXX, no. 120, dec.

Fuà, G. (1980), Problemi dello sviluppo tardivo in Europa, Il Mulino, Bologna.

Fuà, G. (1983), L'industrializzazione nel Nord-Est e nel Centro, in: Fuà, G. and Zacchia, C. (eds.), Industrializzazione senza fratture, Il Mulino, Bologna.

Garofoli, G. (1981), Lo sviluppo delle 'aree periferiche' nell'economia italiana degli anni settanta, L'Industria, II, no. 3, July-Sept.

Garofoli, G. (1983a), Industrializzazione diffusa in Lombardia, I.R.e.R., Franco Angeli, Milan.

Garofoli, G. (1983b), Aree di specializzazione produttiva e piccole imprese in Europa, Economia Marche, vol. 2, no. 1, June.

Garofoli, G. (1983c), Le aree-sistema in Italia, Politica ed Economia, XIV, no. 11, November.

Garofoli, G. (1983d), Sviluppo regionale e ristrutturazione industriale: il modello italiano degli anni '70, Rassegna Economica, XLVII, no. 6, Nov-Dec.

Garofoli, G. (1984a), Diffuse industrialization and small firms: the Italian pattern in the '70s, in: Hudson, R. (ed.), Small firms and regional development, Institute for Transport, Tourism and Regional Economy, Copenhagen School of Economics and Business Administration, publication no. 39, Copenhagen.

Garofoli, G. (1984b), Barriere all'innovazione e politiche di intervento a livello regionale e sub-regionale, in: Camagni, R., Cappellin, R. and Garofoli, G. (eds.), Cambiamento tecnologico e diffusione territoriale, Franco Angeli, Milan.

Garofoli, G. (1984c), Modelli locali di sviluppo, paper presented at the 25th Scientific Meeting of the Società Italiana degli Economisti, Rome, Nov.

Garofoli, G. (1987), Il modello territoriale di sviluppo degli anni '70-'80, Note Economiche, no. 1.

Garofoli, G. (1988), Modelli locali di sviluppo: tipologia di aree e politiche di intervento, CNR, Progetto Finalizzato 'Economia Italiana', Sottoprogetto IV, Quaderno 7, Pavia, July.

Granados, V. (1984), Small Firms and Rural Industrialization in Spain: Some Results from an OECD Project, in: Hudson, R. (ed.), Small Firms and Regional Development, Institute for Transport, Tourism and Regional Economy, Copenhagen.

Guinjoan, M. and Santacana, F. (1985), PME y desarrollo endogeno. Las areas de industrializacion endogena en España, II encuentro de economistas italo-españoles, Politica economica y territorio, Girona, October.

Hadjimichalis, C. and Vajou, D. (1987), Changing Patterns of Uneven Regional Development and Forms of Social Reproduction in Greece, Environment and Planning D: Society and Space, vol. 5, pp. 319-333.

Houssel, J.P. (1984), L'industrialisation spontanée face à la crise de 1973 en Europe Occidentale, Revue Géographique de Lyon, vol. 59, no. 4.

Kafkalas, G. (1984), Small Firms and the Development of a Peripheral Region: the Case of Thraki, Greece, in: Hudson, R. (ed.), Small Firms and Regional Development, Institute for Transport, Tourism and Regional Economy, Copenhagen.

Lewis, J.R. and Williams, A.M. (1984), The formation and role of small and medium size industrial enterprises in the regiao centro, Portugal, paper presented to the 24th European Congress of Regional Science Association, Milan, August.

Lipietz, A. (1980), The Structuration of Space, the Problem of Land and Spatial Policy, in: Carney, J., Hudson, R. and Lewis, J. (eds.), Regions in Crisis, Croom Helm, London.

Mifsud, P. (1984), National and Regional Planning in a Competitive International Environment: Comments on a Reform, Papers of the Regional Science Association, vol. 55, pp. 13-24.

Momigliano, F. (1982), Determinanti ed effetti dell'attività innovativa: revisione di teorie e implicazioni di politiche pubbliche per l'innovazione industriale, Economia e Politica Industriale, no. 35.

Momigliano, F. (1984), Revisione di modelli interpretativi delle determinanti ed effetti dell'attività innovativa, della aggregazione spaziale dei centri di R & S e della diffusione intraindustriale e territoriale delle innovazioni tecnologiche, in: Camagni, R., Cappellin, R. and Garofoli, G. (eds.), Cambiamento tecnologico e diffusione territoriale, Franco Angeli, Milan.

Monck, C.S.P., Porter, R.B., Quintas, P.R., Storey, D.J. and Wynarczyk, P. (1988), Science Parks and the Growth of High Technology Firms, Croom Helm, London.

Pecqueur, B. (1986), Quelles politiques industrielles locales?, mimeo, IREP-D, Université des Sciences Sociales de Grenoble, Grenoble, June.

Pecqueur, B. (1987), De l'espace fonctionnel à l'espace-territoire. Essai sur le développement local, Thèse de Doctorat d'Etat és Sciences Economiques, Université des Sciences Sociales de Grenoble, Grenoble.

Piore, M. and Sabel, C. (1984), The Second Industrial Divide, Basic Books, New York.

Rosenberg, N. (1982), Inside the Black Box: Technology and Economics, Cambridge University Press, Cambridge.

Sabel, C. (1982), Work and Politics, Cambridge University Press, Cambridge.

Sabel, C. (1983), La new Italy della piccola impresa, Politica ed Economia, XIV, no. 7-8, July-August.

Sabel, C. (1987), The Reemergence of Regional Economies, paper presented at the International Symposium Nouveaux Regards sur l'Industrialisation, Tunis, December.

Sabel, C. and Zeitlin, J. (1982), Alternative storiche alla produzione di massa, Stato e mercato, no. 5, August.

Silva, M.R. (1987), Développement industrial et éspaces productifs locaux: étude sur la Vallée de l'Ave (Portugal), paper presented at the International Symposium Nouveaux Regards sur l'Industrialisation, Tunis, December.

Stöhr, W.B. (1978), Center-Down-and-Outward Development versus Periphery-up-and-Inward Development: A Comparison of Two Paradigms, I.I.V.R.S., University of Economics, Vienna.

Stöhr, W.B. (1981), Development from below: the Bottom-up and Periphery-Inward Development Paradigm, in: Stöhr, W.B. and Taylor, D.R.F. (eds.), Development from above or below? The Dialectics of Regional Planning in Developing Countries, J. Wiley & Sons Ltd., Chichester.

Stöhr, W.B. (1984), La crise économique demande-t-elle de nouvelles stratégies de développement régional?, in: Aydalot, Ph. (ed.), Crise & espace, Economica, Paris.

Thompson, W. (1968), Internal and External Factors in the Development of Urban Economies, in: Perloff, H.S. and Wingo, L. (eds.), Issues in Urban Economics, Johns Hopkins Press, Washington.

V.A. (1987a), Areas rurales con capacidad de desarrollo endogeno, MOPU (Ministerio de Obras Públicas y Urbanismo), Instituto del Territorio y Urbanismo, Madrid.

V.A. (1987b), Proceso de formulación de las politicas de desarrollo local. La experiencia española, MOPU (Ministerio de Obras Públicas y Urbanismo), Instituto del Territorio y Urbanismo, Madrid.

Vasquez Barquero, A. (1983), Industrialization in Rural Areas: The Spanish Case, OECD, Intergovernmental Meeting, Senigallia, June.

Vasquez-Barquero, A. (1987), Local Development and Regional State in Spain, Papers of the Regional Science Association, vol. 61.

Vasquez-Barquero, A. (1988), Local Development Initiatives under Incipient Regional Autonomy: the Spanish Experience in the eighties, paper presented at the workshop Depressed Regions in the Mediterranean European Countries and Endogenous Development, Scilla (Reggio Calabria), June.

Zeitlin, J. (1985), Distretti industriali e struttura industriale in prospettiva storica, in: Innocenti, R. (ed.), Piccola città & piccola impresa, Franco Angeli, Milan.

17 Technology transfer and regional policy

R. PREMUS

1. Introduction

Regions with government laboratories have a potentially fertile source of technology to stimulate their economies. Recent changes in federal law have lowered barriers to laboratory technology transfer to the private sector. In spite of these new opportunities, the US federal laboratories system remains largely an untapped source of new technology for national and regional economic development.

This paper will explore the potential of technology transfer as a mechanism of regional economic development. The main thesis of the paper is that a greater co-mingling of the technology systems of government and private industry may be necessary to fully integrate technology into regional policy. Regional policy has a role in developing institutional mechanisms to identify promising laboratory technologies and help area businesses acquire and apply the technology.

The paper begins by examining recent federal policies to encourage technology transfer. A number of topics in the innovation literature are also discussed in terms of their implications for regional technology transfer policy.

Next, a step-by-step analysis of the process of moving technology from the government laboratory into the private sector is discussed. A number of intervention points for regional policy to facilitate the process are identified. The paper then examines some survey evidence on the technology transfer process in a regional context. The Dayton area was chosen for the survey because of the presence of seven major federal laboratories in the region.

A discussion of regional policy implications concludes the paper. A major conclusion of the paper is that establishing effective linkage mechanisms between technology-push and technology-pull programs may be necessary to more successfully commercialize government technology.

2. Legislative and Literature Review

Concern over US competitiveness has prompted the administration and the

Congress to take a number of legislative initiatives to forge closer links between the nation's research establishment and US industry.[1] One response was to increase US industry's access to federally-funded technology. Also, scholarly research has raised a number of issues in the innovation literature of direct importance to regional technology transfer policy. This section addresses both of these concerns.

Legislative initiatives

The Stevenson-Wydler Act of 1980 took the first step to improve the technology transfer process by establishing an Office of Research and Technology Applications (ORTA) in government owned and operated laboratories. The primary mission of ORTA is to disseminate information on government technologies to industry and local governments.

A more recent step was taken with the passage of the National Technology Transfer Act of 1986.[2] This act expanded the authority of the Stevenson-Wydler Act by giving lab directors greater flexibility and discretionary authority to negotiate licensing agreements and enter into collaborative research efforts with industry and universities. One of the purposes of the paper is to access the regional policy implications of this unique opportunity to stimulate regional growth through the application of government technology.

Other provisions of the 1986 Act are also important to discussions of regional technology transfer policy. Laboratories are required to include technology transfer into their mission statements. Employee job descriptions, promotion policies, and evaluation criteria must also include reference to technology transfer activities. These mandatory provisions provide incentives and rewards for technology transfer activities.

Financial incentives are also provided in the 1986 Act. Innovators are to be given up to 15 percent of the royalties (not to exceed $100,000 annually) from the commercialisation of their research. The balance of royalties is to be retained for use by lab management.

Other complementary legislative actions also help to promote technology transfer. The Patent and Trademark Amendments of 1980 allow the government to transfer patent rights from federally-funded research to universities and the private sector, reserving for the government an unrestricted royalty-free license to use the technology for its purposes.[3] Antitrust policies were also amended to make it easier for private firms to enter into collaborative research agreements with less fear of antitrust litigation.

Collectively, these legislative changes have altered the "rules of the game" for generating greater interest in technology transfer. They primarily represent technology-push stimuli to the commercialization of government laboratory research. A major question in this chapter is whether or not they are sufficient to stimulate the technology transfer process within the regions.

Innovation issues

In this section, a number of issues in the innovation literature are discussed in terms of their theoretical and practical implications for regional technology transfer policy.

Technology-push vs. technology-pull. One important issue centers on the relative significance of technology-push versus technology-pull factors in industrial (an by implication regional) innovation. Schmookler, in a major study of innovation, stressed the importance of market demand in stimulating innovative activities.[4] He implicitly assumed the knowledge base of industry to be malleable or easily adaptable to new uses. Entrepreneurial talent was also assumed to be in ample supply to move the technology quickly in response to changes in market opportunities.

Rosenberg argues that Schmookler may have overstated his case for the dominance of technology-pull factors.[5] While patent activity did follow a surge in industrial output in the capital goods sector, as Schmookler maintained, Rosenberg found that increased patent activity also preceded important advances in transistor and circuit board technology. The main point of the argument is that growth in market opportunities could not have lead to a surge in patent activity had not advances in technology been made.

Scherer reaches a similar conclusion by extending Schmookler's model to include all manufacturing industries and materials innovations.[6] Like Rosenberg, he found advances in technological frontiers to be a significant stimulus to innovation.

Nelson explained the link between knowledge and technological innovation in terms of search theory.[7] His model views R&D as a search process whereby search costs and time are lowered by advances in fundamental knowledge, or developmental research, underlying particular inventions. Thus, by reducing R&D costs, advances in knowledge can stimulate innovation.

To summarize, the major conclusion that emerges in the innovation literature is that both technology-pull and technology-push factors must be taken into account in explaining rates of industrial innovation. By implication, this finding also applies in a regional context. Route 128 and the Silicon Valley are reminders that a strong knowledge base is an important factor in the emergence of regional high tech complexes.

Middle ground technology. A closely allied topic is the importance of "middle ground technology' stressed by Nelson, Peck and Kalachek[8], and Eads and Nelson[9]. Middle ground technology is defined as an improved understanding of the fundamental knowledge specific to a particular technology. For example, a better understanding of DNA structure can be helpful in applying gene splicing technology to develop new biotechnology products. Both applied and a basic research are important to middle ground technology, but equally important is the ability to turn ideas into practical results.

It is the lack of emphasis on middle ground technology, or applied science, that Mansfield[10] and Baily and Chakrabarti[11] claim explains why the US lags Japan and Western Europe in productivity growth. For example, Mansfield finds that Japan devotes a large portion of its R&D resources to developmental research and process technologies. In contrast, the US stresses basic research and new product development.

Small versus large businesses. Gellman Research Associates[12] and Edwards and Gordon[13] provide convincing evidence that small businesses are major contributors to industrial innovation, particularly for novelty innovations. Yet, industrial labs are generally associated with large businesses.

The issue of the role of small business versus large business in innovation is far from settled. Nevertheless, a fair assessment is that both large and small firms have an important role to play in regional innovation policy. This suggests that a regional innovation policy ought to include the small business sector as well as large firms.

3. Technology Transfer Process

Technology transfer was described in Joint Economic Committee hearings as a long, laborious process beginning with the identification of technology in the government laboratory and ending when the technology is introduced into the marketplace.[14] There are several stages to the technology transfer process, each with its own associated risks and costs. Specifically, the adopting firm must acquire technical information, adapt it for internal uses, and commercialize the results.

Acquisition stage

An information barrier to technology transfer exists to the extent that potential users must incur costs to obtain the technology and information about its potential commercial applications. Acquisition costs involve search costs, license fees, royalty payments, and any transactions costs in technology transfer. Transactions costs include legal fees, negotiating costs, and learning costs. Also, national security issues are most likely to emerge, imposing additional costs on the technology transfer process. Other things equal, high acquisition and transaction costs are likely to deter "spin off" activity from government laboratory research.

An essential feature of technology transfer is that it involves the diffusion of technology across organizational boundaries.[15] As a result, information networks that cross boundaries of the particular technology systems are critical to the technology transfer process. Information networks establish the necessary linkages between the donor and the recipient of technology transfer.

A number of linkage mechanisms are available for transmitting technical information among technology systems. Technical briefs, engineering abstracts, seminars, and press releases are frequently used channels of technology information dissemination from federal laboratories. In the US, each major federal laboratory is required to prepare technical notes, or briefs, on their labs' research products to be submitted to the National Technical Information Service (NTIS) in Washington, DC. The NTIS maintains an active data base of technical information stored in federal laboratories across the United STates. The Center of Technology Utilization, within the NTIS, promotes the use of this information for commercial purposes.

Studies have shown that technical information systems, in and of themselves, are inadequate conduits of technology transfer. The reason can be found in the fundamental nature of applied research outputs. Applied research involves detailed engineering applications of the ideas of basic research (and other sources). The outcomes of applied research are technology products which embody scientific principles and a great deal of engineering and technical "know how" specific to the technology being developed. Much of the practical information is "person-embodied" in lab employees. Technical notes and published research are inadequate mechanisms for transmitting this "know how".

Another bridging mechanism would be to co-mingle lab engineers and scientists with industry engineers and scientists in applying the laboratory technology to a commercial setting, as permitted by the National Technology Transfer Act. This type of close interaction could be a significant factor in facilitating technology transfer, but it raises a number of legal and ethical concerns. Should government lab scientists be on loan or available as consultants to industry? Would the lure of profits divert their interest from their primary laboratory mission?

A proprietary issue emerges as well. Would industry trust government scientists to be privy to development of federal technology in their labs for commecial use? Experienced lab scientists and engineers would become aware of industry secrets and market opportunities for their research. Since they are likley to also deal with competitor firms, technical and trade secrets could be lost? Imitators will probably find these government lab employees to be productive consultants in their drive to invent around patents of their competitors. If industry perceives this type of third party arrangement to be a problem, they will have an incentive to shy away from federal laboratory technology.

To a large extent, companies have a similar problem of confidentiality with university consultants who want to publish the results of proprietary research. The main difference is that university professors will be largely involved in solving technical problems wich are linked to basic science. The detailed engineering knowledge and experience necessary to develop and apply the new technology will remain with the firm.

R&D stage

The dissemination of technical information is only the first step in the technology transfer process. Potential beneficiaries of this information must use the information to solve technical problems within their organizations. Technology transfer can be said to have occurred when the receiving organization utilizes the information to solve internal problems, develop new products, or improve production processes. For this to happen, the information received from the donor organisation generally must be modified and refined to meet the needs of the adopting organisation. Thus, the adoption of technology at this stage involves R&D expenditures.

If technology is fully malleable (i.e., it can be arranged in many different forms and uses at low cost), and information and transaction costs are near zero, R&D barriers to technology transfer will be largely nonexistent. Government research labs would be multiproduct operations with the technology products being applied jointly in the private and the government sectors. Under these ideal conditions, demand for commercial applications would drive the technology transfer process. On the other hand, if the technology is not perfectly malleable, and if information and transactions costs occur, technology transfer will require the recipient firm to engage in R&D and other expenditures prior to the initial introduction of the innovation.

To be profitable, the innovation must generate a future net revenue stream, which, when capitalized at the appropriate discount rate, allows for an adequate return on the innovative activity. Several factors are important in determining the decision to invest in R&D activities. First, the future stream of benefits is uncertain, creating risks that cannot be envisioned or adequately hedged. Second, a large share of the future benefits of the innovation cannot be captured by the innovator and thus will likely affect the decision to innovate. Third, capital market imperfections can act as a barrier to innovation for costly revolutionary innovations. The cost of capital differentials among countries (regions) is quite large reflecting complex factors such as saving rates, investment alternatives, and macro fiscal and monetary policies. These differentials will have a major impact on differences in regional rates of R&D investment in innovative activities.

Regional differences in innovation rates will also be determined, to some extent, by differences in the ability of regions to apply existing knowledge. Regions with an advantage in applying technology are likely to be regions that offer high-tech manufacturing capabilities, technically competent labor, and excellent applied science capabilities.

For the industry as a whole, market structures will also be an important determinant of the long-run profitability of R&D investments in innovation. Highly competitive market structures, which imply that imitation of the technology is easy, will quickly erode private returns to innovation. Monopoly markets, however, may offer little incentive to innovate in the first place. Scherer argues that an oligopolistic market structure with numerous smaller competitors is more conducive to innovation than either competition or monopoly.[16]

Commercialization stage

The commercialization stage of the innovation process begins with the initial introduction of the product and proceeds over the life cycle of the product. The innovation begins the technology transfer process but the diffusion of the new product throughout the economy is vital to the commercial success of the innovation.

Schumpeter views the innovation process in three stages: research, innovation, and technological diffusion.[17] Technological diffusion is a broader concept than commercialisation used in this paper. Commercialisation refers to the successful selling and leasing of technology products to users of the technology. Diffusion refers to the spread of technology to other firms and users within an industry whether or not the innovator receives a return on the imitation. Thus, commercialisation is only that part of the technology diffusion process that generates the private returns to the innovator.

Imitation can be vital to productivity growth of a region (nation) but it undermines the incentive for firms to undertake the necessary R&D expenditures to innovate. The presence of imitation drives a wedge between the private and social returns to innovation, thus creating a disincentive for firms to invest in technology transfer.

Another dilemma arises from the innovation-imitation process. R&D activities of the industrial laboratories produce technical advances for the firms to use, but the probability of distribution of successful outcomes for industrial laboratories is not known with certainty. In fact, the probability of success of any one single innovation is low, but the expected return for successful innovations is quite high. Thus, the industrial laboratory must achieve a sufficient return on its successful innovations to not only cover the cost of successful innovations but also the cost of all the other unsuccessful innovations and the research efforts that fail to achieve an innovation. Thus, successful innovations must be "big winners" in order to generate a positive return to all of the lab's research activities. This suggests that the successful innovation must generate "supernormal" profits for the industrial laboratory to justify its existence. Of course, the existence of "supernormal" profits makes imitation all the more profitable.

4. Survey Results

Our attention is now turned to the problem of attempting to understand more about the technology transfer process in a regional setting. The Dayton metropolitan area was chosen as a study site because of the large presence of federal government research laboratories in the region. Dayton houses seven government laboratories employing approximately eight thousand scientists and engineers. Six of the laboratories are Department of Defense laboratories located at Wright-Patterson Air Force Base. These laboratories are government owned and government operated. The other facility, Mound Laboratory, is a government-owned, contractor-operated facility of the Department of Energy. Combined, these laboratories produce a wide range of intermediate and advanced technology products that could be available for commercial use, as well as satisfying mission requirements. Thus, the Dayton region provides an excellent test site to determine the extent of technological "spin offs" from government research, the importance of technology-push and technology-pull factors in the technology transfer process, and any internal and external barriers to the process. Although the survey results are specific to the Dayton region, they should also be interesting to other regions attempting to stimulate technology transfer activities.

Two questionnaire surveys were conducted over the January-June 1987 period. One questionnaire was mailed to 497 government laboratory employees residing in the Dayton region. The objective of the laboratory personnel survey was to gather insider information on how government scientists and engineers view the technology transfer process. Another questionnaire was mailed to 315 executives of advanced technology businesses in the Dayton region. The business survey was conducted to gain outsider information about the technology transfer process from the perspective of potential external users of the information.

Results from within

As discussed previously, the intent of the Stevenson-Wydler and the National Technology Transfer Acts is to stimulate technology transfer by emphasizing technology-push factors. This section provides evidence that technology-push actions alone are not likely to be successful.

One hundred sixty usable responses were obtained from the laboratory personnel survey. Of the respondents, 79 were laboratory directors. The remaining 81 respondents were engineers and technicians employed in the laboratories. About 57 percent of the respondents hold graduate degrees in science and engineering. The remaining 43 percent hold undergraduate degrees in associated engineering and science disciplines. About 40 percent of the sample had 20 or more years experience in government laboratory work. Thus, the sample on which this section is based consists of highly skilled and experienced lab professionals within the federal laboratory system. Their perspectives on technology transfer issues should be particularly valuable in providing insights into the technology transfer process.

Internal barriers. Table 1 lists how the R&D professionals rated potential internal barriers to government researchers wishing to engage in technology transfer activities. The most significant barrier listed by 52 percent of the sample is lack of awareness of project engineers and scientists of their role in the technology transfer process. Most of the

Table 1

Lab Personnel Rating of Potential Within-Laboratory Barriers to Technology Transfer

Rating	Important Barriers	Response Percent
1	Project Scientists/Engineers Lack Awareness of the Role of Technology Transfer	52
2	Classified Nature of the Lab Research	45
3	Lack of Positive Incentives for Lab Employees	41
4	Too Much Red Tape for Lab Personnel	35
5	Lab Management Policies Unclear	33
6	Lack of Adequate Channels of Communication with Industry	26
7	Industry's Lack of Awareness of Lab Technology	23
	Minor Barriers	
8	Lab Management Awareness of the Value of Technology Transfer	19
9	Red Tape for Companies Wanting Access to Lab Technology	13
10	Lab Mission Statement	13
11	Lack of Business Interest in Lab Technology	10

Source: Michael Olson. A Study of Technology Transfer to the Commercial Sector, Master's Thesis, Air Force Institute of Technology, Dayton, Ohio, September 1987.

respondents (48 percent) felt that senior lab management was aware of the value of technology transfer activities, and that the laboratory mission statement clearly articulated this view. Evidently the task of getting project managers behind the technology transfer efforts of the research laboratory remains as challenge to lab management.

A relatively high rating (41 percent) given to lack of incentives to employees to engage in technology transfer activities suggests that this is also an important internal barrier to commercializing government technology. The National Technology Transfer Act specifically addressed this barrier by creating a system of awards and royalty fees for inventors of successfully commercialized government technology. Any positive impact of this legislation would not be picked up in the Dayton survey, since the survey was conducted before many labs had an opportunity to fully implement the provisions of the 1986 Act.

Only a relatively few of the laboratory professionals felt that lack of interest (10 percent) or awareness (23 percent) on the part of American business was a deterrent to their technology transfer activities. However, too much red tape for lab researchers to become involved in technology transfer activities (34 percent) and lack of adequate channels of communication between industry and government scientists (26 percent) were listed as the third and fourth most significant internal barriers to technology transfer.

Lab policies. The lab scientists and engineers were also asked to express their opinions of the relative significance of various policies that management can implement to improve the technology transfer process. A list of the potential policy actions, listed in order of importance as reflected in the survey, is presented in Table 2. For convenience, they have been grouped into three categories: Important, somewhat important and unimportant.

Forming joint research consortium with private industry and regional universities was listed by the lab scientists and engineers as being the single most significant action lab management can take to stimulate technology transfer. Providing monetary incentives to inventors and increased funding for commercial technology transfer activities were also related in the important category of possible management actions to improve technology transfer.

Other actions that would be somewhat helpful include increasing lab management support for employee technology transfer activities, clarifying laboratory technology transfer policies, educating lab personnel on labo-

Table 2

Lab Employee Ratings of Potential Lab Policies
To Encourage Technology Transfer

Scale	High Potential	Medium Potential	No Potential
	10	5	0

Lab Policies	Average Rating
Encourage Joint Research Efforts	7.41
Provide Monetary Incentives for Inventors	6.74
Increase Research Funding for Technology Transfer	6.06
Increase Lab Director and Management Support	5.88
Establish Laboratory Technology Transfer Management Policies	5.75
Allow Consulting with Private Firms	5.74
Educate All Lab Personnel on Technology Transfer	5.33
Support Information Requests from Industry	5.31

Source: Olson, 1987

ratory policies and procedures for technology transfer, and allowing lab employees to consult with industry.

Interestingly, lab scientists and engineers gave little support to the idea of upgrading the role of the Office of Technology and Research Applications (ORTA) in the technology transfer process. In particular, they rated improved internal tech transfer reports, increased funding for ORTA, and increasing ORTA's interaction with senior lab management as low on their priority list of potential stimuli to technology transfer. This finding is particularly important because creating an ORTA office in each major government laboratory and improving the reporting and dissemination of technical information were considered major accomplishments of the Stevenson-Wydler Act.

Provisions of the NTTA. Lab specialists were also given an opportunity to express their views on the main provisions of the National Technology Transfer Act of 1987 to stimulate technology transfer. Specifically, respondents to the survey were asked to rate on a scale of 10 (very high) to 0 (very low), with 5 being an average score, how they felt that each of the initiatives in the NTTA would stimulate technology transfer. These initiatives are listed in Table 3 along with the separate responses of lab management to the questions.

Laboratory scientists and engineers listed financial incentives for **inventors** as the most significant incentive mechanism to stimulate technology transfer. The sharing of the balance of the royalty fees with lab management was rated third in significance by the science and engineering group. They rated the ability of lab employees to patent the results of their research as the second most significant incentive.

Not surprisingly, lab management rated financial incentives to lab **management** (i.e., the balance of royalty fees) as the most significant incentive mechanism. In contrast, they rated financial incentives to lab employees at the bottom of their list of important incentives.

Lab management also felt that their increased flexibility to engage in cooperative research efforts with private industry, and co-mingling laboratory and private sector resources in the process, was an important new authority. Flexibility in arranging licensing agreements to commercialize technology was also listed as an important mechanism by lab management.

Thus, in summary, recent changes in federal law appear to go a long way in providing the right mix of technology-push incentives to stimulate technology transfer. We now turn our attention to the business survey to

Table 3

Lab Employee Views on Major Provisions of the National Technology Transfer Act of 1986 to Stimulate Technology Transfer Activities

Scale	High	Average	Low
	10	5	0

Key Provisions	Engineers Rating	Management Rating
Incentives for Inventor	8.34	4.90
Improve Patent Process	6.73	4.96
Retain Balance of Royalties for Lab Use	5.92	7.56
Increase Lab Authority to License Technology	5.64	5.36
Increase Lab Authority to Enter into Cooperation Research Efforts	5.74	6.11
Allow Use of Lab Resources in Technology Transfer Efforts	5.80	5.82

Source: Olson, 1987

determine to what extent, if any, technology-pull barriers may still be inhibiting technology transfer activities.

Business survey results

As discussed previously, the business community must be willing to accept government technology before transfer can occur. The survey was designed to see how receptive business are in regard to technology transfer activities.

An important finding is that most businesses in the Dayton area have not historically used government laboratory technology for commercial purposes. Only 10 percent of the 119 advance technology companies in the Dayton area that responded to the survey reported that their firm is currently using technology obtained from government laboratories in their commercial operations. The relatively low amount of technology transfer activity in the Dayton area is indicative of what other studies found about technology transfer in other regions with government laboratories.

One explanation for the relatively low level of technology transfer activity is that demand-pull forces are weak. To test this hypothesis, the business executives were asked to rate various barriers to business access to government technology (Table 4).

Apparently, many area business leaders view federal technology as "off limits" for their companies. Overall, 44 percent of the business executives responded that they view the classified nature of government research as a barrier to the commercialization of laboratory technology. Acquisition costs under these conditions are prohibitive.

The cost of obtaining information on available laboratory technology is also a potential barrier to business access to laboratory technology. This hypothesis is also confirmed by the survey. Over 42 percent of the businesses listed lack of information as a barrier. Respondents also indicated that their firms are unwilling to make a significant investment in obtaining the necessary information. For example, on another question, only 7 percent of the companies responded affirmatively that they employ a technology transfer specialist to scout new opportunities.

The cost of molding defense-oriented technology into commercial products is another potential barrier to the commercialization process. About 40 percent of the companies reported they consider nonclassified laboratory technology, primarily in the Dayton area, to be too defense-oriented to be easily adapted for commercial uses. The finding also suggests that technology-push policies to lower R&D costs to the recipients may be a

Table 4

Business Community's Perception of Barriers to Technology Transfer for Dayton Area Government Laboratories

Barrier	Overall Percent Response[1]
Lack of Designated Federal Technology Transfer Specialists	33.9
Lack of Available Information on Available Federal Technology	42.1
Available Federal Technology is Often Classified	44.7
Federal Technology is Too Defense Oriented & Not Easily Adaptable for Commercial Applications	39.7
Obtaining a Patent on Federal Technology is a Formidable Obstacle	33.0

[1]Percent of significant or somewhat significant responses

far more significant factor in technological innovation than Schmookler envisioned, at least in transferring government technology to the private sector.

Finally, a significant number of the business executives, 33 percent, felt that the complex organizational structure of government laboratories makes it difficult to gain access to laboratory technology. The also listed cumbersome regulations in getting technologies patented as another formidable barrier.

To summarize, evidence from the Dayton survey suggests that technology-push constraints to the commercial, or secondary, use of laboratory technologies have largely been removed, although the organisation of the technology transfer function within government laboratories needs some improvement. Evidence was also provided, however, that the business community views government technology as too costly or difficult to obtain and use. The problem appears to be a plethora of attitudinal, institutional, and informational barriers that separate the technology systems of private industry and government laboratories in the Dayton region. While the circumstances may differ, this finding is also consistent with the experience of other regions with technology transfer activities.

5. Summary and Recommendations

The study provided an in-depth analysis of regional problems and barriers to finding commercial uses of government laboratory technology. In particular, the process of obtaining commercial benefits from a high concentration of major government laboratories in the Dayton region was investigated using the survey approach. The purpose of the paper was to enhance scholarly and public understanding of the technology transfer process within a regional context.

Recent changes in US policy to encourage technology transfer were discussed. In the US, government laboratories have been given significant new authority to license technology and enter into cooperative research ventures with communities and local industry to promote technology transfer. It is the intent of federal policy to make the vast national laboratory system an important source of new technology to industry to bolster US competitiveness.

A major finding of the paper is that while many technology-push barriers to technology transfer seem to have been removed, a number of damaging technology-pull barriers remain. It was concluded that regional market forces will not automatically utilize laboratory technology unless complementary technology-pull policies to lower information, acquisition and adaptation costs are in place.

Ergas describes the technology policy choices confronting nations and regions as twofold.[18] One policy would emphasise basic research and the creation of new industries. The other approach would emphasis applied science and the application of existing knowledge. A literature review suggested that the latter approach is more applicable to the regional technology transfer issues addressed in this paper, since technology transfer involves finding applications of existing knowledge. Also, a number of other studies revealed that a strong knowledge base can facilitate technology diffusion in industry and regions. The knowledge base peculiar to a particular industry or technology, in contrast to more general knowledge, is called "middle ground technology." The link between strong middle ground technology and industrial innovation provides a persuasive argument for regions to expand their applied science capabilities as an important component of their overall regional technology transfer strategy.

The results of the survey indicate that communications remain the single most important technology-pull barrier to technology transfer. Both industry and government researchers listed lack of knowledge of the capabilities and research efforts of the other as a serious problem. The communications barrier is often caused by institutional constraints such as the classified nature of military research, a cloud of secrecy that surrounds industrial laboratories and a lack of technology transfer focus of regional policy authorities.

As stated, one way communities can better link the technology systems of government laboratories and industries is to promote developmental research capabilities. In addition, the study suggests that regions with federal laboratory facilities should consider the following additional policies:

1. Establishing a Regional Task Force on Technology Transfer composed of leading scientists, engineers and business executives to identify important technologies available in the region's federal laboratory system.
2. Creating formal and informal networking mechanisms to improve communications between government laboratories and industry.
3. Funding several technology transfer demonstration projects in the region. Demonstration projects are needed to call attention to the existence of technology transfer opportunities and serve as a learning process. The cooperation of the Federal Laboratory Consortium should be

sought in these efforts.

4. Establishing a Regional Technology Transfer Clearinghouse to keep area businesses abreast of trends and opportunities in federal research throughout the federal laboratory system.
5. Establishing a Regional Cooperative Research Program to stimulate major cooperative research efforts involving government laboratories, private industry and universities. As indicated in the literature review and the survey, the creation of cooperative research centers offers a particularly promising regional technology transfer mechanism. A regional policy to provide facilities and management services could be a major factor in stimulating cooperative research efforts.
6. Encouraging communities in the regions with extensive federal laboratory facilities to develop long-range programs and strategies to stimulate technology transfer. Federal policy could be an important factor in stimulating community interest in strategic planning efforts of importance to the respective regions and the nation.

References

1. General Accounting Office, (1988), Technology Transfer Constraints Perceived by Federal Laboratory and Agency Officials, GAO/RCED-88-116 BR, March.

2. US Congress, House of Representatives, Federal Technology Transfer Act of 1986, Conference Report, Report 99-925.

3. General Accounting Office (1987), Patent Policy: Recent Changes in Federal Law Considered Beneficial, GAO/RCED-87-44, April 16.

4. Schmookler, J. (1966), Innovation and Economic Growth, Harvard University Press, Cambridge, Mass.

5. Rosenberg, N. (1974), Science, Invention and Economic Growth, Economic Journal, March, pp. 90-100.

6. Scherer, F.M. (1982), Demand-Pull and Technological Innovation: Schmookler Revisited, The Journal of Industrial Economics, March, pp. 295-237.

7. Nelson, R. (1982), The Role of Knowledge in R&D Efficiency, The Quarterly Journal of Economics, August, pp. 453-470.

8. Nelson, R.R., Peck, M.J. and Kalachek, E.D. (1967), Technology, Economic Growth, and Public Policy, The Brookings Institute, Washington D.C.

9. Eads, G. and Nelson, R.R. (1971), Governmental Support of Advanced Civilian Technology: Power Reactors and the Supersonic Transport, Public Policy, Summer, pp. 405-427.

10. Mansfield, E. (1988), Industrial R&D in Japan and the United States: A Comparative Study, American Economic Review, May, pp. 223-228.

11. Baily, M.N. and Chakrabarti, A.K., Innovation and the Productivity Crisis, Chapter 6, pp. 103-131, The Brookings Institute, Washington, DC..

12. Gellman Research Associates (1982), The Relationship Between Industrial Concentration, Firm Size, and Technological Innovation, prepared for the US Small Business Administration, May.

13. Edwards, K. and Gordon, T. (March 1984), Commercialization of Innovations Introduced in the US Market in 1982, prepared for the US Small Business Administration.

14. US Congress, Joint Economic Committee (1984), Hearings on Government Laboratories and Regional Development, August 7.

15. Doctors, S. (1969), The Role of Federal Agencies in Technology Transfer, Chapter 8, The Dissemination Function, pp. 92-114, The MIT Press, Cambridge, Mass.

16. Scherer, F.M. (1970), Industrial Market Structure and Economic Performance," Chapter 15, pp. 346-378, Rand McNally & Company, Chicago.

17. Schumpeter, J. (1934), _The Theory of Economic Development: An Inquiry into Profits, Capital, Interest, and the Business Cycle_, trans. Redvers Opic Harvard University Press.

18. Ergas, H. (1987), Does Technology Policy Matter, in: Guile, B.R. and Brooks, H. (eds.), _Technology and Gobal Industry_, pp. 191-245, National Academy Press, Washington, DC.

18 The territorial tectonics: the locomotive and the train

Towards a new theory of regional space and local development

C. LACOUR

Prologue

The increasing number of conferences, the range of works and the rapidly increasing number of articles devoted to technological development at the local level bear witness to the researchers' critical approach, confirm their suspicions, and rediscover the practices of the 'chefs d'entreprises'[1]. This increase in number, quality and variety of reflections show how uncertain and hesitant theoretical knowledge and political economy still are. It also shows how they seem to have found an academic consensus.

The reason for this encounter may be most varied:

A widely-shared acknowledgement that the major analyses which comprise the dominant points of reference and the growth of industry between the 1950s and 1970s are no longer of much use. Precise analyses hardly matter[2]. What is really at stake are the methods of production, and above all a mode of thinking which at first seems outdated.

A will resolutely fixed on the future. There is an urgent need to prepare for the future. In this respect, technological development, technological transfers, and the 'local level' may play a major role. To a certain extent the opening of Europe creates new ideas in terms of development. These ideas are linked to a major transformation of the international division of labour. Technology is diffused on a global scale, and the market shares of new products are found on a worldwide scale. Unification puts Europe on a more equal footing vis-à-vis the U.S. or Japan. New cultural forms, the progressive adoption of decentralisation, their acceptance in people's minds, enable us to understand the importance of the local level:

Technological Development (TD)	<->	Local Development (LD)

Nowadays there is an astonishing reversal of models from which the rich and prosperous spaces were created. The key industries, which once included the motor, mechanics, textile and 'heavy' industries and which shaped the regions, often find themselves as growth poles in spheres of conversion[3]. The scale of this phenomenon increases when we introduce the 'traditional' industries into the picture: the steel, mining, textiles and clothing industries are defeated as much by new products and processes as by new modes of consumption, or even the costs of production.

One might note that the industry's large traditional units no longer create jobs, that we pay more attention to the real working of business, that we are rediscovering the potential advantages of the middle innovators, the Silicon Valley model par excellence. Much attention should be focused on small and medium size enterprises (SMEs), the generators of a new model of thought and development. And once again we find technological development at the local level.

TD <——— SME ———> LD

1. A New Epistemological Framework

This framework is based on a revival of old ideas, a renunciation of certain beliefs and the rediscovery of 'new' principles leading to a dual train of thought.

a. A necessary reflection

On the one hand, we must be critical of these new firms in the SME sector, which create, demand and diffuse technology.

From another viewpoint, we attach great importance to the local level itself, the middle innovators and the local environment[4]. Here, the emphasis is on the intrinsic conditions, the particular qualities of the actors, their cultural and historical legacies that might explain their ability to initiate development. This brings the third Italy model to mind[5].

TD <——— Local Milieu ———> LD

Our ultimate aim is to discover a general theory, and above all to gain insight into major macro-economic matters. The local level is born: whether national resolved or even anti-economic[6].

The following list serves to illustrate our argument:

- Prato vs. Mezzogiorno;
- South France vs. the North;
- Technopolisation vs. Polarisation;
- New Technology vs. Traditional Industries.

As necessary as it is to think of development at the local level, it is important to put the subject in perspective[7].

Dangerous constraints

The first constraint is the standard mode of reference to technological development. Do the terms used to describe development i.e., 'unequal', 'technological', 'local' have any real meaning? Perhaps one can conceive of some intellectual model of stabilized growth, but what do we mean by 'unequal development'? Perhaps we should distinguish between different forms, moments and favoured notions, but simultaneously in conceptual and concrete terms. If we are to give technology this distinctive role, we should make it the essence of development[8].

A second, and potentially fatal contraint would be to place too much importance on the local level, as an ultimate aim of development[9]. Policy-makers should look upon the increase in local power favourably, inter alia, recognise inventors' extraordinary potential and also the increasing local mobilisation[10]. However, certain pitfalls should be avoided:

- that of taking the local level as the base of everything: seeing it as a priviliged point of focus, an 'electronic eye' endowed with the quality of knowledge;
- that of retaining a space in itself, half-closed and inward-looking;
- that of looking upon high and low technology as incompatible (as was often done in the late 1970s and early 1980s).

The third restriction that we must be aware of consists either of favouring the SMEs, or putting them respectively in opposition to the large units and groups[11]. It is often useful to employ extreme modes of classification but we must bear in mind the dangers of this type of presentation. Weaver's thesis on the SME syndrome seems more relevant than ever before, even if the roles of domination and sub-contractiong may have changed[12]. Besides, it is unclear whether the functional and environmental dynamics of firms can lead to so durable an antagonism: whether the SME would be condemned to grow, or to die?

To insist further on the largely ignored aspects of the dominant analyses of previous years, one must have an open eye for past changes and present developments: technological development at the local level becomes meaningful and all-important when it forms part of a whole theory, and when it is placed at the centre of our political thought.

b. A new epistemological framework for technological development

From this viewpoint, it seems that recent ideas fit into three relatively new paradigms, whilst rediscovering long-forgotten problems. Metaphorically speaking[13] these are:

- neohistorism and neodeterminism;
- a theorical pseudo-melting-pot;
- new behavioural patterns of actors.

Neohistorism and neodeterminism

We see this paradigm blossom and flourish with the growing success of analyses, born from Schumpeter's and Kondratief's ideas, whatever their distinctive traits: the life-cycle theory, the Vernon product-cycle analysis in spatial relation to the 'regional cycle'. This is the dominant theme that we find with **techno-spatial techniques**, the transportation of ideas on geo-technological proximity[14]. Corresponding to the different phases of a products' life-cycle, we find favoured modes of technological development applied to a specific type of enterprise. Thus the small firms are found at the inventive stages, the larger ones during the ensuing boom period. Then, during the maturity period we again encounter the small firms. They are needed to help the large firms, to improve productivity and management and to breathe new life into the system. In the first and third stages, we see the growing importance of the 'service' society.

The opposition between high and low technology on the one hand, and the small and large firms on the other, should only be entertained at a given time. It is important to know why, at any given time and place, the space can change - or alternatively prove incapable of creating or even putting up with change. Perhaps we should reconsider the so-called processes of rupture-filiation and rupture-bifurcation about which Aydalot had begun to theorize. We would find a certain contradiction between the rediscovery of the past and the temptation to retain only that process which is securely placed within a large temporal framework.

Neo-historism and neo-determinism find expression in a relatively new development, accorded to ancient time and temporality, which is part of the urban structure of Italian towns which were founded in the Middle Ages or of cultural and religious traits exacerbated by the French Revolution, acknowledged as explanatory of contemporary spatial dynamics[15].

Likewise, we find a reasserted importance in the theoretical schemas of towns' sequential development[16]. Another example: in a recent symposim on the 'Creation and Transfer of Urban Technologies', the ambiguity was scarcely raised[17]; is it a matter of easily-invented technologies, transmitted by the towns, with the often implicit response that the large towns diffuse them more than the small ones? Or is it really a question of technologies specifically linked to the town, in the same vein as urban civil engineering, congestion management and urban solar energy?

This entrenchment within history, this recourse to Schumpeter-Kondratief models, give another dimension to theories of de-industrialisation and of the rise of the service society[18]. They make us realise the importance of thinking ahead, rather than simply following the logic and models of technology creation, diffusion and consumption. They show a real need for prudence; when we talk about new technology, or its diffusion, we should emphasise that it is really a matter of the arrival of revolutionary products and processes, of remarkable new mechanisms, of older phenomena and more profound consistencies. Re-reading of Vernon (pp. 108-109, for example) is gripping and throws the issue wide open[19].

An intellectual pseudo-melting-pot

This is to a certain extent a hotchpotch in which the hopes and fears of spatial and regional economists are underlined. We are faced initially with a double disorder.

The fact that two of foundations of spatial economics (i.e., micro-economics and macro-economics) presently appear discredited, is the first problem. On the other hand, the lessons of enterprise localisation need to be revised, re-learnt and enriched.

Microeconomics <——	Spatial	——> Macroeconomics
(localisation theory)	analysis	(State intervention theory)

It is, partly, the significance of everything concerning the works on SMEs and on post-Fordism, rediscovering the highest evocations of middle innovation analysis works to which we shall add, for the other part, the theories and analyses of later phases. Will there not be a tendency to effectively forget the spaces, which would suffer from an inability to 'tap' into development? France would be more likely to favour the conglomeration, whereas the small towns and rural environments play a role in Italy and the

U.K.

On the other hand, the Keynesian macro-economic points of reference, the basis of a public and financial economic theory of a state serving as a model to the regions and sub-regions, seems inadequate, if only partially and temporarily so.

The second disarray is another form of the first. The academic fields have been increasingly divided into development, public and industrial economics[20], although the problems and concepts of each may be used by the others; poles of growth, procedures, the black and grey areas of the inter-industrial matrix, models of gravity and accessibility.

The academic frontiers become blurred, as do those of nations, of regions, in short, those of **territories**[21]. How are we to understand the theory and politics of technological development whilst scarcely concerning ourselves with the mechanisms and conceptions of invention and innovation? We must strive to understand the working of research organizations and the way in which the dynamics of technologies and products are diffused. We have not, as yet, fully appreciated nor assessed the influence of the financial and commercial stakes, the market shares and the political dimensions. These dimensions -favoured and disfavoured spaces, ageing and innovative technologies, SME and large groups- can be both interpreted and ultimately put into perspective. If we see the birth of Third Italy, or of Silicon Valley or even if Northern France and Lorraine become models of a certain industrial chaos, measurable only in terms of lost jobs, the reason lies in the profound changes, that are taking place.

These disorders emphasise the need to rethink the bases of regional science, in the hope of finding a new foundation. **A quest for a regional science that is both more general and more specific.** More general for it must entail more than the poor, or rich spaces, well-known to economists as a basis of reflection. More general, because it must re-visualise the concept of time, the qualitative factors, the weight of institutions (whether they become more rigid or more subtle). The new foundation must take into account the industrial and financial rationalities. We shall return to this subject later. Yet more general, because it must allow for a **theorization of the complexity and diversity of the whole**[22].

At the same time, it must be more specific, since a new and richer theory on the large urban and industrial conglomerations, **a theory less manomanic and ethnocentral** than before, may be the outcome of the variables, the certainties and the improbabilities of individual cases. The new foundation should be more specific and general for it shall not be based solely on those aspects which we so often take as alpha and omega: the Italian miracle, for instance, was already a point of reference in the 1960s, as were Silicon Valley, Fos and Dunkirk. More than ever, we must be our guard against potentially false certainties[23].

The final paradigm which plays a key role, underlining the need to take note of the actors' behavioural patterns -too often disregarded in analyses- is 'Neoactorism'.

'Neo-actorism'

This somewhat barbarous term, derived from the Actor's studio is intended to accentuate the importance of the most recent works, justly accorded to individuals, institutions and entrepreneurs, in the words of Schumpeter's new disciples. They claim as he did (and as Veblen did) that family ties, local culture, the vitality of the social environment, and certain traditional activities (textiles, clothing, agricultural products) can be the basis of a technological develoment which is most advanced at the local level with a market of global dimensions: economic efficiency meets micro-entrepreneurial model reliability[24].

The products -hoisery, shoes, duck liver- appear to resemble the old model, as much by their final nature as by their domestic structure or the small size of units. The difference, however, is total: the norms and processes of production, the market research and production management, the enormous incorporation of technical progress. Small units they may be, but they are largely integrated amongst themselves, plugged directly into international demand. Domestic perhaps, and far from Fordism, but the staff of the highest calibre - and not only in technical matters; in financial and commercial engineering, and information processing too, they display a remarkable level of efficiency. The SMEs are progressively using the quality control methods previously found only amongst the leading firms. They can become, if only for a while, masters of the technological world market which they themselves initiated.

We often speak of the flexibility and adaptability of the SMEs, by which they find their 'niche' within the technological world, and create models of management to be taken up by the large groups, 'the niche may be narrow, but the effects are world-wide.'

It is thus necessary to rethink our analyses relative to the behaviour of the actors, the entrepreneurs and the elected authorities. We must look again at the practical and often out-dated distinctions between private and public activity, between the large sectors of activity dear to Colin Clark and Fourastié, as to the statistical institutions. In analysing the actors' behavioural patterns we should always bear in mind two elementary principles:

- Firms, as public institutions, have recourse to innovative technology only out of necessity.
- Secondly, innovation is stifled as much as possible while other products and producers are still functional, while they command social and political support. Whence the key to the importance of our behavioural study: why do the actors change, create or adapt at this point, whilst at another -endowed with globally comparable factors- they do nothing? To answer this question, we must think in qualitative terms.

'Neo-determinism', 'neo-melangism', and 'neo-actorism': these dominant orientations of our problematic theorem underline the need to reconstruct a theory of regional economics. We can do this only by recognising the variety of situations -the complexities of each individual case- and not by the excessively simple operation of merely oppositng them against each other.

2. Theorizing on a New Environment

The 'new' environment is founded upon the multiplicity of present movements, the continuity of major trends, the super-imposition of industrial and technological strata, and the territorial dynamics which render a clear and definite appreciation more difficult than ever: a system of complements, of cohabitation, rather than of opposition between large groups and small units, or between the local level and anything else. Likewise we see complex propulsions towards the free market intermingle with the maintenance, often even the reinforcement, of the more subtle forms of State control; to this picture one might add, for all its negative or positive effects, the actions of supra-national groups, or those linked to the preparation of the Single Market.

Two synthetic representations may account for the major changes -as much economic as mental or cultural- and for logical trends.

a. Territorial tectonics and the complexities of technological development

Our first representation could be metaphorically entitled the 'spatial IS-LM' schema, by reference to a four-part logic[25].

a) The dominant theses concerning the regional spaces, which quite clearly put the territorial upsets into perspective - **territorial tectonics;** the see-saw actions that moderate the depth of regional and inter-regional overlap between the developed and under-developed spaces, between the favoured and disfavoured regions. To these major changes one might add other, related ones, belonging to a hierarchy that is linked to Spain and Portugal's entry into the EEC[26]. We must consider in greater depth the factors which appear to favour the Southern spaces, but a simple opposition, born of previous industrial revolutions, can no longer be taken as decisive, unlike in urban affairs, where North-East Europe still plays a determinative role.

b) In response to these diagnostics, attention has to be paid to the main political tendencies, directly linked to explicit or non-diagnostic interpretations. In one model, a market preference for State intervention leads directly to jobs within industry, jobs which themselves induce tertiary employment i.e., a preference for the Keynesian model, giving the State a central role, the impulse being top-down, via large decentralized units, public or private: in short, what is called **industrial decentralisation**[27]. In another model, by contrast, private enterprise and new technology are favoured. Consequently, a dual logic progressively emerges: the first being financial, the second founded on the intangible (training, advice, the transfer of technology).

c) Next, we have to recall the general principles of town and country planning, for instance, taking a French example, the extraordinary change in the planning process of the Paris metropolis. The institutional modification dates back to February 1987, and the work of Deferre, the godfather of decentralisation. Yesterday we fought to prevent the situation of a Paris (and in general a French) wilderness; today (and tomorrow) we strove to reinforce Paris' position vis-à-vis the larger metropoles of the world. We must question the modalities which permit the emergence of international European towns, and the creation of new wildernesses.

d) Finally, we address the issue of the principal trends of economic interventionism. This model incorporates three dominant dimensions-conceived, financed, and led essentially by the central States. Solidarity seemed a fundamental priority, justifying a sustained importance on redressing the regional inequalities caused by a multitude of historic, cultural, social and economic factors: poverty and unemployment were both means and ends. A progressive reduction in 'carte-blanche' aid, and an increased support, generally for profit-making sectors and specifically as concerns the regions, has extended to cover all activities and almost the whole national territory.

Finally, local communities -either singly or in concert- have increased their own aid by generally duplicating the central model. An inert system, and somewhat inefficient, but a popular one, created from a widespread consensus.

We can see an ever-increasing movement towards a profound change in economic intervention. States, through political choice or budgetary constraints, prefer to keep intervention to a minimum. Provocative claims are put forward, discussing the efficiency of public policies, those initiated by the central State as much as those taken over by the regions. In contrast, **a more qualitative, less quantitative method of intervention, favouring those firms that lay claim to both the concrete and intangible** is to be preferred. Finally, the dominant philosophies reveal a deliberate trend towards more subtle and efficient, less automatic and bureaucratic modalities: the main reference to specificity, to the individual case, a mode of reference that is easier to decree than to put in concrete form.

b. The complex nature of the compass cord

Our second contention, which we propose to call the Compass Cord, seeks to underline the principal movements, characteristic of contemporary economics, and the perspective in which we must place technological development and the local level. Numerous dimensions are intertwined here, accentuating the complexity, the overdetermination of planning and of regional policies. Examples are:

- The regionalisation since long created by the State (and which should, at least in part, remain as it is), the responsibility for local cultures and specificities taken by the local authorities: organisational centralism and territorial specificity.
- New and renewed relations between the large enterprises and the SMEs.
- The confrontation of financial logic and industrial rationality*, where industrial localisation joins forces with immateriality.
- National or European micro-economic forecast models confronted with regional and local perspectives; when the former seem outdated or undesirable, the latter strive to compensate for this regrettable impotence. Forecasting is indeed necessary, as is futurology; the two are complementary rather than opposites, but we must consider the

---

* By 'Financial Logic' we mean strategies motivated by the maximisation of short-term profit, whilst 'Industrial Rationality' is more concerned with long-term advantages, in terms of jobs or products.

local or localized appreciation of technological development as well.

- The subtle links between public, national firms and the tendency towards privatisation; here again the habitual distinctions between Public and Market Economics are defunct. How are we to replace them? The tendency towards definancialisation certainly leads to a new era of Public Economy.
- The difficulty, as well as the necessity, is conserving these national policies in a doubly ruptured framework - at the top, by the growth of economies into global dimensions (the EEC for example) and, at the bottom, by the increasing autonomy of infra-national authorities: the local authority, the State, and the World.

We could, undoubtedly go on. Where these influences and realities meet, regional policies are created, with a multitude of concurrent and interdependent interferences. Devising local and regional policy becomes more delicate, due to technological development, which continuously muddles the previously accepted distinctions.

Faced with these complexities, and given their urgency, two models may be proposed at the methodological level, to move towards some new theory of technological development at the local level, and, more generally, of the relation between the two.

c. Rudimentary principles of a theory on the complexity of technological development at the local level

Our first reflections may be presented on two tracks:

1) 6 P Logic, or the rediscovery of futurology

Here we have the following components: Past, Planning, Programming, Power, Prospects and Prediction. Rather than placing them in opposition, it is better to link them, according to their continuous and intersecting relationships. The continuous relations give the authorities (cf. Neoactorism) pride of place, including the role of entrepreneur. We find the weight of the past, the legacy of various trends and structures (cf. Neohistorism), affirming the need for strategic planning, whatever its content. All in all, everything materializes according to programmes: the physical, the intangible and, above all, the financial programmes which include risk and decision.

2) The 'modernity' model

Without splitting hairs between mode and model, modernisation and modernity, we must always bear in mind a general theoretical framework which incorporates all the important elements and offers on the whole, a means of understanding the complexities.

The 'modernity' model seems to be the essence of a useful theoretical frame of reference. We shall take, for the time being, two examples.

- As regards **technological development**, we shall continue to use the term, but being careful as to its usual meaning. There are indeed continuous trends and slow evolutions, notably in liasion with product and regional cycles. Certain types of change lead perhaps to the resurgence of Kondratief cycles; we indeed recognise the persistence of far-off and cultural influences. The unconscious imitation of analyses created by technology does not fully answer the question of whether it is a matter of a new product (where, how, and with what, can we make it?), or if, at any given moment, a slow process believed to be of little consequence, can bring about the birth of a new world invisible until that point, but undeniably present and already in action. Formerly we called this the Iceberg theory; it re-emerges today as the problematic theory of Territorial Tectonics.
- As regards the **local level**, we can say much the same. How can we discuss the social environment, without seeking to understand why, here and now, a given space should prove capable of adapting, of reversing its history, of calling to arms all local resources or of tapping into outside currents, eventually leading ot an endogenous dominance? But here we must consider another question: is this not the rediscovery of polarisation? The local level must ensure its integration within a larger environment, the surety of its success, but possibly also of its internal disintegration, even of its death. Technological innovation, furthermore, may also be viewed with this caveat; it enables the conquest, or indeed the defence, of the fortress at any given moment, whilst allowing for a more general dispersion.

3. Managing New Policies

Our model touching on the modification of diagnostics, principles of national and regional development, and of economic intervention is a partial response to this need, even if it is practical to oppose solidarity and efficiency, the dominant intangibles and the continual demand for heavy infrastructures. Generally, however, the policies in favour of technological transfer at the local level appear essentially tri-directional:

a. **The importance of the economic, social, technological and general environment**

Through both pragmatic and ideological will, and through recognising that creation and technological development are not self-proclaiming, the authorities seem quite convinced that the determinant factor of technology is a favourable environment, and that it is wise to encourage the conditions for innovation. It may be a profound change evoked by the ambiguous term, **technopolisation:** the withdrawal of direct support to firms, compensated by an increased interest in urban affairs and the requirements of training. Once again the question of urban crisis and decline -often synonymous with that of industry[28]- is indirectly posed. Yet we still do not really know why some industrial towns are loathsome, whilst others -notably in South France and Britain- are attractive. The trend dictated by organisational liberalism, mobilised local and regional actors and institution, in deliberate tandem with Brussels. Within this general direction yesterday's dominant theme of the quality of life is somewhat dispelled, while that of innovative spaces (the urban environment, wit all its qualities, receptive capacities and reputation for efficiency) is affirmed. What seems very much to have the edge over anything else, is the idea of advance preparation-permitting, of course, the design, propagation and reinforcement of technical allure, the enrichment of information tides. This leaves open the following questions:

- whether spaces, particularly capable of generating or accommodating new techniques, exist. What is a 'distretto technologico', or a 'distretto industriale'?[29] Is it very different from A. Marshall's views? Basically we still lack a stable and robust analysis.
- are there any favoured sizes, places, functions, or actors?[30]

We can distinguish the apparently preferential localisations: the spaces of South France, those rural zones which escape the 'population drain', or the reinforcement of international metropolitan areas such as London, Barcelona, Paris or Milan. But we still do not really know why things differ from one place to another, let alone how to encourage such differences. From the all-State to the all-entrepreneur, there is much room for manoeuvre, all the more so since we know that the questions are no longer couched in these terms.

b. The importance of industrial and financial rationality

Another possibility open to us is a new form of conciliation between industrial and financial rationality; the excessive opposition between them need no longer be retained. However, as regards the strategies of firms faced with technological development, two main courses of action seem to emerge:

a) The first, favouring industrial rationality and the logic of industrial know-how, is distinguished by a dual concentricity:

   - around a craft of product (such as the automobile, heavy chemicals, parent electronics or textile industries).
   - around favoured spaces, born largely from initial implantations, eventually bringing us back to the source of activity, or of delocalised and distant subcontraction. In many sectors, there was and there still is one production unit, one large group, per country. The current tendency is for further concentration - one sole unit for all Europe. Whence the risk of over-centralisation.

   This dual concentricity can only bring about a major redeployment of personnel, with the large-scale injection of technological development.

b) The second modality is marked more by a fettered financial dominance, creating a variety of products, markets and spaces. Here the essential characteristic is knowing how to act, to innovate. Certainly, the SMEs and certain complementary structures are favoured.

   One might thus have a certain preference for maintaining, on the one hand, the old industrial metropolises -which, whilst adapting to new technology, retain their capacity to attract, accommodate and continuously re-convert- and, on the other hand, those places where technological innovation is the dominant rule, pressure and preference as much for their emergence as for their maintenance. We might also notice the major Brownian movements - a greater instability than we have been used to - since the last industrial revolution, greater than we would wish to see, based on habitually transposed models.

   Thus, SMEs would correspond more to the ability to create or classify innovation, than to a definitive status. Likewise, certain spaces, subject to being in a continuous state of evolution and agitation, can have a well-defined local origin, rediscovering the social environment while taking, as a rule, the immediate watch over their worldly but fragmented markets.

c. Specific measures to ensure the funding of the SME and their adaptation to technological development

As regards funding, numerous policies are followed. Some undoubtedly, are born of the more general will to favour an already-formed environment. Others are linked to positive fiscal measures, yet others touch once again on the creation or support of capital-risk systems, participative loans, and improved rates of interest. The creation of a second French stock market was intended as a means of giving the SMEs access to high development technology in the capital market; it is still too early to judge its success. Likewise certain large groups have re-read the regulations as to subcontraction, because -as confirmed by a study carried out by Banque de France for the period from 1979 to 1983- the dominant role is that of the SMEs[31].

- The most active SME suffer from a drop in financial independence, despite a marked increase in self-financement.
- The most profitable SMEs are only as self-financing as is necessary to consolidate their capital stocks, and to better ensure their financial independence.
- In industry, building, civil and agricultural engineering, the most capitalist firms enjoy the highest level of independence.
- An important degree of self-financement, and an increasing tendency towards borrowing is typical of SMEs with a high investment level.
- Despite its favourable impact on self-financing, export does not appear to promote financial independence

The sample chosen (Centrale de Bilans), the period studied, and the fact that no explicit distinction was made as to the SMEs linked essentially to technological development, lead to the need for some caution. However, we can well imagine the following:

- It is somewhat awkward to try to draw a casual link between the local level and the creation of SMEs. Cholet, an often quoted example, undoubtedly relies on its textile and hoisery industry; however, we should not ignore its electronics industry. Innovative SMEs may be found in the Landes deserts, just as in urban Bordeaux. In effect their common characteristic is their location on a technological vein, a world market and a pool of buyers and sellers which itself compels continuous innovation. Their technological originality is immensely varied, their evolution difficult to forecast, but rarely static, in terms of size and products.

This accounts perhaps for the tri-directional policy trends which seem to emerge.

- Actions resolutely aimed at creating a more dynamic social environment, at making it aware of the ongoing technological revolution; at centring everything on the current, and permanent, view of dominant instability.
- Measures aimed at encouraging innovation and technological development per se: think-tanks, 'Euro-counters', innovators' clubs, financial and social aid to public participants, rather than the old method of direct aid to investment and employment. In the same vein we find public-private financial systems, buyer-entrepreneur co-operatives, created in the hope of identifying some specificity of technological development.
- Finally, a widely held conviction, founded on the confident belief that the SMEs will continue to create employment. This induces the conviction, exaggeratedly optimistic and under-informed, that the small units -almost by nature and their very essence- have become doubly privileged by local and technological development. From this viewpoint, a series of comprehensive measures are taken in favour of SMEs, stating that in the small units lie the hopes for employment, the guarantee of all development. A largely admissible confidence, so long as we take it soley on its merits.

Notes

1. C.E.R. Symposium. Aix en Provence in R.E.R.U. 1986, no. 5, Les P.M.I. innovatrices et leur environnement local et économique, Lacour, C., l'Astrolabe et la boussole, pp. 743-755.
Symposium ASRDLF-CEEE-GREMI, 1986, Paris: technologies nouvelles et développement régional, and the symposium GREMI, Barcelona 1989.
Colloque CREUSET, Saint-Etienne in RERU, 1987, no. 3, milieux industriels et développement économique.
Innovations Policies at the local level, Paris I C3E, decembre 1987 (see GREMI, 1987).
See Nijkamp, P. (ed.) (1986), Technological Change, Employment and Spatial Dynamics, Springer Verlag.
Federwisch, J. and Zoller, H. (ed.) (1986), Technologie Nouvelle et Ruptures Régionales, Economica.
It is interesting to note, from an epistemological viewpoint, that our essential learnings come from colloquiums and seminars.

2. Aydalot, P. (ed.) (1984), Crise et Espace, Economica, Paris; RERU, 1986, 5.

3. Langumier, J.F., Laborie, J.P. and De Roo, P. (1987), l'Amenagement du territoire, la Documentation française, Paris.
Lajugie, J., Delfaud, P. and Lacour, C. (1985), Espace Régional et Aménagement du territoire, Dalloz, Paris, (noted LDL 1985).
De Gaudemard, J.P. (1988), Les deux défis majeurs de la politique d'A.D.T.: la crise et l'Europe, Séminaire Decta III, IERSO, Bordeaux; more generally see Piatier, A. and De Gaudemard, J.P. in RERU, 1989, no. 1.

4. See, therefore, the reference to the French Revolution (notably for Vendée and Cholet).... or, again, 'L'Italie des cents cites', so opposed to Paris and its French desert: Modene, Parme, Vérone, Padoue and the example of the Italian model. Yet at the same time, we appear to unconsciously favour the present, the instantaneous: see, below, the so-called Modernity Model.

5. There is a wealth of literature on the subject. Some references: Antonelli, C., op.cit., Ciciotti, E. (1986), Natalita delle imprese e diffusione delle innovazioni di processo in un distretto tecnologico, Tecnocity, F.G. Angelli, Turin.
Camagni, R. and Pattarozi, M. (1984), La diffusione territoriale di una innovazione di processo e di prodotto, L'Industria, no. 4.
Also refer to Beccatini, G. (1979), Amendola, M. (1983), Bagnasco, A. (1977), Garofoli, G. (1983), Brutti, M. (1982), Fua, G. (1985).
For French examples see: Ritaine, E. (1985), Un système industriel localisé en Italie, F.N.S.P.; Prato...., Sociologie du travail, 1987, 2.
Courlet, C. (1987), Developpement territorial et systèmes productifs locaux en Italie, IREP-D, Grenoble.
We may soon hear of a third Spain...
A wealth of small firms thus create HTs (The Lectra system of Bordeaux, for example) and others, all of which use new methods, superior to the traditional products and sectors.

6. Our 'Espace' in L.D.L., (1985).

7. Our contribution in Auriac, F. and Brunet, R. (1986), Espaces, jeux et enjeux, Fayard.

8. Cf., the introduction in Keeble-Wever (1986), or Castells, M. (ed.) (1985), High Technology Space and Society, Sage; Thwaites, A.T. and Oakey, R.P. (eds.) (1985), The Regional Economic Impact of Technological Change, St. Martin's Press, New York, who write: 'We believe that a compilation of such work is timely since the study of the impact of technological change on the prosperity of regional economics, from both theoretical and policy view points, has now become established as a coherent subdiscipline within the broader subject area of regional development studies', p. 4.
This is really a question of impact, while other works are more concerned with the conditions in which a 'milieu' can be at the heart of technological birth. Generally, are there not two dissimilar thought processes here? The first is marked essentially by Anglo-Saxon research (the works of J.B. Godard, for example), the second linked more to French and Italian works, although some 'aberrations' may be found: B. Planque fits closer to the first mould, while D. Maillat is more an illustration of the second. Our contention is that we must try to bridge these two methodologies and take a more global viewpoint; cf., Molle, W. (1983), pp. 22-38.
We refer to R. Bar-El, D. Felsenstein, for whom the relations between technology and regional development must be fundamentally multi-dimensional, in Economic Geography, July 1987, pp. 281-284, in their commentary on B. Van der Knaap-Wever (1987). However, we need analytical tools: see Mignolet, M. (1986). Can we, furthermore, see another difference between those who advocate new high technology, and those who seek to integrate new technology into the traditional industries? Nonetheless there are more general ways of looking at things, linked to de-industrialisation (Bluestone, B., Harrison, B., 1982), or to the evolution of economic systems (Massey, D., 1984, or Scott-Storper (ed.), 1986).
Van der Knaap-Wever, B. (1987), New Technology and Regional Development, London, Croom Helm.
Mignolet, M. (1986), Innovation technologique et redeploiement industriel, ERESA.

9. Pecqueur, B. and Silva, M.R. (1988), Industrialisation diffuse et développement, IREP D, Grenoble.
Greffe, X. (1984), Territoires en France, Economica, Paris.
Greffe, X. (1988), Décentraliser pour l'emploi. Initiatives locales developpment, Economica, Paris.

10. Courlet, C., Pecqueur, B. and Rousier, N. (1988), Etude sur les politiques industrielles locales dans le cadre de la promotion des PME. IREP D., April. Lacour, C. and Puissant, S. (1987), La tectonique des territoires, op.cit. Lacour, C., Lhert, J., Peyrefitte, M.L. and Puissant, S. (1988), Evaluation des politiques en matière de développement local, Ministre de l'Emploi, Paris, May.

11. Nijkamp, P. (1987), The role of small and medium size enterprises at the local and regional level, GREMI.
Swyngedouw, E. (1987), Social innovation, production organization and spatial development, RERU, no. 3.
Puissant, S. (1987), Savoir faire et faire savoir, Aquitaine Eco, Sud-Ouest.

12. Weaver, C. (1985), in: Boisvert, M. and Hamel, P. (eds.), Redeploiement industriel et planification régionale, Montreal.
It seems that the all-important thing is reasoning in terms of time for financial, industrial and technological rationality. A discussion with B. Planque shows that the distinction between financial logic/rationality and industrial logic may be illuminating. In future, we should deepen this distinction, perhaps replacing it with logic in which field economies of

scope prevail, opposite a logic prioritizing economies of scale.

13. Probably because we are living in an unstable period, as regards theoretical knowledge, we notice major empirical production and a tendency to the use of metaphors. This is really no solution: cf., the Modernity model and the Compass card. In a symposium at Aix (July 1988) we swing between precise definition and metaphorical explanation. At the Barcelona meeting (GREMI) (1989) Perrin, J.C. shows some ways.

14. An essential theme 'les milieu...' (op.cit.) heatedly debated by D. Maillat in 'Regional Strategies, 1987; above all, we have searched for the conditions that would enable the social environment to be more innovative than the spatial or high technology.
An interesting presentation by Bramanti, A. and Senn, L. (1987) in <u>GREMI</u>. In particular, the synoptic table of policies supporting innovation is a most useful synthesis.

15. Whence the term neodetermonism, or neodeterminism for Emilie Romagne, Prato, Cnolet, where one cites, the French revolution, the Middle Ages, and, for Spain, the Franco era. The references, more evoked than analysed in depth, merit a quantum of thought, and also of caution (see notes 11 and 4). There is still one fundamental dispute; D. Maillat, B. Pianque and J.C. Perrin go further than this sole, circular-suggestion. For our part, the important factor is the interest -and the danger- of retaining one sole point of entry, as, for example, 'PME' and Exportation (see the debate at Aix, above).

16. The Drewett-Klaassen revisited: see, for example
Suarez, L. (1987), Evolution métropolitana, <u>Estudios territoriales</u>, 23, pp. 155-181.
Montanari, A. (1987), Crescita economica e abusivismo urbano nell' Europa mediterranéa' (1945-1985), <u>Rassegna Economica</u>, no. 6, pp. 1033-1052. Lacour, C., Puissant, S., Economie urbaine, to be printed in 1989.

17. May-June 1988, if the relations between regional development and the local level, or between local and regional development are already unclear, we shall see further complexity. The role of urban actors defies coherent definition.

18. P. Aydalot, in a doctorate seminar at Bordeaux, January 1985, strongly insisted on this point: how can we reconcile a return to Kondratief, and a change in the nature of Society as a result of the 'new technologies' when the dominant factor appears to be the multiplicity and concomitance of structures, technologies and changes? A question also raised by M. Castells, P. Hall, A. Markusen.

19. R. Vernon, Metropolis 1958, Doubleday 1963.
How can we tell if it is really a matter of new phenomena or merely of the disguised repetition of history?

20. These divisions are becoming more blurred: cf. the works of ASRDLF at Marrakech, RERU 1985, no. 5 and also our contribution 'Espace et développement': p. 847. See also the relecture 'Commerce International...', 1988. The final cut will perhaps be the 1989 ASRDLF-IERSO symposium at Bordeaux on the theme 'The regional science and the State'. However, these concentricities are organised, the question of Regional Science's status is passed with the same acuteness.

21. Territorial tectonics, op.cit., or what other studies call the upsets of spaces. See also Brunet, R. and Sailois, J. (1986), Les dynamiques du territoire, Gip.

22. Lung, Y. (1987), which makes a clear and simple synthesis of the essential reading: l'ABC des ruptures de l'analyse spatiale, Pub, Bordeaux.

23. Planque, B. (1985), RERU, 5.
Gordon, R. and Kimball, L.M. (1987), Beyond industrial Maturity: Planning the future of Silicon Valley, GREMI.
Hall, P. and Markusen, A. (eds.) (1985), Silicon Landscape, Allen Unwin.

24. Territorial tectonics, op.cit., and C. Weaver, op.cit.

25. Effectively a metaphor, but one which underlines the essential role of Keynesian thought and policy in model 1. Whilst it may have lost some of its poignancy, the model is by no means negligible. Allen, K. (ed.) (1988), European Regional Incentives.

26. 3rd period report of the EEC, May 1987.

27. Langumier, J.F., Laborie, J.P., De Roo, P. (1987), op.cit.; LDL, (1985), op.cit.

28. Antonelli, C. (1986), op.cit., p. 158, points a, b, c.

29. In particular, how to reconcile the phenomenon of technopolisation, implying a certain volume, and a certain geographical preference, with the industrial or technological districts? How to theorize on the medium-sized Italian towns and the regional metropoles of France?
Bernardy, M. and Boisgontier, P. (1985), Grains de technopoles..., Pug.
Sometimes I have the feeling it is a French dispute!

30. Le Monde, A., 21-03-1987.
In particular, concentricity and large sizes must not be forgotten for certain activities: chemical, automobile. By contrast, the computer industry must come out of its market niche.
Likewise, the PME make as much effort as the large groups, as concerns commercialisation, innovation, training and logistics: the ethereal accounts for 40% of FBCF, see Le Monde, 13-02-1986, quoting the SOFARIS colloquium on the PME's investment in 'grey matter'.
In territorial tectonics, we created the LEADER model to explain the conditions for success: a Leader (L); a team (Equipe); and orientation Development (D), centered around the Exterior (E), and relying on the support of the Regions (R). L. Senn op.cit., where the local can only succeed by opening onto the international. We must therefore, as underlined by the 'IS-LM spatial', reconsider the new functions of regional actors as the State, the 'policy response to the third industrial revolution', Schmandt, J. (1987), GREMI

31. Banque de France, CEM 1987. As regards exports, the threshold of 50 employees was considered essential, Colloquium at Aix.

19 Technological development and the environment

G. SCIMEMI[1]

1. Diverging Views on the Technology/Environment Interface

Since the beginning of human civilisation, technology and the environment have experienced an uneasy, ambivalent, controversial, often conflicting mutual relationship. This relationship has been considered by analysts within the wider context of the interactions between technology, human action and human conditions, a subject on which viewpoints have widely varied, and still substantially diverge.

Following a somewhat schematic approach, individual positions may be plotted as points on a diagram framed by the two orthagonal axes.

Along the first axis, a deterministic view of the historical development of technologies occupies one pole, whereas, at the opposite pole, the same development is seen as a voluntary and controllable process. At the first extreme technology generation follows a fatalistic pathway; man is subject to the blessing (or perhaps the curse) of what philosophers with a Socratic lineage might wish to call a maieutic role: indeed, man is instrumental in giving birth to **one** new technology after another. But **what** technology will see the light and **when** the birth will take place are matters which escape his control. Whether this is seen to be dictated by some sort of predestination or simply the result of an inescapable chain of cause-and-effect is immaterial to the present discussions. A typical exponent of this school of thought are Jacques Ellul: 'La technique ou l'enjeu du siècle'. One should also mention Sigfried Giedion: 'Machines Take Command'. At the other extreme, man stays in command: he stands alone bearing the entire responsibility of facing the Promethean challenge. This leads to a further dilemma: who should decide? and for whom? Are technologists responsible for technological decisions including their implications to society? Or: should society decide for them, thereby subordinating, in effect, the freedom of innovation to special needs and goals?

Along the second axis, at one pole are the views of those who see the

---

[1] The opinions expressed and the arguments employed in this paper are those of the author and do not necessarily represent those of the Organisation or of its Member countries.

consequences of technological progress as basically benign, whereas those who consider them to be inherently malignant stand at the opposite extreme. These two schools of thought -the technological optimists and the technological pessimists- were particularly vocal at the time when people started talking and worrying about the environment: that is to say around the late 1960s and early 1970s. 'The Limits to Growth', and the world model underpinning its conclusions recognised both the negative and the positive role of technology. The positive aspect would, however, only result in a limited period of grace (a couple of decades) before the inevitable ecological catastrophe. Subsequent critics of the Report of the Club of Rome have usually adhered to the automatic, market-demand-prompted 'technological fix' persuasion: a classical response to the classical Malthusian (and Ricardian) argument founded on the limited quantity of physical resources, hence the unsurpassable limits to growth. H.D.S. Cole, R.C. Curnon and K.L.R. Pavit of Sussex University articulated this critique as early as 1973. Julian L. Simon must also be counted among the outstanding epigones of this school. On the opposite front, Barry Commoner, in his bestseller 'The Closing Circle' also published in the early 1970s, makes a convincing case that the general technological revolution which occurred in the two-and-a-half decades after World War Two had played a major role in the rapid environmental deterioration during that period; indeed, its role was much greater than that of the increase in volume of per capita consumption combined with population growth. Totally new technological processes had been introduced to produce food, clothes, and various other consumer goods, while an unprecedented variety of synthetic substances had been employed to replace the natural raw materials which had previously been used for the same commodities. A similar line of thought had inspired Alvin Toffler, in another bestseller 'Future Shock', published the previous year (1970), to advocate vehemently the immediate establishment of powerful public institutions aimed at controlling and indeed 'taming' technology.

The **OECD Monograph** on the **Promotion and Diffusion of Clean Technologies** submitted as a background paper to this Seminar, obviously reflects the non-deterministic school of thought, and may well fit into the optimistic class within the paradigm I have just described. This does not imply that a positive presumption prevails in OECD work concerning the controllability of the processes of technological development and diffusion; nor does it mean that their potential favourable effects are generally taken for granted. This particular monograph deliberately focuses on a special set of environmentally benign technologies (in fact a sub-set, identified as 'clean' technologies, as distinct from 'add-on', or 'end of pipe' technologies), and includes a brief survey of policies which various Member countries have adopted and applied -with various degrees of success- in order to promote their development and diffusion.

The OECD monograph on clean technologies must be seen within the context of overall OECD work in the domain of technology. OECD has been active for years in this area. One of its Directorates is responsible for science, technology and industry. The so-called Frascati Manual and the recent OECD Reports on Science and Technology Indicators are widely known. Less widely known, perhaps, is an entirely new initiative, a new, bold start, announced to and welcomed by Ministers only a few weeks ago. This initiative, labelled the OECD Society, Technology, Economy Programme, is bound to have major implications for the orientation of the work of the OECD as a whole, at least over the next three or four years.

The idea is to define an integrated and comprehensive approach to the different technology-related issues which are dealt with within the various components of the Organisation: the Environment Directorate being one of these components, and certainly not a minor one.

The general structure of the programme is developed around three major axes. The first axis will strive for a better understanding of the conditions for the production of technology as well as its diffusion. A second axis will analyse the relationship between technological innovation and economic performance: how technology affects productivity -and therefore economic growth- and how technology can be better integrated into macro-economic analysis. The third axis involves the human and social dimensions of technology, including issues such as the structure of employment, training and education, labour-management relations, and of course health, safety and the environment. The new OECD-wide Society-Technology-Economy Programme (STEP) will be oriented mostly towards governmental policies, as is generally the case for OECD work. The issue of governmental responsibilities vis-à-vis the technology/environment interface will have to be looked at in the broad context of the previously examined paradigm and its two dimensions: a) the deterministic versus the governable character of technological development; b) pessimistic versus optimistic views of its effects.

As to the first dilemma: the acceptance of the governability assumption is of course a precondition for a meaningful discussion concerning the involvement of governments as well as intergovernmental agencies. If technological development were an entirely deterministic process, governmental efforts both to direct and/or control it would be entirely futile.

If, on the contrary, technological development can be directed along preferential lines, and deliberately steered away from less desirable ones, then governments, both individually and jointly, have a role to play. And it can be a very significant one.

Concerning this matter, official views of OECD governments differ as widely as those of individual analysts. There is, however, at present a general trend in the governments of OECD countries to get involved increasingly in technological development issues.

Whilst it is recognised that steering funds and research efforts towards particular technologies may stifle breakthroughs and innovations in less predictable technological areas of technological development, and that governments should avoid picking winners and losers, increasing importance is given to technology planning initiatives. Japan, which for many years has invested considerable effort in elaborating lists of key emerging technologies, is a prime example. Government officials, industrial scientists, corporate executives and others are brought together to select the most promising technologies for development. The demonstration effect resulting from the public selection of new technologies is often considerable in influencing private sector decision-making on research priorities. This collaborative process helps produce a common outlook for industry and society. Several other OECD countries, including Germany, Sweden and Canada acknowledge the need for public support of technological development and, at the same time, the inability to support adequately all areas, in view of limited governmental resources. Thus, a certain degree of priority-setting becomes inevitable in governmental technology supporting measures.

Accepting the governability postulate and, indeed, the notion of technology planning does not eliminate the dichotomy between the pessimistic versus the optimistic view of the impact of technologies upon the environment. It is evident that significant public actions can be called for in both directions, i.e., in order to see how best to prevent the potential negative impact of technologies on environmental quality (technology as a problem) and, symmetrically, in order to see how best to enhance their potential positive effects (technology as a solution). Governmental action is not a novelty in either direction; however, at the time of this writing, several significant developments are under way, both on a national and an international scale. In the next sections, I will briefly touch upon both these policy directions in sequence, focussing on what governments are doing, what they could do, what they intend to do, and what the role of international organisations is and could be. I will also try to identify a number of promising avenues, as well as the obstacles which they should expect to meet along the way.

2. The Precautionary Function of Technology Assessment

How can the potential negative impact of technologies be avoided? And perhaps even more radically: how can the very development and diffusion of technologies susceptible of resulting in environmentally detrimental effects be prevented? The principal experience (as concerns government actions) in this respect goes under the rubric of Technology Assessment (T.A.). And by far the widest and earliest and therefore longest experience with technology assessment is that of the United States.

In 1966, the United States House of Representatives Subcommittee on Science, Research and Development presented a bill on the side effects of technological development. Although environmental concerns may not have been a primary determinant behind this initiative, the bill contained a proposal that an early warning system be set up to prevent the adverse effects of technological development. The term 'technology assessment' was used officially for the first time in this bill. As a result, in 1972 the Office of Technology Assessment (OTA) was set up as an agency closely associated with Congress.

According to one of OTA's leading scientists, J. Coates "technology assessment is the name for a class of policy studies which attempt to look at the widest possible scope of impact in society of the introduction of a new technology, or the extension of an establshied technology in new and different ways. Its goal is to inform the policy process by putting before the decision-maker an analysed set of options, alternatives and consequences. T.A. is extremely wide-sweeping in scope; it is not the decision-making process itself, but only one input for that process.

The US/OTA focuses on subjects that are submitted by Congressial committees for assessment. Several observers agree that the OTA activities, while very valuable, are 'reactive' in nature rather than 'anticipatory'. They provide documentation for political decisions rather than 'early warning'. The same observers also note that the prevailing perspective is not a long-term one, since Congress does not require long-term decision data. The assessments are usually technology-based, that is to say: given a specific technology, their object is to answer the question whether such a technology meets expectations and whether or not it would be advisable for Congress to support its development. These features of technology assessment in the United States and the advantages and limitations attached to them are closely related to the special constitutional structure of the U.S. and the typical U.S. manner of dialogue (including occasional conflict) between Congress and the Executive.

Technology assessment in Europe in a systematic, institutionalised form, started much later than in the United States, and has to some extent followed in the footsteps of the American model. In some cases, technology assessment institutions have been established by national parliaments. It has been much more common in Europe, however, for technology assessment functions to be placed in the framework of government offices. Under a parliamentary system of government it may be assumed that decisions concerning technology policies will by and large be adopted by the parliamentary majority. Negotiations do not take place between the executive and parliament, but between the parliamentary majority and minority, which is why the European technology assessment institutions are normally found in government offices and independent research institutes.

Europe's recent experience with T.A. includes the following developments:

An Office Parlementaire d'Evaluation des Choix Scientifiques et Technologiques (OPECST) has been operating in France for some time.

The new Netherlands Organisation for Technology Assessment (NOTA) Nederlandse Organisatie voor Technologisch Aspectenonderzoek) was created in 1986. It hosted the first European congress on the subject in 1987, drawing 200 delegates.

In 1986, an Office for Technology Assessment was also established in Denmark, with the mandate 'to start general assessment on the possibilities and consequences of technological developments for individuals and for society, and to promote a public debate on these questions'.

Likewise, in 1986, a Mixed Commission from the Chamber of Deputies and Senate for Scientific Research and Technological Development was created in Spain (Law 13/186 of 14th April).

The United Kingdom parliamentary effort to estalish the practice of technology assessment expects to draw on the expertise of the Royal Society in its early work.

In Germany, too, there have been repeated efforts to set up an OTA-like organism. German officials have closely examined the structure and modus operandi of the American system. These efforts are finally bearing fruit. The European Commission has recently extended its Forecasting and Assessment in Science and Technology (FAST) programme -started in 1978- to run at least until 1992, with a £13 million allocation over five years.

Also Europe-wide is a modest effort in Science and Technology Option Assessment set up in 1987 by the European Parliament, which is about to complete its first 18-month trial period.

But the most instructive and innovative effort today to cope with the problem of the T.A., especially new technology and high technology, is probably underway in Japan. In recent years, technological innovation centering on high technologies has brought about a drastic change in Japan's industrial structure. The Environment Agency set up the Conference on Environmental Technology (chaired by Jiro Kondo, chairman of the Science Council of Japan) within the Planning and Co-ordination Bureau in August 1984 with a view to clarifying how best to ensure the integration between environmental protection and technological innovation, particularly high technologies, by examining the problem from various angles. The conference has studied the problem for three years, and a report was published in April 1988. This report sheds new light on a major issue, which could have important implications for the future practice of T.A., and which might call for a totally new undertaking at the international level.

Indeed, the Japanese analysis suggests that the current manner in which certain specific technologies are selectively identified for an assessment, to the exclusion of all other technologies (that is to say of the large majority of technologies under development, and in some cases already at the stage of application) is totally unsatisfactory. This reflects the difficulty of spotting, detecting and tracing emerging technologies at a sufficiently early stage in their development, a problem which is due, according to the Japanese Report, to the very special characteristics of high technology industry.

High technologies, such as micro-electronics, new materials and bio-technology, differ in nature and degree of industrial maturity. But in contrast with the traditional industries, such as steel, electricity and petrochemicals, they generally share the following characteristics:

(i) they produce a large variety of goods, each in limited quality;

(ii) their sites are scattered in many inland areas;

(iii) their products as well as their production processes are subject to rapid change;

(iv) information on materials used by these industries is largely subject to corporate secrecy;

(v) the wastes which these industries generate are often difficult to dispose of, or they contain a large variety of harmful substances in very small quantities and thus defy separation of those substances; furthermore, some of these substances are not even included in the list of those subject to government control;

(vi) there is, as yet, very limited experience (if any) with environmental pollution caused by such industries.

In other words, new high technologies tend to emerge (or rather to come into existence **without** emerging) in places where one would not look, they tend to be inconspicuous in terms of the volumes produced, extremely unconventional in terms of their chemical components and other inputs as well as the composition of their wastes; their life-cycle tends to be too ephemeral and their mutability too rapid in order to be easily detected and monitored; on top of all that, they are increasingly covered by proprietary rights, corporate secrecy and confidentiality.

For these reasons, high tech industries must be dealt with in a different manner from what has so far been done to prevent traditional industries from causing pollution. Regarding conventional industries, a basic policy has been to prevent a recurrence of pollution through regulatory and other measures on the basis of the past experience (namely, occurrence of pollution, clarification of the cause, establishment of environmental quality standards and emission/effluent control for the purpose of achieving the standards). Along this line of reasoning, there is no ground, of course, to question the good judgment of the American Congress in choosing certain (new) technologies rather than others, for assessment. Similarly, there is no reason to disagree with France for giving priority to nuclear disasters and acid rain, or with the United Kingdom for looking first at animal experiments, genetic engineering, geological surveys and space. Nor with the European Parliament for putting its finger on telecommunications, transboundary chemical pollution and nuclear fusion. All these choices are essentially generated by **post hoc** (and, on occasions, even belated) reactions to developments which have reached a fairly advanced stage (as opposed to early warning elicited at a still exploratory stage); in fact most Technology Assessments thus far have been clearly triggered off by widespread perceptions of the potential risks, prompting governments to respond to mounting public concern.

While recognising the legitimacy of these motivations, there is no doubt that the approaches so far pursued by various countries at the national level only cover the field in a sporadic way. They ar hardly adequate to cope with the vastness, complexity and, I would almost say, inscrutability of the subject, as intimated by the Japanese analysis. Even considered on the international scale, they are far from forming a mosaic which could cover the entire field. On the contrary, there is every reason to expect that, as this sporadic approach continues, there will be increasing probability for large gaps as well as fruitless, costly duplication. This prospect runs counter to the increasing expectation that governments should guarantee the safety and health of their citizens, as well as protect their property, through appropriate surveillance systems. At the same time, it raises industry's concern about the effect of irrational and/or uncoordinated control systems. It has been repeatedly pointed out that the lack of national and international technical standards for many new products and industries is an impediment to development and commercialisation. The development of manufacturing processes for new ceramics and composite materials, e.g. depends on the characterisation of materials properties and reference data bases. Similarly, trade in some advanced electronics technologies has been hampered by the lack of internationally recognised standards and certification systems.

Industry recognises that governments can provide a suitable environment for technology development by helping to create the needed regulatory infrastructure - including standards, measurements and test methods; there is indeed a limit to the extent to which the private sector and industrial organisations can be relied upon to develop systems of standardisation and certification. Moreover, there is ample recognition of a need for greater co-ordination of standards for new high-technology industries and products at the international.

Overcoming this problem is indeed essential in order to dispel the prospect of serious risks for the future of technologies: their development could become blocked if an extensive but blindly-designed maze of regulations within countries starts to proliferate; their international transfer, which is vital for innovation, could be paralysed if a network of bureaucratic, non-tariff trade barriers gets established in each country and if it operates within each country's regulatory system, at enormous national cost, without even exploring the possibility of validating and accepting the results obtained abroad. Clearly, there is a dominant international dimension to this problem. Clearly, intergovernmental organisations face the challenge of its solution.

The OECD has no ready recipe to cope with the problem. OECD has done some important assessment work linked to single specific, well-identified technologies, such as nuclear. But OECD has a vast and successful experience in a field not unrelated to that of assessing **new technologies.** This is the field of assessing **new chemicals.**

As is the case for technologies, chemicals present governments with the double challenge of: (a) protecting the environment from harmful effects, and (b) avoiding stifling industry as well as international trade with redundant, contradictory, costly and time-consuming regulatory burdens. As in the case of technologies, reactions to initial concerns were episodical and were confronted on an ad hoc basis. Concerted regulatory action in OECD was initially taken on certain specific organochlorine compounds, polychlorinated biphenyls and heavy metals (Hg, Cd, Pb). It soon became evident that such a reactive approach would soon prove inadequate. About a thousand new chemicals are produced and marketed every year. Their environmental control as well as their international exchange could not proceed smoothly if it was based on a wide and proliferating variety of disjointed assessment and control operations scattered among 24 of the major producing and exporting-importing countries. Decisive steps had to be taken rapidly at the international level.

In accepting the challenge, the OECD had to face a number of extremely difficult issues. The analogies between these issues, and those raised with respect to the environmental dimension of technology assessment are striking. Two initial questions were be confronted at the very outset:

- Should a supra-national institution be created and vested with the task and the authority of testing every new chemical, before its marketing or export, for its effect on the environment and human health?
- Should an international super-laboratory be established where all the

tests would be implemented, regardless of the nature of the chemical, its origin, its destination?

Based on a number of in-depth consultations, expert advice and practical considerations, this course of action was rejected. However, the alternative option, i.e., co-operation and harmonisation of national requirements, etc., which eventually formed that basis of a Special OECD Programme on Chemicals, opened up a number of other questions. For example:

- How was harmonization of test methods to be achieved if assessment were to be carried out in so many different countries and possibly at different times (taking scientific advancement constantly into account)?
- Could the array of the best available testing methods be identified through co-operative international action, and was it feasible to develop internationally uniform guidelines for conducting the assessment of a new chemical?
- How could one ensure that the result of the tests performed in our country would be automatically received by all others, so as to prevent duplication of efforts; and that it would be automatically accepted by all others, so as to avoid barriers to trade?
- How could countries agree on a priority list of chemicals calling for testing; and having established such a list, how was it possible to distribute the candidates for assessment to various countries, thus minimising the aggregate effort and time requirements and avoiding duplication?
- What criteria could be followed in order to decide on the need to proceed from relatively simple (and inexpensive) type of tests, to much more laborious, costly and time-consuming tests (for mutagenesis, theratogenesis, other long-term effects)?
- How could international confidence be ensured and maintained in a geographically scattered testing system and machinery, concerning the comparability, quality and rigour of national monitoring procedures over local laboratory practices? Was it feasible to develop internationally consistent standards of good laboratory practice including conditions under which studies are planned, performed, monitored, recorded and reported?
- How was it possible to ensure proprietary rights and confidentiality of data and trade secrets in the course of internationally shared and internationally recognised assessment exercises?

Over a period of work of more than ten years, the OECD Chemicals Programme has provided operational answers to this entire array of issues.

Analogies -like technologies- need to be cautiously tested before being operationally applied. Subject to this caveat, the OECD Chemicals Programme may offer some inspiration and some lessons on what is feasible as well as what needs to be avoided in the early assessment of new technologies.

3. The Promotion of Environmentally Sound Technologies

Technology Assessment - as it emerges from the previous notes, and as commonly understood, appears as a very effective instrument for preventing the development, or at least the application and diffusion of technologies detrimental to the environment.[2] From the point of view of the role of governments, this is, so to speak, one side of the coin. The other, equally important side is the promotion of environmentally safe or even favourable technologies.

The previously mentioned OECD monograph on this Semianr, on the 'Promotion and Diffusion of Clean Technologies' falls under this rubric. The OECD monograph is not too sanguine as to the amount of effort which governments dedicate to promoting environmentally sound technologies, nor is it too enthusiastic as to positive results which can be clearly and solely attributed to such efforts. The situation, however, may be improving. One cannot complete even a very cursory overview like this one without mentioning a few significant international initiatives undertaken after the date of the OECD survey, which seem to justify a cautiously optimistic view.

Interestingly enough, several of these initiatives, while definitely oriented toward the **promotion** of environmentally benign technologies (as distinct from the **control** of environmentally detrimental or risky technologies) retain the terms 'technology assessment' as part of their names. In some cases, and in some countries (e.g., Denmark, The Netherlands) phrases such as 'active technology assessment' or 'creative technology assessment' have been introduced, presumably in order to stress the opposition with respect to the 're-active' and 'regulatory' function. The risk of semantic ambiguities, however, still persists, and needs to be kept in mind when one undertakes to list, and briefly delineate, the variety of

---

[2] Beyond the domain of the environment, technology assessment has expanded to a variety of other concerns, such as the potential for new technology to provide jobs and other social benefits.

national and international initiatives which are currently underway.

At the international level, activities within the European Communities, the UN Econòmic Commission for Europe, and the United Nations Environment Programme are worth mentioning.

The European Communities Forecasting and Assessment in Science and Technology (FAST) includes one programme on integrated development of renewable natural resources. Between 1984 and 1987, 23 projects were carried out focussing on environmental monitoring technology, forestry, agriculture, chemistry, energy, biological diversity and other environmental subjects. The European Communities have also collected information on clean technology applied to a variety of industrial sectors. It has recently (1984) allocated a budget of 6.5 Ecus for a 3-year period to support demonstration projects in which more than one member country participates. The projects are meant to help bridge the gap between research prototype stages and the market on clean technologies in various industries, and recycling and re-using waste. Furthermore, in March 1988 the European Communities launched a new Network for Environmental Technology Transfer (NETT) to serve information both to the supply and the demand side of the pollution control technology market.

The UN/Economic Commission for Europe is working on a compendium containing 134 monographs on low-waste and non-waste technologies. This exercise should include, in principle, the development of uniform technology assessment procedures and criteria to be applied by all countries in this area; and eventually, the establishment of an international data bank to gather and disseminate information on low-waste and non-waste technology.

UNEP's Office on Industry and the Environment supplies information to 15 subscribers based on a main register of pollution control technology, covering six heavily polluting industrial sectors.

Even as we sit at this Seminar, not far from here we can find further positive signs. Recently, the European Ministers for Scientific Research, meeting in Luxembourg, decided in favour of a total functional transformation of the European Research Centra at Ispra on the Lago Maggiore. The original aims of the Ispra Centre -focused on joint European nuclear fission research- have increasingly lost ground over the past few years, due to a generalised crisis of nuclear industry, aggravated by lack of consensus on a joint European strategy. Largely in order to avoid the risk of closure, a restructuring of the European Centre has finally been adopted. One of the five research institutes which will compose ISPRA's new structure will address member governments' research needs in the domain of environmental improvement, pollution abatement and control, the prevention of ecological disasters and the safety of hazardous industrial plants.

Concerning national initiatives, one of the most recent, and probably one of the most important, particularly in view of its potentially international implications, was launched in Sweden on December 17, 1987. On that date, the Swedish government authorised the Minister for Environment and Energy to appoint a Committee for the purpose of studying the orientation, scope and organisation of an 'international institute for the assessment of environmentally sound technologies'. Although the proposal is still before the Swedish Parliament, there are indications of its imminent approval: including an initial appropriation of (Swedish government) funds to the level of 25 million Swedish kronor every year, for a period of five years.

The phrase initially used to identify this soon-to-be-created institution (Swedish International Institute for Environmentally Sound Technology Assessment). The Institute's mission is certainly not limited to the precautionary assessment of potentially damaging technologies, but also, and in fact to a much larger extent, 'to research and development of environmentally sound technology for sustainable economic development, and also international diffusion of such technology'. In ohter words, this is clearly a case where the extended, 'active' or 'creative' concept of technology assessment prevails. This does not exclude a very important 'precautionary' function, where assessment is meant in its more traditional connotation.

As this paper goes to print (July 1989), the Swedish initiative has materialised and has started to operate under amore appropriate name: International Institute for Environmental Technology and Management: the 'Stockholm Environment Institute' for short.

The documentation so far produced by the Swedish authorities and summarised in a book produced by well-known experts and government officers, ('SIESSTA: an International Institute for the Assessment of Environmentally Sound Technologies') allows for the highest expectations, both with respect to the assessment and the promotion functions of this new institution.

Ambitions are very high indeed. The establishment of the Institute is seen by the proponents as nothing less than 'an international OTA'; the new institution, as they state, will have to gain the necessary recognition and authority for this role to be accepted by the world community in all its forms, governmental, intergovernmental, non-governmental. Its task will be to provide national and international political decision-makers, as well as industry, with non-partisan advice based on reliable, scientifically-founded, decision-supporting data. It should also provide data for the general public, which will enable people to form an opinion on matters related to environment and technology. Furthermore, it is hoped that it will be able to develop generally accepted methodologies with relevant criteria for uniform approaches to technology assessment (the usefulness of this endeavour, if at all possible, does not need not be emphasized) and for the comparative assessment of technological solutions. Finally, it also intends to work towards the identification of so-called 'environmental indicators of development', intended as an instrument for economic decision-making.

In its fundamental approach, the authors reject the traditional method of technology assessment, which is based on technology as given factor, and only proposes minor modifications to eliminate undesirable effects. The starting point for action will not be 'technology-based' or 'technology-prompted or triggered', but rather the identification of a number of major, well-defined problems of a global nature, the solution of which is of vital interest for the world community. The first step will consist of subjecting these very problems (rather than any particular technology) to assessment.

Four priority areas have already been tentatively identified on which assessment work should start as soon as the Institute becomes operational:

- the greenhouse effect and the risk of climatic change;
- biotechnology and genetic engineering;
- the release of hazardous substances from production and products, including hazardous wastes;
- pollution due to use of fertilizers and pesticides in agriculture and forestry.

Further proposals, in the future, will be submitted by the Institute's Scientific Council.

Starting from there, the Institute will subsequently proceed to identify environmentally-sound technology options and new technologies which can fill the previously assessed environmental needs.

The Institute's approach, largely inspired by the conclusions of the World Commission on Environment and Development, is a rather bold one: its general philosophy seems to be, in a nutshell, that technologies are a major, if not **the** major, cause for today's ecological predicament: but that, nevertheless, technologies are essential, in order to **solve** the problem.

This, however, calls for a radical reorientation of the entire technological research effort: a prerequisite for an equally radical reorientation of government R & D policies. 'Investment in environmentally sound technology must not be restricted to marginal technology development as a subsidiary to mainstream technology development. On the contrary, mainstream technology development, which commands the lion's share of investment and R & D, must be modified to accord with the environmental approach'.

Experience will tell whether an ambition of this magnitude meets the kind and calibre of political will and public support which will be necessary for their implementation.

Bibliography

Cole, H.S.D., Freeman, C., Jahoda, M. and Pavitt, K.L.R. (1973), Thinking About the Future: A Critique of the Limits to Growth, Chatto & Windus, London.

Commoner, B. (1971), The Closing Circle, Jonathan Cape, Thirty Bedorf Square, London; Alfred A. Knopf, New York.

Ellul, J. (1954), La technique ou l'enjeu du siècle, Librairie Arman Colin, Paris.

Giedion, S. (1969), Mechanization Takes Command, Norton, U.K.

Meadows, D., Meadows, D., Randers, J. and Behrens, W.W. (1972), The Limits to Growth: Report for the Club of Rome's Project on the Predicament of Mankind, Potomac Associates Book, Universe Books, New York.

OECD (1981), The Measurement of Scientific and Technical Activities, Proposed Standard Practice for Surveys of Research and Experimental Development, Frascati Manual 1980.

OECD (1987), The Promotion and Diffusion of Clean Technologies in Industry, Monograph no. 9, OECD, Paris.

Simon, J.L. (1981), The Ultimate Resource, Princeton University Press.

Svenson, G., Bolin, B., Forsberg, H.G., Niblaeus, K. and Svensson, U. (1988), SIIESTA: An International Institute for the Assessment of Environmentally Sound Technologies, Ministry of Environment and Energy, Sweden.

Toffler, A. (1970), Future Shock, Random House, New York.